Mathematical Excursions to the World's Great Buildings

Mathematical Excursions to the World's Great Buildings

ALEXANDER J. HAHN

PRINCETON UNIVERSITY PRESS

PRINCETON AND OXFORD

Copyright © 2012 by Princeton University Press

Published by Princeton University Press, 41 William Street,
Princeton, New Jersey 08540
In the United Kingdom: Princeton University Press, 6 Oxford Street,
Woodstock, Oxfordshire OX20 1TW
press.princeton.edu

Library of Congress Cataloging-in-Publication Data

Hahn, Alexander
    Mathematical excursions to the world's great buildings / Alexander J. Hahn.
        p.   cm.
    Includes bibliographical references and index.
    ISBN 978-0-691-14520-4 (hardcover : alk. paper) 1. Architecture—Mathematics.
2. Architectural design. 3. Architecture—Data processing. I. Title.
    NA2750.H325 2012
    720.1'51—dc23

                                                                    2011028356

British Library Cataloging-in-Publication Data is available

This book has been composed in ITC New Baskerville
Printed on acid-free paper. ∞
Printed in the United States of America

10 9 8 7 6 5 4 3 2

Contents

The *Mathematical Excursions* into the world of architecture that this book undertakes are organized around two historical narratives. The primary narrative has a focus on aspects of architectural form (the role of geometry, symmetry, and proportion) and structure (matters of thrusts, loads, tensions, compressions) of some of the great buildings of western architecture from the pyramids of Egypt to iconic structures of the twentieth century. Some of the high points of this narrative are the Parthenon in Athens, the Colosseum and Pantheon in Rome, the Hagia Sophia, historic mosques, great Romanesque, Gothic, and Renaissance cathedrals, Palladio's villas, the U.S. Capitol, the Sydney Opera House, and the Guggenheim Museum in Bilbao. (This narrative is not comprehensive, but instead seeks to illustrate important architectural features with compelling examples.)

A second narrative develops current elementary mathematics from a historical perspective. This includes selected aspects of Euclidean geometry, trigonometry, the properties of vectors, coordinate geometry in two and three dimensions, and (at the very end) basic calculus. It is the raison d'être of this book to intertwine these two stories and to demonstrate how they inform each other. The mathematics provides clarifying insights into the architecture, and, in turn, the architecture is a stage that gives visibility to applications of abstract mathematics. To be clear, the two narratives meet around topical issues and not when they happen to be aligned chronologically. In fact, the chronological alignment between mutually informing architecture and mathematics is rare. (Greek geometry and architecture is an exception.) The reality is that the elementary mathematics that might have clarified the understanding of a complex structure was almost always beyond the reach of the builders of the time.

The collage of historic buildings (all at the same scale) of Plate 1—refer to the section of colorplates after chapter 4—provides a snapshot of this book. It studies many of these buildings and focuses its mathematical analysis on their domes, arches, columns, and beams.

The prerequisites that you will need to bring to this book are a working knowledge of some basic high school math (such as elementary algebra and a little geometry) as well as an interest in learning about architecture and its vocabulary (as the Glossary presents it). You will be able to use the book in different ways. To a large extent, your choices will be influenced by your background in mathematics. If you are not on particularly friendly terms with this discipline, I encourage you to be patient and persistent (and work through the details) when you take up the Euclidean geometry, the few elements of trigonometry, and the basic study of vectors of Chapter 2. You will then have the architectural narratives of Chapters 1, 2, 3, and 5 open to you. A little more patience and persistence will allow you to absorb the basic historical mathematics and the elementary introductions to coordinate systems in two and three dimensions that Chapter 4 presents. My hope and expectation is that your reward will be an engaging journey through the first six chapters of the book. The two sections on perspective that conclude Chapter 5 and the two historical sections on structural engineering in Chapter 6 are a bit technical, but they may be skipped as they do not directly impact the rest. If you wish a mathematically more challenging expedition, take these four sections on and also have a thorough go at Chapter 7. This chapter reviews the basics of calculus and applies its methods to the analysis of domes and arches. Each of the seven chapters concludes with a Problems and Discussion section. Many of the 200-plus problems and 18 discussions focus on particulars of the topics presented; others go off on tangents that expand their scope. For a lighter and quicker journey through the book they can all be skipped. Most of the problems are designed to facilitate the understanding of the mathematics. Some of them are challenging. An instructor of a course using this text should assign them with care.

Acknowledgments

This book has drawn much benefit from a number of my friends and colleagues at the University of Notre Dame. Those of the Department of Mathematics deserve particular mention for the supportive environment that they have provided over the years. A special word of thanks to Timothy O'Meara, mathematician and former provost, for reading much of an earlier version of the book and for providing important suggestions that improved it. Thanks go to Neil Delaney of the Department of Philosophy, for many enriching conversations about the creative structures of Frank Gehry and Santiago Calatrava, and many other topics in which we share an interest. Thanks also to David Kirkner of the Department Civil Engineering and Geological Sciences, who always found the time to answer the questions about structural engineering that I brought to his attention. A word of thanks to Michael Lykoudis, Dean of the School of Architecture, for stimulating my interest in architecture, and to his colleagues Richard Bullene, Norman Crowe, Dennis Doordan, Richard Economakis, David Mayernik, John Stamper, and Carroll Westfall for supplying information about various matters that arose in the preparation of this text. My gratitude also to my friends and colleagues at

Notre Dame's Kaneb Center for Teaching and Learning for always reminding me that teaching must be measured not by the quality of the exposition but by the learning experience of the students.

I am most grateful to Marc Frantz, Department of Mathematics at Indiana University, Bloomington, for his positive response to my text. The critical, but always well reasoned, insightful, and constructive remarks of his review enhanced several discussions of the text. Thanks also to Robert Osserman, director of special projects for the Mathematical Research Institute at Berkeley, for clarifying a point about the Gateway Arch in St. Louis.

A warm word of gratitude to Princeton University Press, especially Vickie Kearn, my editor, and her assistants Stefani Wexler and Quinn Fusting; and to Sara Lerner in production. They were always quick to respond, very helpful, encouraging, and consummately professional.

To my wife Marianne, thank you so much for your love and support, and for not minding too much that my mind always seemed to be on something else. Finally, to my father, Dr. George Hahn, thank you for teaching me the values of commitment and discipline with the example of your life.

Alexander Hahn
Department of Mathematics
University of Notre Dame
September 2011

Humanity Awakening:
Sensing Form and Creating Structures

The earliest ancestors of humans began to emerge about 7 million years ago, and the human species has existed for about 100,000 years. For most of this period humans were preoccupied with acquiring food and securing shelter. They lived in caves, made stone tools and weapons, and hunted and foraged for food. Some 10,000 years ago, an important transition occurred. By that time, the Ice Age that had started about 60,000 years ago was over. The ice sheets that covered Europe and Asia had receded to make room for forests, plains, and deserts. Seeing how plants sprang forth and grew in nature, these humans began to cultivate their own. Over time, they emerged from their caves, built their own primitive dwellings, and scratched an existence from the soil. Remains of some early huts show that they were constructed with skeletons of pine poles and bones and covered with animal skins. In time, villages formed, bread was baked, beer was brewed, and food was stored. The crafts of weaving, pottery, and carpentry developed, and basic goods were exchanged. Words expressed very concrete things and the constructions of language were simple. Copper was discovered, then bronze was made, and both were shaped into tools and weapons. Pottery and woven fabrics began to be decorated with geometrical patterns that reflected numerical relationships. Trading activity increased in radius and languages increased in range. With the continuing development of crafts, food production, and commerce, the need to express "how many?" and "how much?" in a spoken and also symbolic way became increasingly relevant so that a concept of number emerged.

Larger communities were a later development. One of them unearthed in the plains of Anatolia (in today's Turkey) consisted of a dense clustering of dwellings. Access to them was gained across their roofs. Mud-brick walls and timber frameworks enclosed rectangular spaces that touched against neighboring ones to form the town's walled perimeter. Interspersed between the houses were shrines that contained decorative images of animals and statuettes of deities. The settlements that began to develop along the world's great rivers at around 5000 B.C. profited from the arteries of communication and commerce that connected them. They became economically thriving, literate, urban communities. Those in Mesopotamia (in today's Iraq) and those on the upper and lower Nile in Egypt would become the cradles of Western civilization. The fertile plains near these rivers gave rise to a large-scale agriculture that required organization and storage facilities. Irrigation projects and efforts to control flooding drove technological advances. In this environment the practice of mathematics began. It was a mathematics of basic arithmetic. It had almost no symbolism and did not formulate general methods. It computed elementary areas and volumes and was strictly a tool for solving particular practical problems. Architecture developed in response to the requirements of commerce and agriculture, the need to honor the gods on whose good will success depended, and the rulers' insistence on a secure afterlife. Cities had storehouses, sprawling configurations of temples, and elaborate tomb complexes. Depending on their purpose, these structures were built with sun-baked bricks, stone columns and wooden beams, and massive stone slabs. A sense of aesthetics found expression in ornamented glazed tiles, decorative terra cotta elements, and monumental statues of rulers and deities.

Sensing Form and Conceiving Number

As humans began to be more acutely aware of their surroundings, watchful waiting turned into awareness, and fear and instinct became caution and reflection. Humans became curious about the physical world, began to observe similarities in things, and noticed regularities and sequences. They observed the points of light of the night sky, the line drawn by sea and sky at the horizon, the circular shapes of the moon, sun, and irises of eyes. They became conscious of the perpendicularity of the trunk of a tree against a flat stretch of land, the angles between the trunk and its branches, and the triangular silhouettes of pine trees against an illuminated sky. They wondered about the arcs of rainbows, shapes of raindrops, designs of leaves and blossoms, curves of horns, beaks and tusks, spirals of seashells, the oval form of an egg, and the shape of fish and starfish. They looked up at the vast reaches of sky and the moving clouds, sun, and moon within it, and became aware of the spatial expanse of their environment. The cave paintings executed about 30,000 years ago (Plate 2) demonstrate a wonderful ability of early humans to record what they observed. In fact, they show us that they had a heightened sense of their surroundings, a capacity to reflect about what they saw and experienced, and a sense for composition. Humans became aware of nature's organizational structures: the leaf configuration of a fern, the branching patterns of a bare tree in the winter, the arrangement of seeds in their housings, the intertwined form of a bird's nest, the hexagonal repetition of honeycombs, and the netted arrangement of a spider web. They began to gain a sense for basic shapes such as those depicted in Figures 1.1 and 1.2. When humans noticed a common aspect about a group of three trees, three

Figure 1.1. An owl traced in the Chauvet Cave. Photo by HTO

Figure 1.2

Figure 1.3. Bone discovered in the village Ishango, Democratic Republic of Congo, Africa. Photo from the Science Museum of Brussels

grazing zebras, three chirping birds, three mushrooms, and three roars of a lion, they began to gain a sense of number. The earliest records of the practice of counting are from 15,000 to 30,000 years old. The bone shown in Figure 1.3 is an example.

The important transition from the gathering and hunting for food to the cultivation of crops and the domestication of animals began about 10,000 years ago. Humans left their caves, began to erect primitive dwellings, and clustered in villages for protection. With ropes and sticks, they could trace out lines and circles. The living quarters that they laid out, whether circular yurts or rectangular huts, borrowed forms and structures from nature. They baked bread, built granaries, conceived of the wheel and axle, and made carts. Trade began and spoken language grew more sophisticated. When they encountered the need to count objects, estimate distances, and measure lengths, they used fingers, feet, and paces to do so. Elementary pottery, weaving, and carpentry developed. These early efforts of designing, building, and shaping cultivated a sense of planar and spatial relationships. The connection in our language between "stretch" and "straight" and between "linen" and "line" provides some evidence for the links between these early crafts and early geometry. Textiles were decorated with geometric designs of the sort depicted in Figure 1.4. They provide evidence of an increased awareness of order, pattern, symmetry, and proportion.

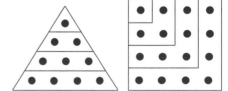

Figure 1.4

Powerful natural images and events such as threatening weather formations, angry thunderstorms, devastating floods, and volcanic eruptions were attributed to supernatural forces. Myths and primitive religions responded to a basic need to explain these phenomena. The realization that seasonal weather patterns and life cycles of plants follow a rhythm that is related to the variation in the height of the midday Sun in the sky gave importance to the tracking of celestial phenomena within the passage of time. Architectural structures organized these beginnings of astronomy. Stonehenge in southern England is an example. Started around 3000 B.C. and added to for a millennium and more, it featured huge stone slabs arranged in vertical

Figure 1.5. Stonhenge, southern England. Photo by Josep Renalias

pairs with horizontal slabs across the top. The slabs were arranged in a careful pattern that included a circular arrangement 100 feet across. The fact that this pattern was aligned with solar phenomena tells us that Stonehenge served as a prehistoric observatory that made the predictions of the summer and winter solstices (the longest and shortest days of the year) possible. Figure 1.5 shows some of the several dozen slabs that remain. The largest weigh as much as 20 tons (1 ton = 2000 pounds). Stonehenge also testifies to the incredible ability of its builders to move and shape huge megaliths (in Greek, *mega* = great, *liths* = stones) and to arrange them for intelligent purposes.

Rising Civilizations

Urban settlements became possible when agricultural surpluses allowed some people to assume specialized roles (priests, merchants, builders, and craftsmen) that were not directly tied to the production of food. Starting in the fifth millennium B.C., more advanced societies evolved along the banks of the great rivers Tigris and Euphrates, Nile, Indus, Huang, and Yangtze. These mighty rivers served as channels of communication and trade and brought raw materials from neighboring uplands. Extensive irrigation systems spread floodwaters to the low-lying, fertile plains, and made it possible to grow an abundance of crops. Large structures such as levees, dams, canals, reservoirs, and storage facilities rose to restrain and regulate the flow of water and to order agricultural production. Elaborate temples were built to appease the gods on whose good will the success of these efforts was thought to depend. Monumental burial complexes were designed and constructed to provide Egypt's ruling pharaohs and other important citizens with a smooth

Figure 1.6. The great pyramids of Giza, Egypt. Photo by Ricardo Liberato

and comfortable transition to and existence in the afterlife. Structures made of wood and earthen bricks did not survive the forces of time and erosion, but there are impressive remains of great stone structures. One of these, a tomb complex built near ancient Egypt's capital Memphis around 2500 B.C., consisted of a step pyramid with a burial chamber, a reconstruction of the pharaoh's palace, courtyards, altars, and a temple, all surrounded by a 33-foot wall laid out in a rectangle with a perimeter of one mile. The pyramid and sections of the wall survive. The greatest pyramids were built around the same time and not far away near today's city of Giza. They are shown in Figure 1.6. Their construction was an amazing feat. Without any machinery beyond ramps, levers, and strong ropes, and no metal harder than copper, thousands of Egyptian laborers cut massive blocks of stone, moved them to the site, and stacked them into precise position. The largest of these pyramids, the pyramid of the pharaoh Khufu (on the far right in Figure 1.6) rises to a height of 481 feet from a square base with 755-foot sides. The tip of the pyramid is almost exactly over the center of its square base. The pyramid was built with 2.3 million blocks weighing from 2.5 to 20 tons. The lowest layer of blocks rests on the limestone bedrock of the area and supports about 6.5 million tons. The building materials used most frequently by the Egyptians in their monuments are limestone and sandstone. Both are formed by sedimentation. Limestone consists of calcium carbonate. Sandstone is usually harder. It consists of sand, commonly quartz fragments, cemented together by various substances. The structural qualities of limestone and sandstone depend on the particular deposit, but both can be carved and cut without great difficulty. The sizable burial chambers deep in the interior of the pyramids were constructed with granite so that they could resist the enormous loads of the blocks above them. Granite, a rock formed by the crystallization and solidification of molten lava from the hot core of the Earth, is much harder and stronger than limestone and sandstone.

Large constructions such as the pyramids necessitated organizational capacity, enhanced technological expertise, and record keeping that needed to be promoted by growing central administrations. All this required a richer symbolic representation of language and an enhanced development of mathematics. The mathematics of these river civilizations originated as a practical science that facilitated the computation of the calendar, the surveying of lands, the coordination of public projects, the organization of the cycle of crops and harvests, and the collection of taxes. In the hands of a class of administrator priests, the practice of arithmetic and measuring and the study of shapes and patterns evolved into the beginnings of algebra and geometry.

The most advanced of the great river civilizations were the people of the "fertile crescent" of Mesopotamia (in Greek, *meso* = between, *potamia* = rivers) between the Tigris and Euphrates (of today's Iraq). They introduced a positional numerical notation for integers and fractions based on 60 and used wedge-shaped symbols to express $1, 60, 60^2 = 3600$, and $60^{-1} = \frac{1}{60}$, and $60^{-2} = \frac{1}{60^2}$. Traces of this system survive to this day in the division of the hour into 60 minutes, a minute into 60 seconds, and a circle into $6 \times 60 = 360$ degrees. The mathematicians of the Babylonian dynasty that followed around 2000 B.C. solved linear, quadratic, and even some cubic equations. In particular, they knew that the solutions of the quadratic equation $ax^2 + bx + c = 0$ are given by the formula

$$x = \frac{-b \pm \sqrt{b^2 - 4ac}}{2a}.$$

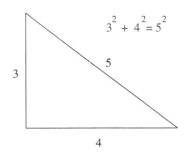

$$3^2 + 4^2 = 5^2$$

5

3

4

Figure 1.7

The Babylonians knew the theorem now called the Pythagorean Theorem. For any right triangle with side lengths *a*, *b*, and *c*, where *c* is the length of the hypotenuse, the equality $a^2 + b^2 = c^2$ holds. A clay tablet cast between 1900 and 1600 B.C. gives testimony to the achievements of the Babylonians. Now referred to as Tablet 322 of the Plimpton collection, it lists triples, in other words threesomes, of whole numbers *a*, *b*, and *c*, with the property that $a^2 + b^2 = c^2$. As such triples represent the sides of right triangles, they provide specific instances of the Pythagorean Theorem. The triple of numbers $a = 3$, $b = 4$, and $c = 5$ is an example (see Figure 1.7). Another is $a = 5$, $b = 12$, and $c = 13$. Some of the very large triples listed on the tablet strongly suggest that the Babylonians had a recipe for generating such threesomes of numbers. The Babylonians had formulas for the areas of standard planar figures and volumes of some simple solids. They also analyzed the positions of the heavenly bodies and developed a computational astronomy with which they predicted solar and lunar eclipses.

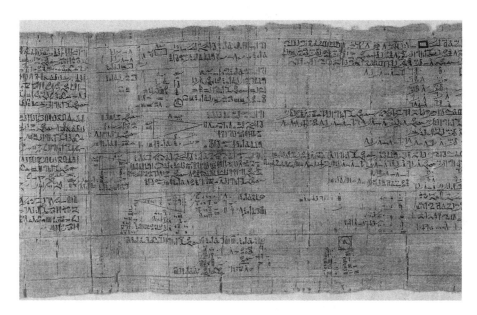

Figure 1.8. The Rhind papyrus, British Museum, London. Photo © Trustees of the British Museum

The Egyptian Rhind papyrus is a long scroll that dates from around 1600 B.C. A portion of it is shown in Figure 1.8. The Rhind papyrus (papyrus is a plant product that served as the paper of that time) gets its name from the Scotsman A. Henry Rhind, who bought it in Egypt in the nineteenth century. Its introduction promotes it to be "a thorough study of all things, insight into all that exists, knowledge of all obscure secrets." But it is simply a handbook of practical mathematical exercises of the sort that arose in commercial and administrative transactions. The 85 mathematical problems it presents and the elaborate theory of fractions on which many of the solutions rely give a good idea of the state of Egyptian mathematics at the time. It also provides the approximation $(\frac{16}{9})^2 = \frac{256}{81} \approx 3.1605$ for the ratio $\frac{c}{d}$ of the circumference c of a circle to its diameter d. (The symbol $\approx$ means "is approximately equal to.") Today, this ratio is designated by π (and better approximated by $\pi \approx 3.1416$). The papyrus also contains some practical advice: "catch the vermin and the mice, extinguish noxious weeds; pray to the God Ra for heat, wind, and high water."

As physical witness to the growing sophistication of thought, architecture advanced as well. The walled city of Babylon with its imposing temples and soaring towers was well known for its architectural splendors. The Greek historian Herodotos traveled widely in the Mediterranean region in the fifth century B.C. and recorded what he saw. About Babylon he wrote that "in magnificence, there is no other city that approaches it." Unfortunately, very little remains. One of the main gates to the inner city was built early in the sixth century B.C. and dedicated to the goddess Ishtar. Its central passage featured a high semicircular arch, walls covered with blue glazed tiles, and

Figure 1.9. The facade of the Great Temple at Abu Simbel, Egypt. Lithograph by Louis Haghe, 1842–1849, from a painting by David Roberts, 1838–1839

doors and roofs of cedar. The middle section of the Ishtar Gate (only a small part of the ancient gate complex) has been reconstructed from materials excavated from the original site. A full 47 feet tall, it stands in the Pergamon Museum in Berlin. A similar reconstruction exists near the original site in Baghdad, Iraq.

Egypt's buildings withstood the challenge of time better than those of Babylon. The rise of the sun god Amun in the middle of the second millennium to the position of primary state god inspired the building of stone temple complexes that were grander and more elaborate than before. An impressive example is the great temple of Amun, built near today's city of Karnak from the middle of the sixteenth century to the middle of the fourteenth century B.C. A succession of pharaohs ordered the construction of monumental entrance gates, obelisks, colossal statues, and grand ceremonial halls. The largest of these halls was built in the reign of the powerful Rameses II. It measured 165 feet by 330 feet and was tightly packed with tall and massive columns that supported the heavy stone slabs of its roof. Plate 3 tells us that enough of the structure is preserved to give today's visitor a sense of its former size and grandeur.

The great temple of Abu Simbel that the same Rameses II had built in his honor in the fourteenth century B.C. in southern Egypt is another example. Figure 1.9 shows the temple carved into a sandstone cliff on the banks of the Nile, its facade dominated by statues of the great man himself in ceremonial pose. These massive statues are 67 feet high and weigh 1200 tons. Smaller figures immortalize the queen and lesser dignitaries. The depiction confirms that sandstone can be brittle. The ancient temple has a modern history. It was rediscovered around 1815 as it emerged from the shifting sands that had buried most of it. When the temple was threatened in the 1960s by the rising waters of the artificial lake created by the dam being constructed near Aswan in Upper Egypt, the United Nations organized a monumental effort to save it. The temple's facade and its elaborate interior (reaching 200 feet into the rock) were cut into sections weighing many tons each, moved carefully block by block, and reassembled, exactly as they had been, on higher ground a few hundred feet away.

The very brief survey of early mathematics and architecture presented above is a story of the developing ability of humans to recognize shape, pattern, and structure in their surroundings and their later efforts to impose shape, pattern, and structure on the activities that impacted their existence. However, the mathematics and architecture of ancient civilizations were driven by different forces. Mathematics arose primarily in response to the practical need to organize and order production, commerce, and their underlying infrastructures. The primary purpose of architecture on the other hand was to give powerful visual expression to the importance and grandeur of the rulers and their gods.

Problems and Discussions

The problems below are related to matters discussed in the text. They provide an opportunity for thinking about and maneuvering through some basic mathematics.

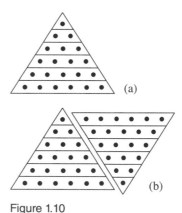
(a)

(b)

Figure 1.10

Problem 1. Consider the diagram in Figure 1.10a. Count the dots from the top down to get $1 + 2 + 3 + 4 + 5 + 6$. Turn to the diagram in Figure 1.10b and notice that

$$2(1 + 2 + 3 + 4 + 5 + 6) =$$
$$(1 + 6) + (2 + 5) + (3 + 4) + (4 + 3) + (5 + 2) + (6 + 1) = 6 \cdot 7.$$

So $1 + 2 + 3 + 4 + 5 + 6 = \frac{1}{2}(6 \cdot 7)$. Use the same strategy to show, for any positive integer n, that $1 + 2 + \cdots + (n - 1) + n = \frac{1}{2}n(n + 1)$.

Problem 2. It seems to be the case that any sum of consecutive odd numbers starting with 1 is a square. For example, $1 + 3 = 2^2$, $1 + 3 + 5 = 3^2$, $1 + 3 + 5 + 7 = 4^2$, and $1 + 3 + 5 + 7 + 9 = 5^2$. The diagram of Figure 1.11 shows that $1 + 3 + 5 + 7 + 9 + 11 = 6^2$. Let n be any positive integer. Consider the term $2k - 1$. Plugging in $k = 1$, $k = 2, \ldots$, $k = n$, provides a list $1, 3, 5, \ldots, 2n - 1$ of the

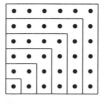
Figure 1.11

first n odd integers. Show that their sum $1 + 3 + 5 + \cdots + (2n - 1)$ is equal to n^2. [Hint: Try the strategy used in the solution of Problem 1.]

Problem 3. Take any two positive integers m and n with $n > m$. Now form the positive integers a, b, and c, by setting $a = n^2 - m^2$, $b = 2nm$, and $c = n^2 + m^2$. This is a recipe for generating numbers a, b, and c that satisfy $a^2 + b^2 = c^2$. Taking $m = 1$ and $n = 2$ gives us $a = 4 - 1 = 3$, $b = 2 \cdot 2 = 4$, and $c = 4 + 1 = 5$. Because $3^2 + 4^2 = 5^2$, the recipe works in this case. Taking $m = 2$ and $n = 3$ gives us $a = 9 - 4 = 5$, $b = 2 \cdot 6 = 12$, and $c = 9 + 4 = 13$. The fact that $5^2 + 12^2 = 169 = 13^2$ tells us that the recipe works in this case as well. Verify the recipe in general, and then use it to list five additional triples (a, b, c) of positive integers that satisfy $a^2 + b^2 = c^2$.

Problem 4. Study the three diagrams of Figure 1.12. A right triangle with sides a, b, and c is given. The diagram at the center is a configuration of two squares arranged in such a way that the four triangular regions that they determine are each equal to the given triangle. Make use of the diagrams to write a paragraph that verifies the Pythagorean Theorem.

Problem 5. The Pythagorean Theorem was also known to the Chinese. The essential information of the old Chinese diagram depicted in Figure 1.13a is captured by Figure 1.13b. It depicts four identical right triangles (each with sides of lengths a, b, and c) arranged inside a square. Determine the size of the inner square and use the diagram to verify the Pythagorean Theorem.

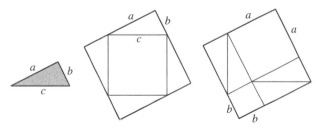

Figure 1.12

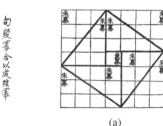
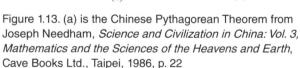

(a)　　　　　　(b)

Figure 1.13. (a) is the Chinese Pythagorean Theorem from Joseph Needham, *Science and Civilization in China: Vol. 3, Mathematics and the Sciences of the Heavens and Earth*, Cave Books Ltd., Taipei, 1986, p. 22

Discussion 1.1. Solving the Quadratic Equation. The solutions of the equation $ax^2 + bx + c = 0$ (with $a \neq 0$) are given by the quadratic formula $x = \frac{-b \pm \sqrt{b^2 - 4ac}}{2a}$. Today's verification of this formula is a consequence of a procedure known as *completing the square*. The procedure consists of several algebraic steps that are illustrated below in the case of the quadratic polynomial $6x^2 + 28x - 80$.

First factor out the coefficient of the x^2 term. So $6x^2 + 28x - 80 = 6(x^2 + \frac{28}{6}x - \frac{80}{6})$. Focus on $\frac{28}{6}x = \frac{14}{3}x$, divide $\frac{14}{3}$ by 2 to get $\frac{14}{6} = \frac{7}{3}$, and square this to get $\frac{49}{9}$. Now rewrite $6(x^2 + \frac{14}{3}x - \frac{80}{6})$ as $6(x^2 + \frac{14}{3}x + \frac{49}{9} - \frac{49}{9} - \frac{80}{6})$. Regroup to get $6[(x^2 + \frac{14}{3}x + \frac{49}{9}) - \frac{49}{9} - \frac{80}{6}]$. Because $(x^2 + \frac{14}{3}x + \frac{49}{9}) = (x + \frac{7}{3})^2$, you now have

$$6x^2 + 28x - 80 = 6\left(x^2 + \tfrac{28}{6}x - \tfrac{80}{6}\right) = 6\left[(x + \tfrac{7}{3})^2 - \tfrac{49}{9} - \tfrac{80}{6}\right] = 6\left[(x + \tfrac{7}{3})^2 - \tfrac{169}{9}\right].$$

Having rewritten $6x^2 + 28x - 80$ as $6[(x + \frac{7}{3})^2 - \frac{169}{9}]$, you have completed the square for the quadratic polynomial $6x^2 + 28x - 80$. Notice that it is now easy to solve $6x^2 + 28x - 80 = 0$ for x. Divide $6[(x + \frac{7}{3})^2 - \frac{169}{9}] = 0$ by 6 to get $(x + \frac{7}{3})^2 - \frac{169}{9} = 0$. So $(x + \frac{7}{3})^2 = \frac{169}{9}$, and hence $x + \frac{7}{3} = \pm\sqrt{\frac{169}{9}} = \pm\frac{13}{3}$. Therefore, $x = -\frac{7}{3} \pm \frac{13}{3}$. So $x = 2$ or $x = -\frac{20}{3}$.

Problem 6. Repeat the steps above to complete the square for $4x^2 - 8x - 12$. Use the result to solve $4x^2 - 8x - 12 = 0$ for x.

Problem 7. Complete the square for the polynomial $-5x^2 + 3x + 4$. Then use the result to solve $-5x^2 + 3x + 4 = 0$ for x. Try the same thing for $-5x^2 + 3x - 4$. [Note: The solution of the equation $-5x^2 + 3x - 4$ requires square roots of negative numbers. Such *complex numbers* will not be considered in this text and we will regard such equations to have no solutions.]

Problem 8. Verify by completing the square that the solutions of $ax^2 + bx + c = 0$ (with $a \neq 0$) are given by the quadratic formula $x = \frac{-b \pm \sqrt{b^2 - 4ac}}{2a}$. What happens when $a = 0$?

Problem 9. Let x and d be any two positive numbers. Study the diagrams in Figure 1.14 and write a paragraph that discusses their connection with the completing the square procedure.

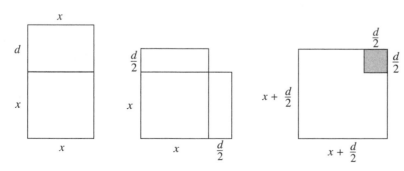

Figure 1.14

2
Greek Geometry and Roman Engineering

By the beginning of the last millennium B.C., the city-states of Greece started to flourish. These trading towns along the Aegean and southern Italian coasts were ruled by an independent, politically aware merchant class. Their growing trade made them wealthy and connected them with other lands on the shores of the Mediterranean and beyond. The new social order that they established spawned a new rational approach. Rather than the acceptance of "mythos," the account that gods and demi-gods controlled nature and unleashed its forces on a whim, there was a realization that observed phenomena operated in accordance with a rational "logos" that reason could begin to sort out and comprehend. Greek statesmen, philosophers, dramatists, sculptors, architects, and mathematicians approached and shaped reality with this mindset.

Greek architecture is the most visible manifestation of this spirit. The design and execution of Greek temples combines a focus on the aesthetics of the important architectural elements with attention to the composition of the structure as a whole. A column has a base, its shaft often has vertical grooves, and it tapers as it rises to meet the decorative element that tops it. A row of such columns holds sculpted horizontal components that carry a triangular section that in turn supports the roof structure. There are general conventions about the size and spacing of the parts and the proportional relationships between them. There is a logic and a discipline that is rooted in a sense of geometry. Some shrines feature circular arrangements of columns and delicately spiraling patterns of floor tiles. Large theaters are banked into hillsides, their seats arranged as ascending and widening semicircular configurations of heavy stone slabs.

From 600 B.C. until about A.D. 200, Greek geniuses, working in Greece and its colonies along the coast of the Mediterranean, laid the foundations of mathematics and science. Many of their answers "all matter is made up of the four basic constituents earth, air, water and fire" were wrong or incomplete, but the important fact is that when they asked "are there basic elements that combine to make up all matter?" they posed the right questions. Mathematics was regarded to be central to the understanding of the natural world and the design of the universe. From about 300 B.C. onward, Alexandria (in today's Egypt) with its great library was the center of this activity. Working in the Museum, a state-supported institute for advanced studies (a "house of the muses" for the arts and sciences), scholars investigated astronomy, mathematics, and medicine.

At about 300 B.C., Euclid put together the *Elements*, a comprehensive treatise that built on the mathematics of the Babylonians and the followers of Pythagoras. It is a tightly structured exposition, driven by logic and organized into 13 books. Ten fundamental statements are given a central position as axioms or postulates. Placed at the beginning, they could be examined critically at the outset. The rest is derived in the form of several hundred propositions about various aspects of plane geometry and properties of numbers. Today's mathematical theories still adhere to this basic structure. Archimedes (287–212 B.C.) was an extraordinary mathematician, a powerful physicist, and a legendary mechanical engineer. He understood the law of the lever, namely that the rotational effect of a force is given as the product of its magnitude times the distance to the axis of rotation. His law of hydrostatics (in Greek, *hydro* = water, *statikos* = at rest) told him that the force with which a liquid pushes up against a floating object is equal to the weight of the volume of the liquid displaced by the object. He calculated the centers of mass (or gravity) of planar regions and solids with methods that anticipated modern calculus. He also designed pulley systems and catapults for military purposes. Apollonius (262–190 B.C.) contributed the *Conic Sections*, a comprehensive study of the ellipse, parabola, and hyperbola. Claudius Ptolemy (around A.D. 150) built on the work of his Greek predecessors to develop a quantitative trigonometry and to devise an elaborate scheme of circles that describes how the Sun, Moon, and planets move from the vantage point of a fixed Earth. This was the accepted theory of the motion of the heavens until the discoveries of Copernicus, Galileo, and Kepler placed the Sun at the center of the universe and told us that the curves of Apollonius describe how the heavenly bodies move around it.

The Roman civilization flourished from about 600 B.C. to A.D. 400 side by side with the Greek. By the first century B.C., Rome had extended its empire to include the entire Mediterranean world. Roman, Greek, and Syrian engineers, armies of workers, and construction machinery driven by men and animals provided the infrastructure. Harbors were dredged, docks were constructed, swamps drained, an extensive network of durable roads and bridges was built, large heated public baths were erected, and underground sewers were dug. Aqueducts were constructed that brought water to the cities from distant springs tens of miles away. To maintain the elevation and grade that the even flow of water required, these channels cut through mountains and soared over valleys. In response to Roman enthusiasm for public spectacles, architects constructed theaters, amphitheaters, and stadiums. The architecture of Rome was greatly influenced by Greek principles and designs. Vertical columns, horizontal elements, and triangular components were common features. However, Roman architecture also made extensive use of curves, and spanned space with arches and vaults. The Pantheon, a temple with massive cylindrical walls, an expansive hemispherical dome, an elegant portico, and a classically ornamented interior, combines the power of Roman engineering with the aesthetic of Greek forms. The discovery of concrete was fundamental. Easily poured, molded, and shaped, it set to achieve the strength and resilience of stone. *The Ten Books of Architecture*, a treatise by the Roman architect Vitruvius from the first century B.C., is the only work on the architecture of classical antiquity that has come down to us. It is an important source of information about Greek and Roman design, methods of construction, and fundamentals of city planning.

Roman mathematics was limited to rudimentary arithmetic and practical geometry. The Romans were aware of the legendary Archimedes and his exploits. We know the story of Archimedes's "Eureka" moment as well as his method of discovery from one Vitruvius's ten books. (This famous tale recalls how Archimedes, sitting in his bath, realized that his law of hydrostatics solved the problem of the crown, and how, in celebration of his insight, he ran naked through the streets shouting "I have found it.") Given the scale and complexity of the public buildings and infrastructure that Roman engineers designed and executed, it is surprising that they did not appear to be interested in the potential value of either Greek geometry and trigonometry or the applied mathematics of Archimedes.

Greek Architecture

Athens, the city of democratic Greece, reached its great flowering between the years 500 and 350 B.C. This is the age of the great statesmen Themistocles and Pericles, the great thinkers Socrates, Plato, and Aristotle, the great dramatists Aristophanes, Sophocles, and Euripides, and the great sculptors Praxiteles and Phidias. The architecture of Athens was a visual expression of the city's greatness.

The Acropolis (in Greek, *acro* = high, *polis* = city) of Athens consists of a cluster of temples that rose on a rocky hill in the center of Athens in the fifth century B.C. It was the sacred place of Athena, the patroness of the city and the goddess of peace, wisdom, and the arts. The most important of the temples is the Parthenon. It was one of the largest built in classical Greece. As Plate 4 shows, the Parthenon dominates the hill. Built with the finest marble (a form of limestone) from a nearby quarry, it measures 110 feet by 250 feet with 8 columns at the front and rear and 17 columns on each side. A great marble statue of Athena, clad in ivory and gold, dominated the inner, sacred space of the temple. Its exterior is depicted in Figure 2.1. The structure is an example of the Doric order (an architectural style named after the Dorian Greeks who developed it). It features powerful columns that support marble slabs. Just above the slabs is a frieze, a row of horizontally arranged sculpted marble segments. The frieze supports a triangular section called pediment. The marble of the pediment was decorated with a relief of triumphant scenes from Greek mythology. Only a few of the images remain. Above the pediment was a tile roof supported by heavy timbers. The columns that bear the heavy loads consist of stacks of cylindrical sections, or drums, that are crafted with such precision that there are virtually no gaps between them. The seams between other marble components are executed with similar exactness. The columns are closely spaced. They are thicker at the bottom and taper gradually as they rise. Remarkably, the taper of the columns is guided by intentionally delicate and identical curves. Parallel vertical grooves called flutes enhance the appearance. The sharp ridges formed by successive grooves move gently toward each other as they follow the narrowing columns upward. The two columns at the ends of the front row are slightly thicker and more tightly spaced. This provides added strength at the corners of the structure. It also counters the thinner appearance that these columns have when light flows past them through the corners. The rectangular marble floor of the Parthenon is not flat. It is highest in the middle and slopes gently but visibly down to the sides. This means that the columns rest on a base that curves from a high point at the center to low points at the ends. Had the columns been set on this base without correction, they would lean outward. The architects compensated for this by making the lowest drums of the columns higher on one side (by about three inches for some drums) than on the other. In fact, they overcompensated to give the columns a slight

Figure 2.1. The Parthenon of Athens.
Photo by Onkel Tuca

inward lean. This provides better support for the heavy loads they carry. Historians who have studied the temple have suggested that the architects of the Parthenon introduced these gently curving and leaning elements to give the temple a less rigid and more dynamic look. Although reconstructing what was in the minds of the architects is impossible, the structure itself testifies to the fact that they proceeded with the same creativity and insistence on perfection as the Greek geometers.

The Acropolis has another impressive shrine, the Erechtheion, named after Erechtheos, an early king of Athens. It can be seen just to the left of the Parthenon in Plate 4. The roof of its portico, a porch structure featuring columns and usually attached to a building, is held up by graceful statues of six female figures. Historians tell us that these six caryatids bear the heavy load as symbolic punishment for the fact that the state of Caryae had supported the Persians in a war against the Greeks. The Erechtheion combines diverse styles and scales, including Ionic and Corinthian elements, into a wonderfully cohesive classical composition. The Ionic and Corinthian architectural orders, introduced by the Greeks of Ionia and Corinth, are ornate and more slender variations of the Doric.

The Acropolis hosted a yearly festival that celebrated Athena's mythological birthday. An eyewitness account tells of thousands of joyous Athenians, including women in saffron and purple robes, horsemen, warriors on chariots, and champion athletes, some carrying ceremonial torches and some silver trays with offerings, in ceremonial procession up to the hilltop. The architects of the Acropolis intended for the celebrating throng to experience an architectural drama of unfolding visual experience of its structures as it proceeded along the designated route. Similar spatial considerations influenced the layout of the Athenean agora (in Greek, *agora* = market), the civic and commercial heart of the city at the foot of the Acropolis. General conventions regulated the height, length, and width of a structure, as well as the spacing of columns and the proportions of the components to the whole. Careful attention was given to city planning and the spatial relationships between public and civic buildings, shrines and temples, monuments, fountains, and the long rectangular colonnaded stoas (large porticos used as a place to meet or promenade).

Two structures in Epidaurus (south of Athens on the Peloponnese peninsula) from the fouth century B.C. illustrate the connection between Greek architecture and geometry. The Tholos of Epidaurus was a round colonnaded structure of unknown purpose. (The Greek *tholos* refers to many different forms of such classical circular buildings.) Surviving fragments from the foundations and floor were uncovered and reassembled. The lower half of Figure 2.2 shows the intricate geometric pattern of the tiles of the floor (and the upper half depicts parts of the ceiling). The two circular configurations of dark discs represent the positions of columns. The outer circle had a diameter of about 72 feet. The diameter of the round wall inside it was about

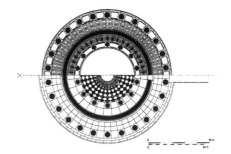

Figure 2.2. The Tholos of Epidaurus, Greece

Figure 2.3. The Theater of Epidaurus, Greece. Photo by Olecorre

45 feet. The Theater of Epidaurus, constructed in 360 B.C. with a seating capacity of 14,000, is depicted in Figure 2.3. The stone blocks that serve as the seats were set into a hillside in an expanding and rising pattern of semi-circles. The largest of these has a radius of about 200 feet. To this day, the theater provides a dramatic setting for a summer program of performances of ancient Greek plays.

Greek builders handled the challenge of moving and placing the heavy components of their structures in several ways. They moved round segments of columns by inserting iron pins at their ends and using pack animals to roll them along. They moved rectangular slabs in a similar way after constructing heavy wooden wheels around their ends. Heavy slabs were dragged up ramps made of sand or loose earth that were removed after the slabs were in position. Later, the Greeks developed cranes with systems of ropes, drums, pulleys, and winches that could hoist and position heavy items. These deployed mechanical devices that were developed for the machinery of war that was used to lay siege to walled cities.

The effect of centuries of intermittent wars and indifference of civic institutions brought damage and neglect to the splendid Greek structures and the Parthenon in particular. At the time of the conversion of the Parthenon to a Christian church in the fifth century, the statue of Athena and the rows of columns of the interior were removed. After the fall of Athens to the Turks in the fifteenth century, the Parthenon was turned into a mosque. During their war with the Venetians in the seventeenth century, the Turks used it to store munitions. Gunfire from the Venetian fleet caused an explosion that ripped out the interior walls. Toward the end of the Turkish occupation at the beginning of the nineteenth century, the British Lord Elgin obtained authorization from the Turks to purchase some of the friezes. Cut away and shipped to England, these exquisite works of art can now be admired in the British Museum in London. Traces of original paint discovered in protected corners tell us that the friezes were painted in vibrant colors and had an appearance quite different from the elegance in white marble that we experience today. What humans and time did not destroy is now being threatened by automobile exhaust that damages the marble surfaces. Extensive efforts are being made to repair and save the structure. What remains of the Parthenon is but a shadow of its former self and yet it continues to remind us that Western civilization has its roots in ancient Greece. The many buildings in the style of the Parthenon that exist in today's cities give it the distinction of being the most imitated and admired temple of antiquity.

Gods of Geometry

Basic geometric forms are evident everywhere in Greek architecture. In the construction of their buildings the Greeks tied ropes to pegs and stretched and rotated them to lay out the straight lines and circular arcs of their designs. Greek geometers abstracted this practice and developed it into the study of straightedge and compass constructions. This study is laid out in Euclid's *Elements*. Let's begin with a look at two very basic constructions from Book I.

Start with Figure 2.4a. Use a straightedge (a ruler without the markings that allow lengths to be measured) to draw a line segment *AB*. Place the point of a compass at *A*, stretch it to *B*, and draw a circular arc as shown. Then draw a circular arc with the same radius with center *B*. Let *C* be the point of intersection of the two arcs. So the equilateral triangle *ABC* has been constructed with straightedge and compass. Why are the angles at *A*, *B*, and *C* all equal to 60°? (The answer is a consequence of the conclusions of Problems 1 and 2 at the end of the chapter.) Now turn to Figure 2.4b. Let *AOB* be any angle. Place a compass at *O*, draw a circular arc through the angle, and let *C* and *D* be the points at which the arc intersects the segments *OA* and *OB*. Then place the point of the compass respectively at *C* and *D* and draw two circular arcs of the same radius. Let *E* be the point of intersection of the two arcs. The line from *O* through *E* cuts the angle *AOB* in half. (See Problem 3.) So the angle *AOB* has been bisected with straightedge and compass.

In Book VI Euclid explains that "a straight line is cut in 'extreme and mean ratio' when the whole line is to the greater segment as the greater is to the less." Let's have an algebraic look at what Euclid is saying. Figure 2.5 shows a line segment that has been cut into two pieces. Their lengths are *a* and *b* with *a* ≥ *b*, so that the length of the segment is *a* + *b*. Euclid says that

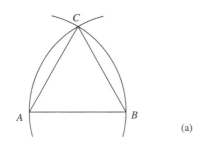

(a)

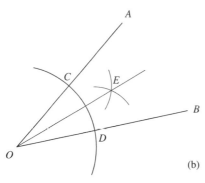

(b)

Figure 2.4

Figure 2.5

the segment is cut in "extreme and mean ratio" if $a + b$ is to a as a is to b, therefore, if $\frac{a+b}{a} = \frac{a}{b}$. Observe that if this is so, then

$$\frac{a}{b} = 1 + \frac{b}{a} = \left(\frac{a}{b}\right)^{-1} + 1.$$

Multiplying this equation through by $\frac{a}{b}$ shows that $\frac{a}{b}$ is a root of the polynomial $x^2 - x - 1$. So by the quadratic formula, $\frac{a}{b} = \frac{1 \pm \sqrt{1+4}}{2} = \frac{1 \pm \sqrt{5}}{2}$. Since $\frac{a}{b}$ is positive, we can conclude that $\frac{a}{b} = \frac{1+\sqrt{5}}{2}$.

In today's terminology, such a cut is a golden cut or golden section of the segment and the ratio $\frac{a}{b} = \frac{1+\sqrt{5}}{2}$ is the golden ratio. (This ratio would also become known as the divine proportion.) It is common practice in mathematics to denote the golden ratio by the Greek letter phi. This is often claimed to be a tribute to the great Greek sculptor Phidias, who is said to have made use of it. There certainly is something intrinsically compelling about a proportion that is specified by the requirement that the whole is to the larger part as the larger part is to the smaller.

We will follow tradition and use a lowercase Greek ϕ to denote the golden ratio. So

$$\phi = \frac{1+\sqrt{5}}{2}.$$

Because ϕ is a root of the polynomial $x^2 - x - 1$,

$$\phi^2 = 1 + \phi \quad \text{and} \quad \phi^{-1} = \phi - 1 = \frac{\sqrt{5}-1}{2}.$$

A calculator tells us that $\phi = \frac{1+\sqrt{5}}{2} = 1.618033989\ldots$

If the lengths of the sides of a rectangle are in golden ratio to each other, then the rectangle is a golden rectangle. A golden rectangle is constructed with a straightedge and compass as shown in Figure 2.6. Let b be any length and start by constructing a square with b the length of the sides. The required 90° angles are obtained by bisecting 180° (given by a straight line and a point on it) with the procedure described in Figure 2.4b. Now construct the midpoint of the base of the square, place the sharp point of your compass there, and stretch the leg of the compass along the dashed line to the upper corner of the square. An application of the Pythagorean Theorem shows that this dashed line has length $\sqrt{\frac{b^2}{4} + b^2} = b\frac{\sqrt{5}}{2}$. Swing the circular arc downward until it intersects the extension of the base of the square. Take the segment that this point of intersection determines and complete it to a rectangle as shown in the figure. Let a be the base of this rectangle and notice that $a = \frac{b}{2} + \frac{\sqrt{5}}{2}b = b\frac{(1+\sqrt{5})}{2}$. Because

$$\frac{a}{b} = \frac{(1+\sqrt{5})}{2} = \phi,$$

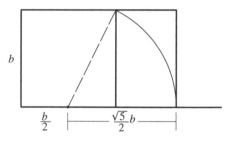

Figure 2.6

this rectangle is golden. Consider the base of the golden rectangle. The right side of the square cuts it into a longer segment of length b and a shorter segment of length $a - b$. Because

$$\left(\frac{b}{a-b}\right)^{-1} = \frac{a-b}{b} = \frac{a}{b} - 1 = \phi - 1 = \phi^{-1},$$

it follows that $\frac{b}{a-b} = \phi$. So this cut is a golden cut of the base of the rectangle.

This construction also informs us where to "place the knife" for a golden cut of any segment. Let AB be a segment and attach the construction of Figure 2.6 to it as shown in Figure 2.7. Draw the line from the right endpoint D of the base of the golden rectangle to the right endpoint B of the segment. Construct a parallel to this line through the point P. (How this is done is pursued in Problem 6.) This parallel line determines a point C on the segment. By a basic property of similar triangles (reviewed in Problem 16 and the preamble to the problem), $\frac{AB}{AC} = \frac{a}{b} = \phi$. Because $\frac{CB}{AC} = \frac{AB-AC}{AC} = \phi - 1 = \phi^{-1}$, $\frac{AC}{CB} = \phi$. Therefore $\frac{AB}{AC} = \frac{AC}{CB}$, and the cut at C is golden.

The constructions illustrated above can all be carried out with only a compass for drawing circular arcs—this abstracts what a peg and a string can accomplish—and a straightedge for connecting points—this is analogous to stretching a string between two pegs. The "only" is the requirement that needs to be met for any straightedge and compass construction.

A polygon consists of points in the plane that are connected by straight line segments in such a way that the collection of line segments forms a closed loop. Figure 2.8a provides a typical example. The points V_1, V_2, V_3, etc., are called vertices and the line segments connecting them are referred to as edges. Because it has 10 vertices the polygon of Figure 2.8a is called a 10-gon. If a polygon has n vertices (or edges), it is an n-gon. If all of the connecting edges have the same length and all the interior angles $\alpha_1, \alpha_2, \alpha_3$, etc., are equal, then the polygon is called regular. Figure 2.8b shows a regular 7-gon. It is obtained by spacing the points $V_1, V_2, \ldots, V_7$ evenly around a circle. Because the circle is divided into 7 equal sections, each of the angles at the center has $\frac{360}{7}$ degrees.

We turn to the question as to which regular polygons can be constructed with a straightedge and compass. Figure 2.4a told us how an equilateral triangle can be constructed. The construction of the golden rectangle began with the construction of a square. Therefore, the regular 3-gon and the regular 4-gon can be constructed. The most direct approach to the construction of a regular n-gon is the construction of the angle $\left(\frac{360}{n}\right)^\circ$. If this angle can be constructed, it can be used to mark off n equally spaced points on a circle. Connecting consecutive points on the circle with a line segment completes the construction of the regular n-gon. For example, the construction of the equilateral triangle provides the construction of a 60° angle. Marking off 60° angles consecutively provides the points H_1, H_2, H_3, H_4, H_5, and H_6 in

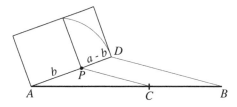

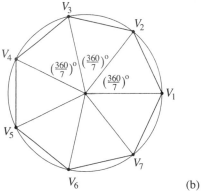

Figure 2.7

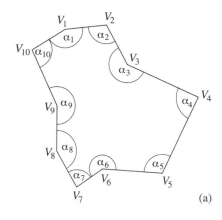

Figure 2.8

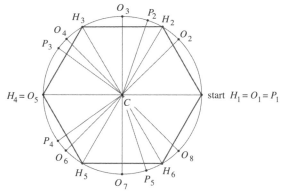

Figure 2.9

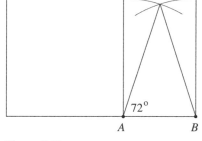

Figure 2.10

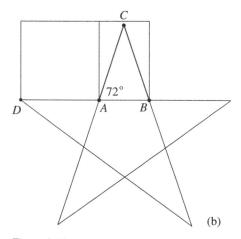

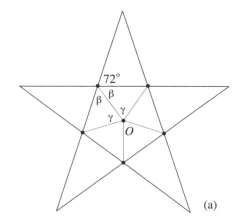

Figure 2.11

Figure 2.9 and hence the construction of a regular 6-gon, or hexagon. The bisection of a 90° angle constructs a 45° angle. Marking off the angle 45° repeatedly determines the points $O_1, O_2, \ldots, O_8$ of Figure 2.9 and hence the construction of a regular 8-gon, or octagon.

The regular 5-gon, or pentagon, can be constructed as well. This can be done by starting with the construction of the golden rectangle of Figure 2.6. In Figure 2.10 two circular arcs of radius b are drawn with centers at A and B respectively. The point C is their point of intersection. It turns out that $\angle CAB = 72°$. Therefore, the angle 72° can be constructed. Marking this angle off five times provides the points P_1, P_2, P_3, P_4, and P_5 of Figure 2.9. Because $5 \cdot 72 = 360$, the construction of the regular pentagon is complete.

But how do we know that $\angle CAB = 72°$? The details of the answer are complicated but the essence is this. Start with the regular pentagon at the center of Figure 2.11a. (It can't be assumed, and it isn't, that this pentagon can be constructed, else you'd be assuming what needs to be verified.) The point O is the center of the circle on which the five vertices of the pentagon lie. The interior angle γ is equal to 72° because $5\gamma = 360°$. A basic property of isosceles triangles tells us that $\beta = 54°$. So $2\beta = 108°$. Now complete the pentagon to the five-pointed star. Turn to Figure 2.11b and label the points A, B, C, and D as indicated. Notice that the angle $CAB = 72°$. Now comes the hard part. This is to show that the rectangle obtained by constructing a square over DA and extending it by using the segment AB is golden! Discussion 2.2 describes in detail how Euclid does this in his *Elements*. The rest is easy. The fact that the segments DA and AC have the same length tells us that the triangles ABC in Figures 2.10 and 2.11b are obtained in the same way. Therefore, the angle CAB in Figure 2.10 is indeed equal to 72°.

We have now seen that the regular polygons of sides 3, 4, 5, 6, and 8 can all be constructed with straightedge and compass. This fact raises an interesting but very difficult question: for which positive integers n can a regular polygon of n sides be constructed with straightedge and compass? It took mathematicians until the nineteenth century to make significant progress toward the answer. Only after the geometric question was converted into a question of higher abstract algebra (and in particular to one about roots of polynomials with coefficients in the rational numbers) was it completely solved. It turns out, for example, that of the regular polygons with sides ranging from 3 to 1002, in other words the first thousand such polygons, those with sides numbering

> 3, 4, 5, 6, 8, 10, 12, 15, 16, 17, 20, 24, 30, 32, 34, 40, 48, 51, 60, 64, 68, 80, 85, 96, 102, 120, 128, 136, 160, 170, 192, 204, 240, 255, 256, 257, 272, 320, 340, 384, 408, 480, 510, 512, 514, 544, 640, 680, 768, 771, 816, 960

can be constructed but *none* of the others can be. A count shows that only 52 of these 1000 regular polygons are constructible. Because 7 is not on the list, the regular 7-gon of Figure 2.8b cannot be constructed. Because 9 is not on the list, the regular 9-gon cannot be constructed. Notice the widening gaps in this table of numbers. These and related facts are far beyond the scope and purpose of this text, but Discussion 2.3 does give a glimpse at them.

As has been noted, Greek building practices clearly influenced the direction of Greek geometry. In turn, the geometry of the Greeks—triangles, semicircles, circles—finds expression in their great structures. But how deep did this go? Did Greek architects attempt to adhere to precise mathematical ratios or strict geometric relationships in their architectural designs or their execution of them? The Parthenon is often used as an example to show that they did. The evidence is provided by diagrams such as the one in Figure 2.12. Does the array of golden rectangles superimposed on the facade of the Parthenon tell us that the Greeks used the golden rectangle as a template? Consider how the squares and rectangles are chosen. How relevant is the choice, how is their placement determined, and how good is the fit? Is this solid evidence that the golden rectangle was used intentionally to shape the design? Or were the golden rectangles simply imposed on the facade after the fact? The golden ratio is something precise, either as number or by construction. In particular, it is not equal to $\frac{3}{2} = 1.50$ or $\frac{5}{3} \approx 1.67$, nor is it equal to $\frac{8}{5} = 1.60$. There is no conclusive evidence that Greek builders followed precise geometric relationships in their architectural designs or their execution. In fact, the plan of the Tholos of Epidaurus in Figure 2.2 suggests that they did not. Notice that this plan features an inner circular arrangement of 14 columns and an outer circular arrangement of 26 columns. The centers of the circles that mark the positions of these columns are the respective vertices of a 14-gon and a 26-gon. Because neither 14 nor 26 appears on the earlier table of numbers, we know today that it is impossible to construct either a 14-sided or a 26-sided regular polygon. If exact execution had been important to the Greeks, would they not instead have made use of regular 12-gons

Figure 2.12. Parthenon with the golden ratio. Photo by Padfield

and 24-gons in the configuration of these circular arrays of columns? They knew how to lay these out precisely with pegs and ropes by starting with the regular hexagon and doubling the number of vertices twice.

Mathematical systems of proportion were developed and used beginning in the fifteenth century by the architects of the Renaissance. They codified, in terms of fixed numerical ratios, the relationships between the various dimensions of the components of their buildings, including the thickness and length of columns, the bases that supported them, and in turn the elements that the columns supported. This is taken up in the section "Alberti, Music, and Architecture" in Chapter 5 and in Discussion 5.1.

Measuring Triangles

The ingenious Greeks were able to squeeze a lot of information from the study of basic triangles, a study that they called trigonometry (in Greek, *trigono* = triangle, *metrein* = to measure). Let's suppose that an inquisitive Greek traveler to Egypt comes upon a pyramid. He is interested in its size and determines the dimensions of its base by pacing off its sides. He then turns his attention to the height of the pyramid. On this bright sunny afternoon, he paces off the distance from the side of the pyramid to the tip of the shadow that it casts and estimates that the tip of the shadow is 310 paces from

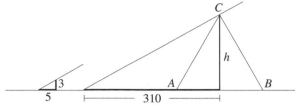
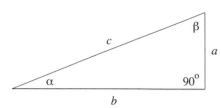

Figure 2.13 Figure 2.14

the center of the pyramid's base. Our traveler knows that he is 3 paces tall and measures the length of his shadow as 5 paces. Organizing the information he has, he draws the diagram of Figure 2.13 in the sand. The triangle *ABC* represents the pyramid and *h* its height. The two slanted lines depict the light rays from the sun that determine the shadows. Our Greek traveler knows about similar triangles and sees that $\frac{h}{310} = \frac{3}{5}$. So he concludes that the height of the pyramid is approximately equal to $h = \frac{3 \cdot 310}{5} = 186$ paces. This traveler could have been the same Herodotos who described the splendors of Babylon. His chronicles recall his visit to the great pyramid of Khufu (see Figure 1.6) in the fifth century B.C. and report some of its dimensions.

Figure 2.14 considers a right triangle with its acute angles α and β. The Greeks had their own terminology and notation, but they understood the relevance of the ratios sine, cosine, and tangent, that are today defined by $\sin\alpha = \frac{a}{c}$, $\cos\alpha = \frac{b}{c}$, and $\tan\alpha = \frac{a}{b}$. They developed standard trigonometric identities. For instance, because $\sin\beta = \frac{b}{c}$, $\cos\beta = \frac{a}{c}$, and $\tan\beta = \frac{b}{a}$ and $\beta = 90° - \alpha$, they knew that

$$\sin(90° - \alpha) = \cos\alpha, \quad \cos(90° - \alpha) = \sin\alpha, \quad \text{and} \quad \tan(90° - \alpha) = \frac{1}{\tan\alpha}.$$

They also noticed that

$$(\sin\alpha)^2 + (\cos\alpha)^2 = \frac{a^2 + b^2}{c^2} = 1$$

follows directly from the Pythagorean Theorem.

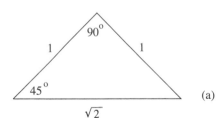

(a)

Consider the two triangles in Figure 2.15. By the Pythagorean Theorem, the base of the triangle in Figure 2.15a is $\sqrt{2}$. It follows that $\sin 45° = \cos 45° = \frac{1}{\sqrt{2}}$. Again by the Pythagorean Theorem, the height h of the equilateral triangle in Figure 2.15b satisfies $1^2 = h^2 + (\frac{1}{2})^2$. So $h^2 = \frac{3}{4}$, and $h = \frac{\sqrt{3}}{2}$. It follows that $\sin 30° = \cos 60° = \frac{1}{2}$ and $\sin 60° = \cos 30° = \frac{\sqrt{3}}{2}$.

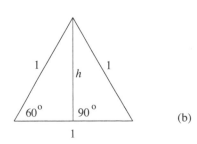

(b)

Let a triangle be given and let α, β, and γ be its angles. If $\beta + \gamma < 90°$, then $\alpha > 90°$. But the definition of the sine, cosine, and tangent that Figure 2.14 provides does not apply to such an angle. To give meaning to the sine, cosine, and tangent of any angle α with $0° \leq \alpha \leq 180°$ proceed as follows. Let α be any such angle and place it into a semicircle of radius 1. If $0° \leq \alpha < 90°$, proceed

Figure 2.15

as shown in Figure 2.16a and define $\sin\alpha = h$ and $\cos\alpha = b$. This agrees with what we already know. If $90° \le \alpha \le 180°$, proceed as in Figure 2.16b and define $\sin\alpha$ and $\cos\alpha$ by

$$\sin\alpha = h \quad \text{and} \quad \cos\alpha = -b.$$

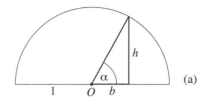

The tangent is given by $\tan\alpha = \frac{\sin\alpha}{\cos\alpha}$ in either case. Let α satisfy $0° \le \alpha < 90°$. So $180° - \alpha > 90°$. Consider Figure 2.16a for α and Figure 2.16b for $180° - \alpha$. Study the two diagrams and convince yourself that the two right triangles that arise are similar and hence that

$$\sin(180° - \alpha) = \sin\alpha \quad \text{and} \quad \cos(180° - \alpha) = -\cos\alpha.$$

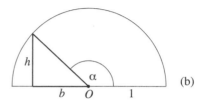

Figure 2.16

It follows, for example, that $\sin 120° = \sin(180° - 60°) = \sin 60° = \frac{\sqrt{3}}{2}$ and that $\cos 120° = \cos(180° - 60°) = -\cos 60° = -\frac{1}{2}$.

Greek contributions to mathematics are nothing short of astonishing. They axiomatized geometry; that is, they presented it as a mathematical structure that starts with a few central definitions and statements and sets out everything else in a cohesive, rigorous, logical way. They studied the ellipse, parabola, and hyperbola, and developed the basic properties of these curves with approaches that are closely linked to coordinate geometry and modern calculus (as we will see in Chapter 4, "Remarkable Curves and Remarkable Maps"). They also developed a practical astronomy. By making daily shadow measurements (similar to those made by our Greek traveler), the Greeks determined the days of the summer and winter solstices and measured the year to have $365\frac{1}{4}$ days. By stretching imagined triangles between the Earth, Moon, and Sun and applying their trigonometry, they gained a sense of the sizes of these bodies and the distances between them. Modern mathematics and astronomy rose from Greek foundations.

Dealing with Forces

It is time to turn to the analysis of the forces on and within a structure. Think of a force as the kind of push or pull encountered in everyday experience, and in particular, a push or pull that one component of an architectural structure exerts on another. Every architectural structure is subject to forces. The force of gravity—in the form of the weight of the elements of a building—is an ever present and central factor, but forces generated by wind, heat, and earthquakes can be of major consequence as well. The architects of antiquity were aware of forces and had an intuitive understanding of their effects, but they were not able to deal with them in the explicit and quantitative way that this section describes. The conceptual challenges that the analysis of forces involves were beyond the reach of Greek builders and Roman engineers.

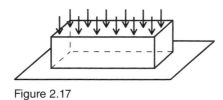

Figure 2.17

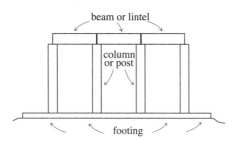

Figure 2.18. Column and beam or post and lintel construction with footing

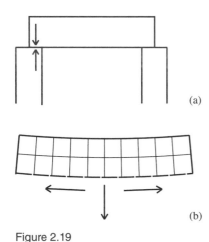

(a)

(b)

Figure 2.19

Return to Figure 1.6. Consider a typical stone block from any of these pyramids. The collective weight of the blocks above it push down on the upper face of this block and put the block under compression as indicated in Figure 2.17. This compression is resisted by the internal structure of the stone. Such internal forces that counter the loads on a structure are called reactions. The bottom row of blocks of a pyramid puts the underlying ground under pressure. Any shifting in the structure that results is called settlement. Uneven settlement can lead to critical dislocations and failure of the structure. This did not occur for the pyramids of Figure 1.6 because they were built on a foundation of natural limestone. Such rock can support loads of 100 tons per square foot. However, other Egyptian pyramids settled unevenly during and after construction and suffered major damage.

A basic structural configuration used by the Egyptians and Greeks in their temples is the column and beam, or post and lintel. This consists of two vertical columns (or posts) that support a horizontal beam (or lintel). See Figure 2.18. This basic structural device gave us the Parthenon and other splendid structures of Greece and Egypt. Stonehenge (depicted in Figure 1.5) provides an earlier example. Beam and column constructions have limitations, especially when executed in stone. There is generally no problem with the columns. The ground on which they rest can be reinforced by stone or masonry footings that distribute the load evenly to the subsoil. The downward push of the beam on the column is matched by the upward reaction of the column on the beam. See Figure 2.19a. The weight of the beam puts the column under compression, but stone can withstand large compressional forces. But what about the beam between the columns? There is nothing to oppose the weight of the beam except the internal resistance of the stone. The beam is pulled downward by its own weight, even if only very slightly, and in the process, its bottom edge is stretched (and its top edge compressed) as indicated Figure 2.19b. Such a stretching force is a tensile force and the component on which it acts is in tension. While stone can resist compression extremely well, it is much less able to deal with tension. In other words, much more force is required to crush a stone slab than to pull it apart. (The collapse of the torso of one of the pharaohs depicted in Figure 1.9 provides some evidence for this fact.) If the supporting columns are too far apart, then the tension on the bottom edge of the beam will be too great for the stone to resist. The stone will crack and the beam will fail. Therefore, columns and beams in stone can support large and heavy roofed structures only if the columns are placed in close proximity to each other (as in Plate 3). It follows that the construction of large clear spans is not possible with columns and beams. We will soon see that the Romans changed all that with the deployment of the arch. But before we discuss arches, let's refine the way we think about forces.

A force has both a magnitude and a direction. The magnitude and the direction together determine the force. In structural studies, forces are represented by arrows called vectors. When a vector is used to represent a force, the vector points in the direction of the force. See Figures 2.17 and 2.19, for example. To represent the magnitude of a force, both a unit of length (the inch, foot, or meter, for example) and a unit of force (the pound or ton, for instance) need to be given. (Incidentally, the common abbreviation "lb" for pound has its origin in *libra*, the name of an ancient Roman unit of weight.) A force of *x* units in magnitude is represented by a vector of *x* units in length. So the magnitude of the force is *numerically equal to* the length of the vector representing it. For instance, in Figure 2.20a, the vector *A* represents a force of 1000 pounds acting horizontally and to the left. Similarly, the vector *B* in Figure 2.20b represents a force of 10 pounds acting horizontally and to the right. Longer and shorter vectors represent forces of larger or smaller magnitudes in a proportional way. Suppose that the beam in Figure 2.19a pushes down on the column with a force of 2000 pounds and that the column reacts with an upward force of the same magnitude. The two vertical vectors depicting this in Figure 2.20a take both directions and magnitudes into account. Figure 2.20b shows various situations of forces of different magnitudes. For instance, the vertical vector might represent the gravitational force exerted on the ground by a woman weighing 120 pounds. The second diagram in this group might involve an object weighing 50 pounds that is subject to a diagonally upward force of 30 pounds. Can you think of a situation that might be represented by the circular situation of 5-pound forces pushing against a point? The directions of the forces that we will discuss will be apparent from the context, so that we will use the capital letters F, F_1, F_2, or P, P_1, P_2 primarily to denote the magnitudes of forces. So when a force F is said to act in this or that way, then the direction of the force will be apparent and F will usually refer to its magnitude.

Consider two forces with magnitudes F_1 and F_2 acting at the same point. What can be said about their combined effect? If the two forces act in the same direction, then this is a force with magnitude $F_1 + F_2$ in that direction. If they act in directly opposed directions, then their combined effect is a force acting in the direction of the larger force. If, say, F_1 is larger, then the magnitude of the combined force will be $F_1 - F_2$. In general, the combination of two forces is determined as follows. Position the vectors that represent the two forces in such a way that their initial points coincide. As Figure 2.21a shows, this configuration determines a parallelogram and the diagonal of the parallelogram from the common initial point determines a vector. This vector represents a force, and this force—both in direction and in magnitude—is the combined effect, or resultant, of the two given forces. This fact is the *Parallelogram Law of Forces*. Figures 2.21b and 2.21c show that the resultant is also obtained by placing the two vectors end to end. Figure 2.21d provides a numerical example. The fact that the vectors of magnitudes 40 and 75 are perpendicular means that the magnitude of the

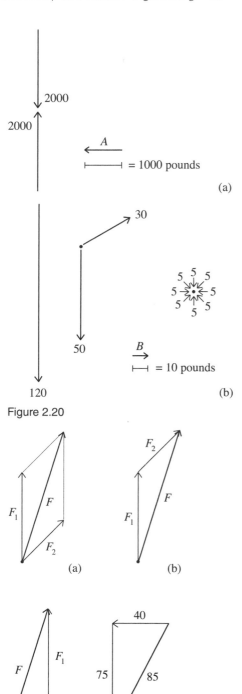

Figure 2.20

Figure 2.21

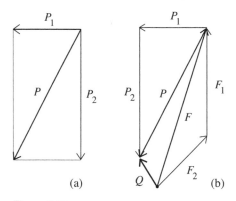

(a)

Figure 2.22

resultant is given by the Pythagorean Theorem as $\sqrt{40^2 + 75^2} = \sqrt{7225} = 85$ pounds.

If several forces act at the same point, then their resultant is determined by applying the parallelogram law to two forces at a time. Take the four forces F_1, F_2 and P_1, P_2 considered in Figures 2.21a and 2.22a. The resultant of the first pair is the vector F in Figure 2.21a and that of the second pair is the vector P in Figure 2.22a. By the parallelogram law, the resultant of F and P is the vector Q in Figure 2.22b. The resultant of any number of vectors can also be obtained by placing them end to tip in any order. The perimeter of the diagram of Figure 2.22b shows how the resultant Q of F_1, F_2, P_1, and P_2 is obtained in this way.

The important fact is that the magnitude of the resultant is always numerically equal to the length of the vector that represents it. This means that the representation of forces by vectors is much more than a convenient way to think about forces: it provides a fundamental insight into the way forces act.

Let a force F be given. Figure 2.23a specifies a direction with a dotted line and denotes by θ with $0° \le \theta \le 90°$ the angle between the dotted line and the direction of the force. Figure 2.23b drops a perpendicular from the tip of the force vector F down to the dotted line and puts in the vector that the perpendicular determines. This vector represents a force called the component of F in the direction θ. If l is the length of this vector, then $\cos\theta = \frac{l}{F}$. It follows that the magnitude of the component of F in the direction θ is $F \cos \theta$. To remember that the magnitude of the component is given by the cosine (rather than the sine), think of the fact that the component is, in a way, "adjacent" to the original force. In the same way, the component of the force in the direction $90° - \theta$ has magnitude $F \cos(90° - \theta)$. By a basic trigonometric identity of the earlier section "Measuring Triangles," this is also equal to $F \sin \theta$. Figure 2.23c tells us that these two components of F determine a rectangle that has F on its diagonal. By the parallelogram law the resultant of the two components is equal to the original force.

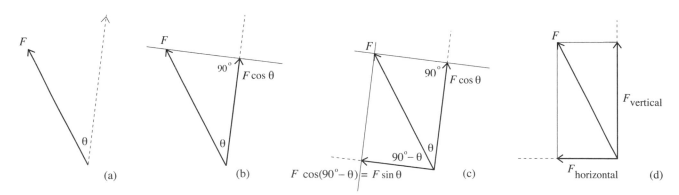

Figure 2.23

The Decomposition of a Force into Components. Let a force with magnitude F and a direction θ with $0° \le \theta \le 90°$ be given. Then the component of F in the direction θ and the component of F in the direction perpendicular to θ have magnitudes

$$F\cos\theta \quad \text{and} \quad F\cos(90° - \theta) = F\sin\theta$$

respectively, and the resultant of these two components is the original force. Figure 2.23d illustrates this decomposition in the most important instance: the situation where the components are the vertical and horizontal components of F.

A pyramid is structurally stable only if each one of its stone blocks pushes back against the weight pressing down on it with a force of the same magnitude as this weight. Similarly, for a column and beam combination to be sound, the column needs to push up with a force equal to the load that pushes down on it, and the internal structure of the material of the beam has to resist the force with which gravity pulls on it. These examples are special cases of a general principle that we call the *First Principle of Structural Architecture*. For a structure and the components of a structure to be stable, *the combined effect of the forces acting at every point of the structure* must be zero. If this is not the case, then there is a nonzero net force acting on some point and this point will move. Consequently, there is movement in the structure and the structure may fail. Even if there is only the slightest imbalance, the excess force will move things. The forces in question are not only external forces (such as gravity) but internal forces (such as reactions and compressions) of the materials of the structure. This "stability only if nothing moves" principle is actually too restrictive. After all, buildings move due to the action of wind, heat, and earthquakes and remain intact. Skyscrapers are designed to respond to strong winds by pushing back in such a way that a controlled equilibrium of movement is achieved. Structures expand and contract in response to changing temperatures without suffering damage. And newer buildings are generally engineered to roll with the action of earthquakes without failing. However, given its focus on rigid masonry buildings, this text regards such effects to be secondary and makes important use of the "stability only if zero net force at every point" principle in its analyses of architectural structures. The vertical downward force of gravity is fundamental. It is for this reason that the resolution of forces into their horizontal and vertical components is a central strategy in such analyses.

Let's close this section with an illustration of the First Principle of Structural Architecture. A ladder at a construction site is supported by a horizontal floor and leans against a rigid vertical wall at an angle of 60°. A worker has reached the top of the ladder. His weight, that of the supplies that he carries, and the weight of the ladder add to 240 pounds. Figure 2.24a depicts a cross section of the ladder. The total load of 240 pounds is regarded to act vertically downward at the point A. Also regarded to be acting at A is the reactive

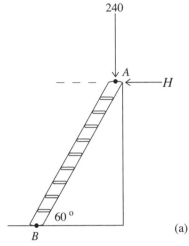

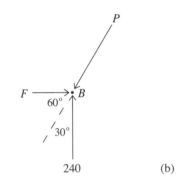

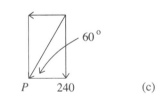

Figure 2.24

force of the vertical wall against the ladder. The vertical wall is assumed to be very smooth, so that the upward frictional force of the wall on the ladder is negligible. Therefore the reaction H is horizontal and the entire vertical load of 240 pounds is supported at the base B. Figure 2.24b represents the three forces acting at the point B. They are the reactive upward force of the floor, the downward slanting push P of the ladder, and the force of friction F necessary to keep the ladder from sliding. We will assume that the system of the ladder and the forces acting on it is stable. It follows from the First Principle of Structural Architecture that the three forces acting at B are in balance. Therefore the vertical component of the push P is equal to 240 pounds and the magnitudes of F and the horizontal component of P are equal. Figure 2.24c informs us that $P\cos 30° = 240$ and $P\cos 60° = F$. Because $\cos 30° = \frac{\sqrt{3}}{2}$ and $\cos 60° = \frac{1}{2}$, it follows that

$$P = 240 \cdot \frac{2}{\sqrt{3}} = \frac{480}{\sqrt{3}} \approx 277.1 \text{ pounds} \quad \text{and} \quad F = \frac{480}{\sqrt{3}} \cdot \frac{1}{2} = \frac{240}{\sqrt{3}} \approx 138.6 \text{ pounds.}$$

Notice that the slanting push P of the ladder at B is larger than the load of 240 pounds that the ladder supports. This does not seem possible. Shouldn't this slanting push be simply equal to the component $240\cos 30°$ of the load on the ladder in the direction of the slant? Not quite, because the component $H\cos 60°$ of the reaction of the wall down along the ladder needs to be considered as well. Adding these two components tells us that $P = 240\cos 30° + H\cos 60°$. From this we can compute H, getting $\frac{1}{2}H = \frac{480}{\sqrt{3}} - 240 \cdot \frac{\sqrt{3}}{2}$, and hence

$$H = \frac{960}{\sqrt{3}} - 240 \cdot \sqrt{3}\frac{\sqrt{3}}{\sqrt{3}} = \frac{960}{\sqrt{3}} - \frac{720}{\sqrt{3}} = \frac{240}{\sqrt{3}}.$$

Notice therefore that $H = F$. This is not surprising. One would expect the two horizontal forces F and H on the ladder to be in balance. A similar analysis of the forces at the point A leads to the same results about P, H, and F.

Figure 2.24b is an abstract representation of the point of contact of the ladder with the floor together with all the forces acting at that point. Because this point has been "freed" from all the other parts of the structure, such a force diagram is called a *free-body diagram*.

The fact that a horizontal frictional force is necessary to keep the ladder from sliding out at the base illustrates a very important general point: Unless the horizontal component of the push of a slanting element of a structure is counteracted by the structure as a whole, the structure will not be stable.

The Roman Arch

Roman builders made principal use of the stone arch. Figure 2.25 shows an ornate example. The arch and related forms, the vault and dome, had been used by other civilizations, but in Roman hands they became the basic element of a new type of construction. In taking advantage of the capacity of stone to resist compression, the arch makes it possible for large spaces to be spanned. Semicircular arches were especially easy to lay out and the Romans used them extensively to build their bridges, aqueducts, arenas, temples, and villas.

Consider a typical semicircular arch in the abstract. Its shape and the wedge-shaped pieces that comprise it are determined by the upper halves of two concentric circles and a number of their radii as shown in Figure 2.26a. A wedge is called a voussoir (a French word with the Latin root *volvere* = to turn). The voussoir at the top is the keystone. The imposts are the elements between the arch and the supporting columns. The two springing points are the points on the inner surface from which the arch begins its rise. The springing line is the line that these points determine. This is usually the horizontal diameter of the inner semicircle of the arch. These elements along with the span of the arch are illustrated in Figure 2.26b.

Figure 2.27a illustrates how the Romans built arches. A wooden support structure called centering kept the voussoirs in place during construction. With the keystone in place, the other voussoirs can no longer fall inward, are locked into place, and the centering could be removed. The gravitational forces that the weights of the voussoirs and the structure that the arch

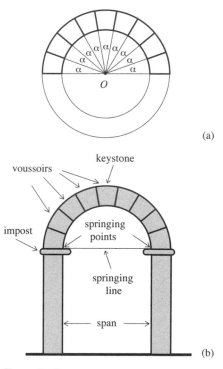

(a)

(b)

Figure 2.26

Figure 2.25. Arch from Hadrian's Temple, Ephesus, in today's Turkey. Photo by Evren Kalinbacak

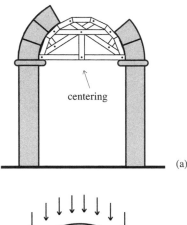

centering

(a)

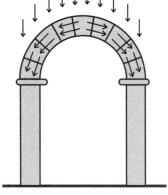

(b)

Figure 2.27

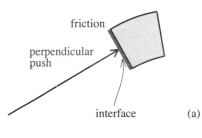

friction

perpendicular push

interface

(a)

interface

resultant upward push

friction

perpendicular push

(b)

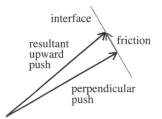

Figure 2.28

supports generate act vertically downward. These forces are redirected by the sloping arch to the supporting columns and from there to the ground as indicated in Figure 2.27b. The stable transmission of these forces relies on the capacity of stone to withstand compression. But the stable transmission of the flow of forces down the sloping arch also requires that the outward horizontal components of these forces be contained. Only after this is attended to does the arch attain stability.

We now turn to the study of the forces generated by the weights of the voussoirs with a focus on their outward horizontal components. It will be assumed that the arch is stable but that it supports only itself and carries no additional loads. As in the situation of Figure 2.26a, we'll suppose that the voussoirs are identical. So all voussoirs have the same weight W and the two sides of every voussoir determine the same angle α.

The keystone is held in place by the two voussoirs immediately below it, and the typical voussoir is held up by the single voussoir below it. The push of a lower voussoir on an upper voussoir has two components, both depicted in Figure 2.28a. There is the upward push perpendicular to the interface between the two voussoirs and there is the force of friction along their interface. Figure 2.28b depicts these two forces as well as their resultant. This resultant is the total force with which a lower voussoir pushes on an upper voussoir. In most cases, Roman architects did not use mortar in the construction of their arches. They relied instead on the precise shaping of the voussoirs. We will therefore regard the sides of the voussoirs to be smooth enough so that the effect of friction is secondary. So the analysis that follows will ignore friction and assume that the push from a lower voussoir on an upper one is perpendicular to the common interface between the two voussoirs. We will consider each voussoir separately, one at a time, starting with the keystone.

By the symmetry of the situation, we can assume that the magnitudes of the two forces that push up against the keystone are equal and we'll let this magnitude be P_0. See Figure 2.29. Because the vertical components of these two forces together support the keystone's weight, it follows that each of these vertical components is equal to $\frac{W}{2}$. Now turn to Figure 2.30a. The point C_0 is the center (of mass) of the keystone. Figure 2.26a provides the point O and the angle $\frac{\alpha}{2}$ at O. The point A is chosen so that segment AC_0 lies along the line of the upward push. The point B is taken so that AB is horizontal and hence perpendicular to C_0O. Notice that the two right triangles with hypotenuse OC_0 and AC_0, respectively, share an angle at C_0. Because the interior angles of any triangle add to 180°, it follows that $\angle C_0AB = \frac{\alpha}{2}$. Let H_0 be the horizontal component of P_0. The equality of angles depicted in Figure 2.30b, and the fact that the vertical component of the push P_0 is $\frac{W}{2}$, provides the force diagram of Figure 2.30c. The decomposition of forces into components tell us that $H_0 = P_0 \cos \frac{\alpha}{2}$ and $\frac{W}{2} = P_0 \sin \frac{\alpha}{2}$. Therefore, $P_0 = \frac{W}{2} \cdot \frac{1}{\sin \frac{\alpha}{2}}$ and

$$H_0 = \left(\frac{W}{2} \cdot \frac{1}{\sin\frac{\alpha}{2}} \right) \cos\frac{\alpha}{2} = \frac{W}{2} \cdot \frac{1}{\tan\frac{\alpha}{2}}.$$

This is the horizontal component of the force that keeps the keystone in place.

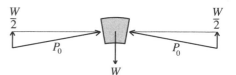

Figure 2.29

Now that the keystone has been attended to, we regard it to be removed from the arch and turn to the forces that hold up the voussoirs just below the keystone. Figure 2.31 considers the voussoir to the left of the keystone. Let P_1 be the magnitude of the push that holds up this voussoir—just this voussoir—and let H_1 be its horizontal component. The vertical component of P_1 is equal to the weight W of the voussoir. In Figure 2.31a, C_1 is the center of the voussoir, and A is chosen so that AO is horizontal and AC_1 lies along the line of the push. The fact that AC_1 is perpendicular to the side of the voussoir tells us that the angle at A is $\frac{3\alpha}{2}$. The equality of angles depicted in Figure 2.31b provides the force diagram of Figure 2.31c. The decomposition of forces into components tells us that $H_1 = P_1 \cos\frac{3\alpha}{2}$ and $W = P_1 \sin\frac{3\alpha}{2}$. Therefore, $P_1 = W \cdot \frac{1}{\sin\frac{3\alpha}{2}}$ and

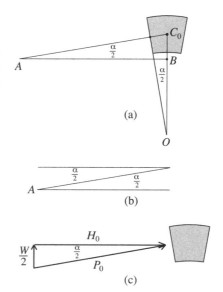

$$H_1 = \left(W \cdot \frac{1}{\sin\frac{3\alpha}{2}} \right) \cos\frac{3\alpha}{2} = W \cdot \frac{1}{\tan\frac{3\alpha}{2}}.$$

This is the horizontal component of the force P_1 that holds up the voussoir to the left of the keystone.

Considerations similar to those illustrated in Figure 2.31 inform us that the horizontal component of the force necessary to hold up each voussoir of the second pair below the keystone is $H_2 = W \cdot \frac{1}{\tan\frac{5\alpha}{2}}$. For each voussoir of the third pair this horizontal force is $H_3 = W \cdot \frac{1}{\tan\frac{7\alpha}{2}}$. Do you see a pattern in the equations for $H_1, H_2,$ and H_3? It is a pattern that continues for subsequent pairs. (The formula $H_2 = W \cdot \frac{1}{\tan\frac{5\alpha}{2}}$ is taken up in Problem 31.)

Suppose, for example, that $W = 300$ pounds and $\alpha = 20°$. The fact that $9 \times 20° = 180°$ tells us that the arch has nine voussoirs. Plugging the values for W and α into the formulas derived above shows (after rounding off what a calculator gives) that $H_0 = 851, H_1 = 520, H_2 = 252,$ and $H_3 = 109$ pounds. So the horizontal forces that these voussoirs generate in each direction total $851 + 520 + 252 + 109 = 1732$ pounds. This is considerable, especially when one considers that the total weight of all nine voussoirs of the arch is $9 \times 300 = 2700$ pounds. Of course, if the arch has to support a load, then the horizontal forces that are generated are greater still.

Figure 2.30

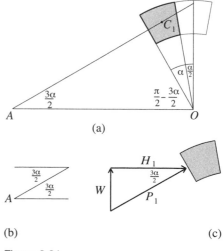

Figure 2.31

The horizontal components of the forces required to hold up the voussoirs of the arch are diagrammed in Figure 2.32. For the arch to be stable, these forces need to be supplied by the structure of which the arch is a part. An alternative perspective is provided by the First Principle of Structural Architecture. It tells us that the voussoirs of a stable arch push out against the structure with forces of these same magnitudes (but opposite in direction)

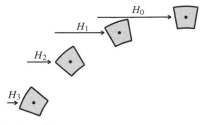

Figure 2.32

and that these outward forces need to be responded to by the structure as a whole. Such an outward force is known as thrust. In the case of the arch in Figure 2.25, the horizontal extensions of the lowest voussoirs push inward and this push is transmitted upward one voussoir at a time to counteract the horizontal thrust.

The analysis of the arch undertaken above considered each voussoir separately. It studied the slanting upward push that supports the isolated voussoir as well as the horizontal component of this push. Figure 2.32 summarizes the results. By the symmetry of things, the same conclusions apply to the voussoirs of the other side of the arch. The actual dynamic within the arch is different. The two voussoirs of each matching pair hold up the entire part of the arch that lies between them and not just the pair of voussoirs immediately above them. The assumption that the voussoirs are very smooth limits the impact of the friction at their interfaces. But within the dynamic of the arch, friction does play a role. Consider an interface and notice that friction pushes the upper voussoir up and the lower voussoir down. While the simplifying assumptions made in our analysis mean that its conclusions are only estimates, the analysis does capture essential behavior of the arch. It is important to keep in mind that any quantitative assessment of a complex structure—no matter how sophisticated it is—must make simplifying assumptions and can therefore only provide estimates. For such an assessment to be useful, it must capture essential features of the structure in spite of the simplifying assumptions that are made. In the context of the analysis of the semicircular arch, it is of interest to note that the magnitudes of the actual horizontal thrusts are less than those computed by ignoring friction. Hence an arched structure designed to respond to the horizontal thrusts computed by this analysis is provided with a margin (or factor) of safety. (See Problem 32.)

The books of Vitruvius are an extensive record of Roman architectural practices. With regard to the current discussion, the comment

> When there are arches . . . the outermost piers must be made broader than the others, so that they may have the strength to resist when the wedges, under the pressure of the load of the walls, begin to . . . thrust out the abutments

confirms that Roman architects knew that their structures needed to respond to the thrusts that arches (as well as vaults and domed structures) generate. However, Roman architects and engineers would not have been able to deal with forces in quantitative terms. The idea that forces could be separated into components and that these could be estimated by using trigonometry was beyond them. They built with trial and error, relied on their experience, and modified what had been done before.

The Romans brought fresh water into some of their cities from springs in hills and mountains in the vicinity with systems of channels constructed

Figure 2.33. The Roman aqueduct Pont du Gard in southern France. Photo by Hedwig

Figure 2.34

from stone, brick, and concrete. The flow of the water relied only on gravity and was not aided by pumps. This required channels that sloped gradually downward from springs in the elevated areas to reservoirs at the terminal points in the city. Along the way, hills had to be tunneled through and valleys had to be bridged. The aqueduct is a structure that made this possible. It consists of rows of semicircular arches, called arcades, that are stacked in tiers. The Roman aqueduct over the Gard river in southern France built in 20–16 B.C. is the striking example depicted in Figure 2.33. The highest row of semicircular arches carries the channel through which the water flows. It rises to a height of 160 feet over the valley and is 882 feet long. Its arches have spans of 20 feet. The semicircular arches of the lower two rows have spans of 60 feet except for the 80-foot spans of the arches at the center. The arches of the bottom row were made broader so that they could support a bridge. Such a broad arched structure is called a vault or, more precisely, a barrel vault. Figure 2.34 illustrates how the horizontal thrusts of the arches in an arcade are counterbalanced by the compression-resistant masonry materials that fill the triangular gaps between the arches.

Another important innovation of the Romans was their use of concrete. They discovered that when a certain volcanic powder is mixed with lime, sand, fragments of stone and masonry, and subsequently with water, the mix hardens to a substance that has stonelike consistency. Before it hardens, the mix can be shaped and molded. Vitruvius described the volcanic powder as "a kind of natural powder from which natural causes produce astonishing results." His treatise discusses the composition and application of Roman concrete. The Romans recognized the strength of concrete and used it in much of their construction. Because concrete was not attractive enough for some applications, the Romans became expert at the use of brick, stucco, marble, and mosaic finishes. Roman concrete had a thick consistency. It was layered by hand around chunks of stone and masonry fragments. It had good compressive strength, but comparatively weak tensile strength. Roman concrete did not offer a significant *structural* advantage over conventional masonry construction of stone and brick, and was used largely because of its *constructional* and economic advantages. It was relatively easy to build with and it was relatively inexpensive. Modern concrete differs from Roman concrete in an important respect. While Roman concrete had a thick consistency, modern concrete is fluid and homogeneous. Not only can it be poured into forms, but reinforcing steel rods and mesh can be embedded in it to give it significant additional strength, in particular tensile strength.

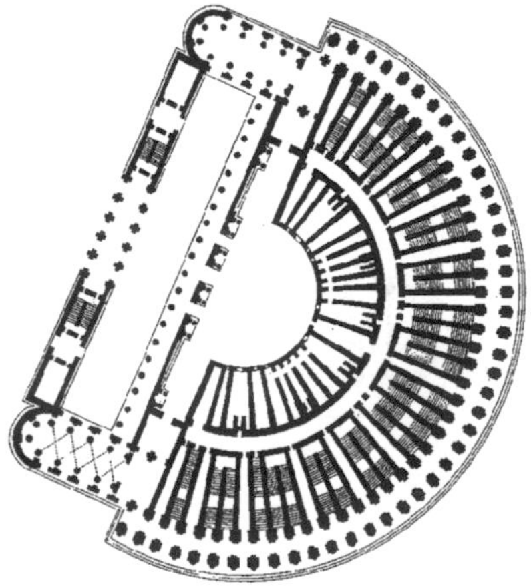

The Colosseum

The Romans made extensive use of arcaded structures. Unlike the Greeks, who carved their great theaters into hillsides that could support the sloping semicircular arrays of seats (as Figure 2.3 shows), the Romans constructed their theaters and arenas in the cities on flat terrain. This required extensive substructures of concrete, masonry, and stone. The Romans relied on configurations of arches and barrel vaults to support the seating sections and to provide access for the spectators. The arcaded exterior walls that often surround Roman theaters and arenas are in essence aqueducts arranged in semicircles or ovals. The Theater of Marcellus built in Rome between 13 and 11 B.C. is an example of what has been described. Its plan is shown in Figure 2.35. It had a seating capacity of about 11,000. The surrounding semicircular arcade had a radius of about 200 feet. Figure 2.36 shows a part of the structure that has survived.

A few decades later, the Romans embarked on a much larger and more ambitious project. Its design starts with an aqueduct like the one of Figure

Figure 2.35. Plan of the Theater of Marcellus, Rome

Figure 2.36. Remaining wall structure of the Theater of Marcellus, Rome. Photo by Joris van Rooden

Figure 2.37. The Roman Colosseum. Photo by Marcok

Figure 2.38. Aerial view of the Colosseum. From Michael Raeburn, editor, *Architecture of the Western World*, Rizzoli, New York, 1980. Marquand Library of Art and Architecture, Princeton University Library, Barr Farre Collection

2.33. But it is longer, bent into an oval shape, and adjoined at the two ends. Inside this oval configuration of arches another one is placed, parallel but not as high. Inside that a third, lower yet, and so on, and then a final lowest oval is set to surround the field of action. Then the spaces between these arcaded ovals are connected with sets of arches and barrel vaults both along the ovals and radially from the outer to the inner oval. What has been designed is the core of a Roman arena and the essence of the Roman Colosseum. The barrel vaults serve as entry passages for the spectators and as support structures for the sloping sections of seats. Figures 2.37 and 2.38 show what remains of the Colosseum today. Figure 2.37 depicts the outer oval and two of the inner ovals. Notice that the smooth exterior finish of the outer oval is intact. The missing finish from the inner oval reveals the rough concrete of its construction.

How exactly the Romans laid out the parallel sequence of ovals for the Colosseum has been a matter of scholarly debate. However, it is likely that they were laid out, one by one, as combinations of circular arcs. What is known is that the oval structures of some earlier and smaller Roman amphitheaters were laid out with circular arcs according to the following scheme. On a flat and horizontal stretch of ground lay out an isosceles triangle *T*, as shown in Figure 2.39a. Its base has length *b*, its height is *h*, its vertices are labeled 1, 2, and 3, and the sides from 1 to 2 and from 3 to 2 have the same length *s*. Attach an identical copy of this triangle to form the diamond shape with vertices 1, 2, 3, and 4. Now extend the four sides of the diamond as

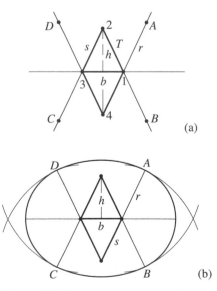

Figure 2.39. Roman construction of ovals

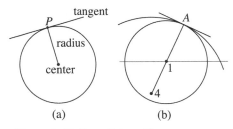

Figure 2.40. Smooth transitions

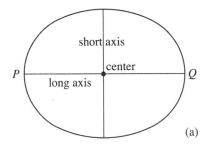

(a)

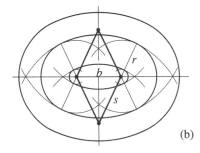

(b)

Figure 2.41. Roman ovals

shown and select some distance r. Put pegs at the points 1, 2, 3, and 4. Tie a rope to the peg at point 1, mark off a length r to the point A, and draw a circular arc from A to B. Next, tie a rope at peg 2, stretch it to point B, and draw a circular arc from B to C. Then do similar things at pegs 3 and 4 to draw circular arcs from C to D and from D to A. Taken together, the four circular arcs form the oval shown in Figure 2.39b. The oval is perfectly smooth at the points of transition A, B, C, and D from one circular arc to the next. This follows from the fact (illustrated in Figure 2.40a) that for any point P on a circle, the tangent at P is perpendicular to the radius from the center to the point P. Applying this fact to the point A and the radius from 1 to A and again to the radius from 4 to A in Figure 2.40b tells us that the tangents at A of the two circular arcs coincide. So the transition at A is smooth. The transitions at B, C, and D are smooth for the same reason. The oval is shown again in Figure 2.41a together with its long and short axes. The other two ovals shown in Figure 2.41b are obtained by repeating the construction just described with the same triangle T but different lengths r. Ovals obtained from the same triangle T are parallel to each other, but of different shape. If r is large relative to the size of the triangle, then the oval is close to a circle. If r is small relative to the size of the triangle, then the oval is flat. By varying the shape of the isosceles triangle, the Romans could lay out an unlimited number of different ovals.

It follows from Figure 2.39b that the long axis of the oval has length $b + 2r$ and that the distance from the center of the oval to the top of the circular arc DA is $s + r - h$. So the length of the short axis of the oval is $2s - 2h + 2r$. By the Pythagorean Theorem, $s^2 = h^2 + (\frac{b}{2})^2$, and hence $h = \sqrt{s^2 - \frac{b^2}{4}} = \frac{1}{2}\sqrt{4s^2 - b^2}$. Therefore, the length of the short axis is $2s - \sqrt{4s^2 - b^2} + 2r$.

Could Roman builders have laid out the outer oval of the Colosseum with the method described? Possibly. The lengths of the long and short axes of the outer oval of the Colosseum are known to be approximately 615 and 510 feet. Take $s = b$ (so T is equilateral) and use the formulas for the lengths of the axes to obtain the equations

$$b + 2r = 615 \quad \text{and} \quad 2b - \sqrt{4b^2 - b^2} + 2r = (2 - \sqrt{3})b + 2r = 510.$$

Solving for b and r shows that $b = 143$ feet and $r = 236$ feet. So the oval obtained with $s = b = 143$ feet and $r = 236$ feet has axes of the required lengths. (Problem 35 explores the matter further.)

The Colosseum was completed in A.D. 80 and is the largest monument of Roman antiquity. Estimates of its seating capacity range from 50,000 to 80,000. It was the arena in which the Roman emperors entertained the public with bloody spectacles. Spectators watched as gladiators fought to the death and as Christians were devoured by lions and tigers. After these "games" were discontinued in A.D. 400, the Colosseum suffered from neglect, vandalism, and earthquakes. During the rebirth of Rome that began in the Renaissance, it became a quarry for new construction in the Eternal City. The dismantling of the Colosseum stopped in the middle of the eighteenth century when the pope proclaimed it to be a sacred monument to honor the Christians who suffered martyrdom there.

The Pantheon

One of the most impressive structures of Roman antiquity is the Pantheon (never to be confused with the Parthenon in Athens). Its supervising architect was the Roman emperor Hadrian. It was built between A.D. 118 and 128 and dedicated to all Roman gods (in Greek, *pan* = all, *theon* = of the gods). Its elevation, the term for a representation of a facade, is shown in Figure 2.42. The entrance hall is a portico in the Greek Corinthian style. The portico fronts a large cylindrical structure that is capped by a hemispherical dome. The exterior of the cylindrical structure consists of flat Roman bricks, carefully laid out row upon row. The dome of the Pantheon was regarded to be a shape of ideal perfection that conveyed both beauty and power. Plate 5 shows the spectacular interior. Its circular floor has a diameter of 142 feet and the

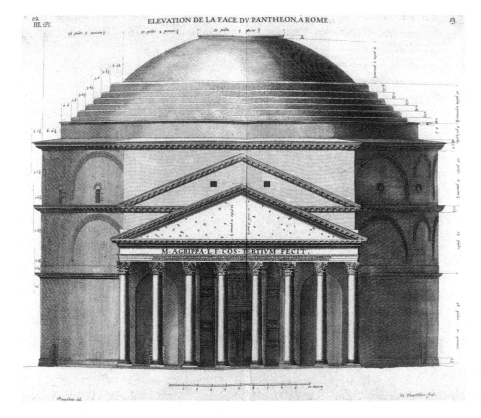

Figure 2.42. Antoine Desgodetz, engraving of the elevation of the Pantheon. From *Les edifices antiques de Rome*, Claude-Antoine Jambert, Paris, 1779

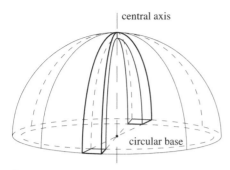

Figure 2.43. The shell of a dome as a composite of arches

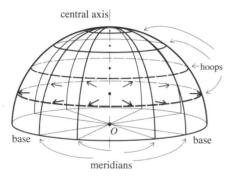

Figure 2.44. Hoop stress on the shell of a dome

interior of the dome rises to a height of about 142 feet above the floor at its highest point. At the top is a circular opening, or oculus (in Latin, *oculus* = eye), with a diameter of about 24 feet to let in light and air. Other than the entrance, the oculus is the sole source of natural light for the interior. Notice the circular arrays of rectangular indentations, or coffers, in the ceiling. The attention to detail in the composition of the interior with its framed statues and sets of fluted Corinthian columns and pilasters (the rectangular vertical elements flanking pairs of columns) stands in contrast to the Greeks' greater focus on external space and form. Before the Romans could build the massive dome of the Pantheon, they had to have some understanding of the structural challenges that a dome presents and they had to respond to them.

The shell of a dome is its structural part. Classically, shells of domes are made of masonry or concrete. Figure 2.43 shows how the shell can be thought of as a composite of arches obtained by slicing the dome vertically through its central axis. These arches, just like those studied earlier, generate outward horizontal forces. Because the arches of the shell taper as they rise, they weigh less comparatively, so that the magnitudes of these horizontal forces are less than before. However, unlike the situation of the aqueduct of Figures 2.33 and 2.34, there is only the base of the dome and the internal resistance of the shell to contain them. The vertical cross sections of a dome through its central vertical axis are called meridians and the horizontal circular cross sections are called hoops. They are depicted in Figure 2.44. A hoop is a ring-shaped, horizontal slice of the shell. Two forces are at work on the hoops. One is the push of the base of the dome propagated up along the rigid shell. The other is the weight of the dome above the hoop pushing down. Toward the top, the weight is not a factor and the push of the structure from below dominates. This puts the hoop under compression, in the same way that the keystone of an arch is under compression. But farther down, the weight of the shell above exceeds the upward push from below. The outward push of the horizontal components of this excess force puts the hoops (those from the base to about two thirds of the way up the shell) under tension. See Figure 2.44. This tension is called hoop stress. The First Principle of Structural Architecture tells us that unless the shell is able to resist this tension, the shell will expand along the hoops, so that cracks will develop along some meridians (that can lead to structural failure of the dome in extreme cases).

The Romans built the Pantheon out of concrete. Concrete was a building material that was relatively easy to build with, but like brick or stone, it had little tensile strength. So it facilitated the construction of the Pantheon, but it had relatively little capacity to contain the hoop stress within the shell of the dome. However, concrete could be made lighter or heavier simply by adding lighter or heavier aggregate, meaning stone and masonry materials, to the mix. The Romans used this to advantage. By making concrete with the light volcanic rock pumice for the upper sections, they reduced the weight of the dome and therefore the hoop stress. The concrete of the dome weighs about 81 pounds per cubic foot for most of the shell and about 100 pounds per cubic foot for the section of the shell above the supporting cylindrical wall. The circular oculus at the top not only supplies light and air, it reduces the dome's weight further. (Chapter 7, "Volumes of Spherical Domes," applies basic calculus to estimate the weight of the dome.) To contain the outward thrusts of the dome at its base, the Romans made the supporting cylindrical wall up to 20 feet thick with concrete that increased in weight from 100 pounds per cubic foot near the top of the wall to 115 pounds per cubic foot at the bottom. The aggregate of the concrete in the cylindrical wall includes the dense, resilient volcanic stone basalt. The rigid concrete shell propagates the push from the base upward. The inward component of this push counters the tensile stress on the hoops in much the same way as the hoops of a barrel keep its wooden slats (or staves) together. The cylindrical wall of the Pantheon rests on a substantial foundation. The concrete used in the foundation also contains basalt and weighs 140 pounds per cubic foot (close to the 150 pounds per cubic foot of standard modern concrete).

The coffering on the inside of the shell shown in Plate 5 is shallow, serves no structural purpose, and has essentially no impact on the weight of the dome. But the vertical and horizontal configuration of ribs that the coffers suggest does resemble the ribbed elements that would play an important structural role in later domes and vaults. The Romans often placed masonry and concrete masses on top of the lower, outer sections of arches and vaults. These masses were intended to increase the stability of such structures. The Romans may have intended for the step rings they built into the lower part of the dome of the Pantheon (see Figures 2.42 and 2.45) to serve such a function and to contain hoop stress. However, recent studies have indicated that the step rings seem to play no significant role in this regard. The step rings may also have been put in to facilitate the construction work on the shell. The dome was constructed with the use of centering. An elaborate forest of timbers reached upward from the floor of the Pantheon to support the growing shell until the construction was completed and the concrete had set.

Figure 2.45 depicts half of a central vertical section of the Pantheon. It shows the structure of the shell, a section of the cylindrical supporting wall, the oculus, and the step rings. The vectors flowing down the shell represent the downward transmission of the weight of the dome. Their horizontal components generate the hoop stress already discussed. The upward-pointing vectors represent the support of the shell from its cylindrical base. Their horizontal components counteract hoop stress. The arc that is highlighted lies on a circle of radius $\frac{1}{2} \cdot 142 = 71$ feet. Its center C is the point on the centering from which the builders of the Pantheon stretched ropes in all upward directions to guide the spherical shape of the shell during construction.

oculus

steprings

springing line

C

Figure 2.45. Section of the Pantheon from Andrea Palladio's *I Quattro Libri dell' Architettura* (*The Four Books of Architecture*), Venice, 1570. Marquand Library of Art and Archaeology, Princeton University Library

In spite of the efforts to control it, the hoop stress in the dome of the Pantheon did lead to extensive cracks along some meridians of the dome. The distribution of cracks generally corresponds to openings within the upper parts of the cylindrical wall (some of these are shown in Figures 2.42 and 2.45). These openings increase the hoop stress in the parts of the shell that rise near them. Nonetheless, the fact that the Pantheon has remained standing for almost 1900 years tells us how well the Romans succeeded. The Pantheon is one of the most important buildings in the history of architecture. Roman design and construction practices have been very influential. It would be hard to imagine today's construction without arches, domes, and the use of concrete.

Problems and Discussions

The first set of problems deals with basic Greek geometry as it can be found in Euclid's *Elements*. The next set considers trigonometric questions, the set thereafter studies forces, and most of the remaining problems consider arches. The solutions of some problems depend on conclusions from previous problems. The three discussions that end this section are closely related to topics taken up in the chapter.

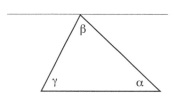

Problem 1. Use Figure 2.46 to show that the sum of the interior angles of a triangle is 180°.

Figure 2.46

The next two problems make use of congruent triangles. Two triangles are congruent if one of them can be moved on top of the other so that they coincide. The two triangles in Figure 2.47 are congruent. How does the one on the left have to be moved so that it coincides with the one on the right?

Figure 2.47

Problem 2. An isosceles triangle *ABC* is given with base *AB* and equal sides *AC* and *BC*. Show that the angle at *A* is equal to the angle at *B*. [Hint: Extend Figure 2.48a to Figure 2.48b so that *CD* = *CE* and show that the triangles *CAE* and *CBD* are congruent. Deduce that triangles *ABD* and *BAE* are congruent.]

Problem 2 is the fifth proposition in Book I of Euclid's *Elements*. It is known as the *pons asinorum*, Latin for "bridge of asses." There are two explanations for the name, the simpler being that the diagram used in the proof (Figure 2.48b) resembles a bridge. The more common explanation is that the fifth proposition in Book I of the *Elements* is the first real test of the intelligence of the reader and serves as a bridge to the harder propositions that follow. Whatever its origin, the phrase *pons asinorum* has been used to refer to any critical test of ability or understanding that separates the quick mind from the slow.

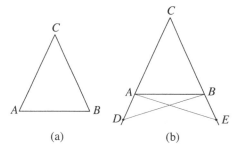

Figure 2.48

Problem 3. Refer to Figure 2.4b of the text and verify that ∠*AOE* = ∠*BOE*. [Hint: Use the result of Problem 2 twice.]

Problem 4. Consider a triangle inscribed in a circle in such a way that one of its sides is a diameter of the circle. Use Figure 2.49 to show that the triangle is a right triangle.

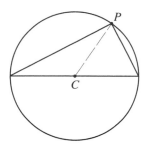

Figure 2.49

Problem 5. Let *R* be any rectangle, and let *a* and *b* with $a \geq b$ be the lengths of its sides.

   i. Form a rectangle R_1 by attaching a square to the longer side of *R*. What are the lengths of the sides of R_1? Show that if *R* is golden, then R_1 is golden.

ii. Place a square inside R so that one side of the square coincides with one of the shorter sides of the rectangle. Let R_2 be the rectangle that remains. What are the lengths of the sides of R_2? Show that if the rectangle R is golden, then R_2 is golden.

[Hint: Use the fact that the golden ratio ϕ satisfies $\phi^{-1} = \phi - 1$.]

Refer to the image of the Parthenon in Figure 2.12 and consider the surrounding golden rectangle. Notice that the other golden rectangles in the figure are generated by applying procedure (ii) of Problem 5 several times.

The next few problems are exercises in Euclidean straightedge and compass construction.

Problem 6. You are given a straight line L and a point P not on L. Construct a line through P that is parallel to L. [Hint: Start by constructing a perpendicular to L through P.]

Problem 7. A unit of length and a line segment of length 1 are given. Execute the constructions called for in (i) and (ii) below and explain how to execute those of (iii) and (iv).

i. Construct a line segment of length 3. Then take any line segment and divide it into three equal pieces. [Hint: Use a construction similar to the one illustrated in Figure 2.7.]
ii. Construct a segment of length $\frac{5}{3}$.
iii. Let n be any positive integer. How can a segment of length n be constructed? How can a segment be divided into n equal pieces?
iv. Let $\frac{n}{m}$ be a positive rational number. How can a segment of length $\frac{n}{m}$ be constructed?

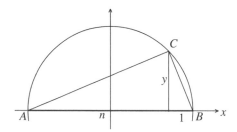

Figure 2.50

Problem 8. A unit of length, a segment of length 1, and a positive integer n are given. Construct a segment AB of length $n + 1$ and the circle that has this segment as diameter. Construct the segment perpendicular to the diameter shown in Figure 2.50 and let y be its length. Use the Pythagorean Theorem three times to show that $y = \sqrt{n}$. Therefore a segment of length $\sqrt{n}$ is constructible for any n. [Hint: Use the conclusion of Problem 4.]

The transfer of an angle to another location is a useful construction. The diagrams in Figure 2.51 illustrate how the angle α is transferred from its location at A to another location at D using a straightedge and compass.

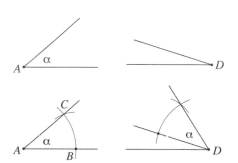

Figure 2.51

Problem 9. Given the angles α and β in Figure 2.52, construct the angles $\alpha + \beta$ as well as $\alpha - \beta$.

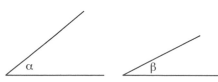

Figure 2.52

Problem 10. Construct an angle of 15° with a straightedge and compass.

Problem 11. Construct angles of 36°, 18°, 54°, and 24°.

Problem 12. Describe how to use the information given in Problems 10 and 11 to construct a regular 10-gon, a regular 15-gon, a regular 20-gon, and a regular 24-gon.

Claudius Ptolemy's comprehensive treatment *The Almagest* contains a development of Greek trigonometry. While Greek trigonometry has a different "look and feel" than our modern version, it is equivalent to it. Problems 13 to 16 consider basic trigonometry.

Problem 13. Turn to Figure 2.10 and suppose that the side of the square has length 1. The triangle ABC of this figure is depicted in Figure 2.53. Show that the base of the triangle is equal to $\frac{\sqrt{5}-1}{2}$ and that its height is $h = \sqrt{\frac{5+\sqrt{5}}{8}}$. Check that $\sin 18° = \cos 72° = \frac{\sqrt{5}-1}{4} \approx 0.31$ and $\sin 72° = \cos 18° = \sqrt{\frac{5+\sqrt{5}}{8}} \approx 0.95$.

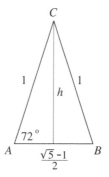

Figure 2.53

Problem 14. Consider any triangle, let α, β, and γ be its angles, and denote the lengths of the sides opposite these angles by a, b, and c respectively. Verify the Law of Sines, namely the equality

$$\frac{\sin \alpha}{a} = \frac{\sin \beta}{b} = \frac{\sin \gamma}{c}.$$

[Hint: Let α and β be two angles of the triangle that are less than 90°. Use Figure 2.54b to show that $\frac{\sin\alpha}{a} = \frac{\sin\beta}{b}$ for the triangle in Figure 2.54a. Conclude that the Law of Sines holds if all angles are less than 90°. If one of the angles of is 90° or more, let it be γ and consider Figure 2.54c. Use Figure 2.54d to show that $\frac{\sin\gamma}{c} = \frac{\sin\alpha}{a}$.]

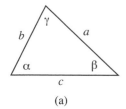
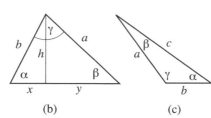
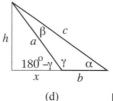

(a)   (b)   (c)   (d)   Figure 2.54

Problem 15. Consider any triangle and let γ be any one of its angles. Let the side opposite γ have length c, and let a and b be the lengths of the other two sides. Verify the Law of Cosines, namely the equation

$$c^2 = a^2 + b^2 - 2ab \cos\gamma.$$

[Hint: If $\gamma \leq 90°$, use a figure similar to Figure 2.54b and apply the Pythagorean Theorem twice. If $\gamma > 90°$, use Figure 2.54d and again apply the Pythagorean Theorem twice.]

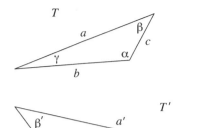

Figure 2.55

Two triangles T and T' are similar if their angles match up, in other words, if there is a correspondence between the angles of T and T', in such a way that matching or corresponding angles are equal. The two triangles in Figure 2.55 are similar because the corresponding angles α and α' are equal, β and β' are equal, and γ and γ' are equal. (Why is it enough for only two corresponding angles to be equal?)

Problem 16. One of the most important and useful facts in all of geometry says that if two triangles are similar, then the ratios of corresponding sides are equal. (Two sides correspond if they lie opposite corresponding angles.) In Figure 2.55 the corresponding sides are $a \to a'$, $b \to b'$, and $c \to c'$, so that

$$\frac{a}{a'} = \frac{b}{b'} = \frac{c}{c'}.$$

Use the Law of Sines to verify this property of similar triangles.

The remaining problems deal with forces, ladders, arches, ovals, and matters involving straightedge and compass constructions.

Problem 17. The weights of the limestone blocks of a pyramid generate considerable compression on the blocks that lie below them. Consider 10 such blocks, each weighing approximately 12 tons.

   i. Stack the blocks vertically and estimate the compressive force on the bottom block.
   ii. Stack the blocks in a triangular configuration. Put down a horizontal row of four, then stack a row of three symmetrically over this row, then stack two more on top of the row of three, and then put the tenth block on top of the two. Estimate the compressive forces on each of the four blocks of the bottom row. [Hint: Regard all the blocks to be cut vertically. Is any of the realism of the situation lost with this assumption?]

Problem 18. The stick figure of Figure 2.56 depicts a human body. The figure is symmetrical and the person it depicts is standing perfectly still. Draw a force diagram for each of the five points singled out. Write a paragraph that describes the action of the forces at each of the five points.

Figure 2.56

Problem 19. The vectors in Figure 2.57 represent forces of the indicated magnitudes and directions. In each case, draw in the vectors that represent their horizontal and vertical components and compute the magnitudes of these components.

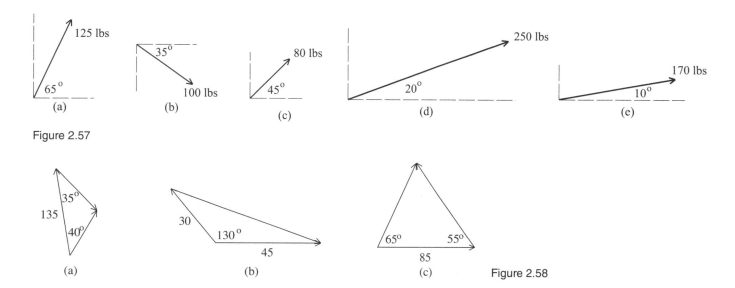

Figure 2.57

Figure 2.58

Problem 20. Each of the diagrams in Figure 2.58 represents two forces and their resultant. Identify the resultant in each case. The magnitudes of some of the forces (and some of the angles between them) are given. Use the Law of Sines or the Law of Cosines to compute the magnitudes of the forces that are not specified.

Problem 21. Refer to Figure 2.59. The vector *A* represents the gravitational force acting on some object. The vectors *B* and *C* provide a decomposition of *A* into components. The vector *D* is the horizontal component of *B*. Figure 2.59 appears to show that the downward force *A* has a nonzero horizontal component (namely *D*). But this can't be since the original *A* acts vertically. Explain the apparent contradiction.

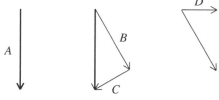

Figure 2.59

Problem 22. Return to the ladder of Figure 2.24. Assume that it still leans against the smooth wall at an angle of 60°, but that it supports a load of only 190 pounds. Compute the magnitudes of the forces *F*, *P*, and *H* in this case.

Problem 23. Suppose that the ladder of Figure 2.24 supports 240 pounds but that it leans against the smooth wall at an angle of 72° instead of 60°. Compute the forces *F*, *P*, and *H* in this case. Compute the components of these forces in the direction of the ladder and show that they are in balance. [Use the conclusions of Problem 13.]

Problem 24. Figure 2.60 shows a ladder supported and kept in balance by the vertical force *U* of the floor together with the pull *V* of a rope in the vertical direction. The angle of lean of the ladder is *β* and *α* is the angle between

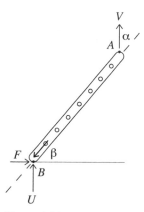

Figure 2.60

the ladder and the vertical. Let W be the weight of the ladder. Why is $W = U + V$? Show that the push of the ladder down in direction of the ladder is $W \cos \alpha - V \cos \alpha = U \cos \alpha$. Conclude from this that the friction F at B must be zero.

Problem 25. Investigate the origin of the semicircular arches at the foot of the Acropolis depicted in Plate 4.

Problem 26. Study the lowest sequence of arches of the aqueduct of the Pont du Gard. Describe several ways in which this structure might fail.

Problem 27. Figure 2.61 shows an arch from the Roman ruins in Palmyra. This is today's Tadmor in the heart of the Syrian desert. What information in the section "The Roman Arch" provides a plausible reason for the slippage of the keystone of the arch?

Problem 28. The arch depicted in Figure 2.62 stands in the Greek-Roman city of Ephesus. It has a span of about six feet. Assume that the voussoirs are made of sandstone weighing 140 pounds per cubic foot. Consider an ideal version of this arch in which all five voussoirs are identical. Estimate the weight of a voussoir of the arch. Then estimate the horizontal components of the outward forces that the top three voussoirs generate on the bottom pair of voussoirs. [Hint: Estimate the volume of a voussoir by regarding the face of the arch to be the difference between two concentric semicircles. Supply rough values for the relevant dimensions.]

Problem 29. Consider the Roman arch with $W = 300$ and $\alpha = 20°$ studied in the section "The Roman Arch." Assume that the arch supports only itself and carries no additional loads. Recall that the estimate of the horizontal thrust generated by the top three voussoirs is $H_0 + H_1 = 1371$ pounds in each direction. In Figure 2.63, the top three voussoirs are cemented together into one larger keystone weighing 900 pounds. Estimate the horizontal thrust of this larger keystone in each direction. Why is this much less than the earlier 1371 pounds? Before you answer, think about the situation where the semicircular arch is cast in one solid piece. Figure 2.64 shows such solid arches. What horizontal forces does such an arch generate on the two supporting columns on which it rests? Does gravity generate horizontal forces down along the arch? If yes, how does the arch deal with them?

Problem 30. Study the arch depicted in Figure 2.65. Do you think it possible that its voussoirs are not cemented together and that the arch is held in place only by the forces at the interfaces between the voussoirs? In this case, explain how the outward horizontal thrusts could be counteracted.

Figure 2.61. A Roman arch in Palmyra, in today's Syria. Photo by Odilia

Figure 2.62. A Roman arch in the city of Ephesus, in today's Turkey

Figure 2.63

Figure 2.64. An arcade on the grounds of Hadrian's Villa, Rome

Problem 31. Return to the discussion of the forces on the voussoirs of the arch. Figure 2.66 is similar to Figure 2.31. It applies to the second voussoir below (and to the left of) the keystone. Use the figure to verify the formulas $P_2 = W \cdot \frac{1}{\sin \frac{5\alpha}{2}}$ for the upward push P_2 on this voussoir and $H_2 = W \cdot \frac{1}{\tan \frac{5\alpha}{2}}$ for its horizontal component.

Problem 32. The analysis of the horizontal thrusts generated by the voussoirs of an arch ignored the effects of friction. Consider a modification of this analysis that takes the action of friction along the interfaces between the voussoirs into account. Would this modified analysis have resulted in larger or smaller estimates of the magnitudes of these horizontal thrusts? [Hint: Study Figure 2.28b.]

The next three problems deal with the construction of ovals described in the section "The Colosseum."

Problem 33. An oval with a short axis of 150 feet and a long axis of 200 feet is to be laid out with T an equilateral triangle. What lengths should be taken for $b = s$ and r?

Problem 34. Let L and S be the lengths of the long and short axes of an oval determined by the triangle construction. Show that all the ovals constructed with a fixed triangle have the same $L - S$.

Problem 35. The long and short axes of the outer oval of the Colosseum are estimated to be 615 feet and 510 feet, respectively. For the inner oval that bounds the arena within the Colosseum, the estimates are 287 feet for the long axis and 180 feet for the short axis. Could both ovals have been laid out with the same equilateral triangle ($b = s = 143$ feet)?

Problem 36. Find pictures of the New York Stock Exchange in New York City and the Jefferson Memorial in Washington, D.C. What buildings studied in this chapter do they remind you of?

Discussion 2.1. The Columns of the Parthenon. We learned in the section "Greek Architecture" that the builders of the Parthenon gave the columns of the facade a slight inward lean. Serious architectural literature mentions that this inward lean is so precise, that if the central axes of the columns of the facade are extended upward, they meet at a point high above the Parthenon. The stated height of this point varies. One account lists it as 6800 feet, another as 16,200 feet. Figure 2.67 depicts a corner column of the facade and shows the extensions of the central axes of both corner columns to the point in question. The figure makes use of the fact that the facade of the Parthenon is 110 feet wide (but it is not drawn to scale). The angle α is

Figure 2.65. An arch from the Roman ruins of Volubilis, in today's Morocco.

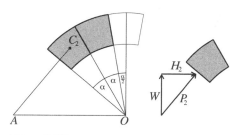

Figure 2.66

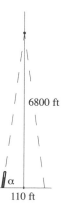

6800 ft

110 ft   Figure 2.67

the angle of lean of the column. Suppose that the Greek builders of the Parthenon intended to achieve the convergence of the extensions of these axes as described. Because $\tan \alpha = \frac{6800}{55}$, they would have needed to take α very close to 89.54°. (To be exact, $\alpha = 89.5366°$ is required.)

Problem 37. Discuss the consequences of even the slightest error in any effort to set the column at the angle α. Consider $\alpha = 89.55°$ or $\alpha = 89.53°$, for example, and study the impact of this error on the question of the convergence of the axes. Is it plausible that the architects of the Parthenon could have succeeded in getting the axes of the massive columns to converge in the way described? If they did succeed, is it possible that the axes of today's Parthenon converge in this way? What event in the history of the Parthenon suggests that this is improbable?

Discussion 2.2. The Golden Rectangle and the Pentagon. This section supplies the proof from Euclid's *Elements* of the fact that the rectangle determined by the five-pointed star shown in Figure 2.11b is a golden rectangle. This is the information needed in the verification that the angle 72° can be constructed with straightedge and compass. Figure 2.68 shows a regular pentagon. All of its five vertices lie on a circle and the lengths of its five sides are the same. The length of a side is s and the length of a diagonal is d. The point O is both the center of the circle and the center of the pentagon. Consider the segments from the center O to two successive vertices. Notice that the angle between the two segments is $\frac{360°}{5} = 72°$. The triangle that the two segments determine together with the side of the pentagon is isosceles. It follows that the other two angles of the triangle are 54° each. Therefore the angle between two successive sides of the pentagon is 108°. Label the vertices of the pentagon A, B, C, D, and E. Let M be the intersection of the two diagonals AC and BE and consider Figure 2.69. Because the angle at B is 108° and $\triangle ABC$ is isosceles, it follows that $\angle BAC = 36°$. So $\angle BAM = 36°$. By the same argument, $\angle ABM = 36°$. This implies that $\angle AMB = 108°$. Because there is an equality of corresponding angles, the triangles ABC and AMB are similar.

Let's study the quadrilateral $MEDC$. Euclid shows first that $MEDC$ is a parallelogram. We know that the angles of the quadrilateral at M and at D are both equal to 108°. The angles at C and E are both $108° - 36° = 72°$. So these angles are also equal, and $MEDC$ is a parallelogram, as asserted. Because the adjacent sides ED and DC are equal, it follows that all four sides of the parallelogram $MEDC$ are equal. Because s and d are the lengths of a side and a diagonal of the pentagon, Euclid knows that $MC = ED = s$ and $AM = d - s$. From the similarity of $\triangle ABC$ and $\triangle AMB$ he can conclude that

$$\frac{d}{s} = \frac{AC}{AB} = \frac{AB}{AM} = \frac{MC}{AM} = \frac{s}{d - s}.$$

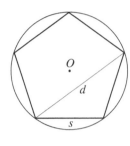

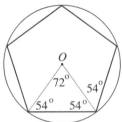

Figure 2.68

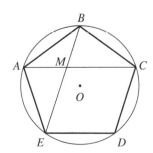

Figure 2.69

Notice that $\frac{s}{d} = \frac{d-s}{s} = \frac{d}{s} - 1$. It follows that $\frac{d}{s} - (\frac{d}{s})^{-1} - 1 = 0$ and hence that $x = \frac{d}{s}$ is a positive root of the polynomial $x^2 - x - 1$. So $\frac{d}{s}$ is the golden ratio $\frac{1+\sqrt{5}}{2}$. Euclid has established that the ratio of the diagonal of a pentagon to its side is golden. So $\frac{MC}{AM} = \phi$ and $\frac{AC}{MC} = \frac{AC}{AB} = \phi$. In Figure 2.70, N is the intersection of the diagonals AC and BD. By the symmetry of things, $AN = MC$ and hence

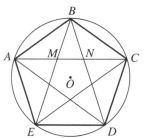

$$\frac{AC}{AN} = \phi = \frac{AN}{AM}.$$

Figure 2.70

It now follows that the rectangle with shorter side AM and longer side AN is a golden rectangle. The only thing left to verify (and this is routine) is that the small pentagon at the center of the figure is regular. Therefore the star with corners A, B, C, D, and E is the star determined by extending the sides of a regular pentagon. The proof is now complete.

Focus on the diagonal AC and notice that N is a golden cut of AC and that in turn M is a golden cut of AN. Because one of the diagonals behaves like any other, this is true for each of the five diagonals of the pentagon. In particular, the five-pointed star that the sides of the pentagon determine is full of golden ratios.

This five-pointed star is in wide use today. It adorns national flags, it is a symbol of military power, and it is attached to a variety of commercial products. It is also known as the pentacle or pentagram, especially when it is depicted with its surrounding circle. The pentacle has also become—for reasons beyond the interest of this text—a symbol used by cults of witch and devil worshippers.

Discussion 2.3. More about Straightedge and Compass Constructions. We saw in the section "Gods of Geometry" that the regular n-gons with $n = 3, 4, 5, 6$, and 8 can be constructed with straightedge and compass. However, the regular n-gon for $n = 7$ and $n = 9$ cannot be constructed. The regular 10-gon can be constructed because the regular 5-gon can. Table 2.1 summarizes facts about the constructibility of a regular n-gon for $n \leq 30$. The

Table 2.1. The constructibility of regular n-gon for $n \leq 30$

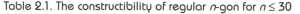

7-gon	8-gon	9-gon	10-gon	11-gon	12-gon	13-gon	14-gon	15-gon	16-gon	17-gon	18-gon
No	Yes	No	Yes	No	Yes	No	No	Yes	Yes	Yes	No
	45°		36°		30°		7-gon	24°	22.5°	Gauss	9-gon

19-gon	20-gon	21-gon	22-gon	23-gon	24-gon	25-gon	26-gon	27-gon	28-gon	29-gon	30-gon
No	Yes	No	No	No	Yes	No	No	No	No	No	Yes
	18°		11-gon		30°		13-gon		14-gon		24°

second row provides reasons why they can or cannot be constructed. For instance, the regular 14-gon is not constructible because the regular 7-gon is not constructible. But the regular 20-gon is constructible because the angle 18° is constructible.

The question as to which constructions are possible and which are impossible is difficult. The problem was solved only after it was translated into the realm of abstract algebra and number theory. The great German mathematician Carl Friedrich Gauss (1777–1855) knew what the basic facts were. In reference to the table, he knew that the regular 7-gon, 9-gon, 11-gon, 13-gon, 19-gon, 23-gon, and 29-gon are not constructible, but that the regular 17-gon is. One such construction is depicted in Figure 2.71. (The points J, L, and K are vertices of the 17-gon and the remaining points can be marked off using them.) Gauss knew the facts, but he did not supply mathematical proofs for all of them. Only when the French mathematician Pierre Laurent Wantzel (1814–1848) and the German Carl Louis Ferdinand von Lindemann (1852–1939) filled in the last important pieces of the puzzle was the solution complete. The solution included the answers to the three famous constructibility problems of Greek antiquity:

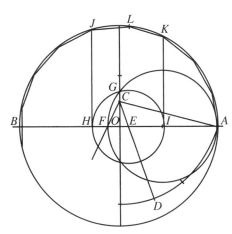

Figure 2.71. A construction of the regular 17-gon

i. Can an angle be divided into three equal parts? This is the question of the "trisection of an angle." Lots of angles can be subdivided into three equal angles by using a straightedge and compass. Because the angle $60° - 45° = 15°$ can be constructed, 45° can be trisected. Because the angles 18°, $54° - 30° = 24°$, and 12° can be constructed, the angles 54°, 72°, and 36° can all be trisected. But can *any* angle be trisected using a straightedge and compass?

ii. For a given circle, is it possible to construct a square that has the same area as the circle? This is the question of "squaring the circle." Since the area of a circle of radius 1 is π, this would give the construction of a segment of length s such that $s^2 = \pi$, or $s = \sqrt{\pi}$. Possible?

iii. For a given segment of length a is it possible to construct a segment of length b so that $b^3 = 2a^3$? This is the question of "duplicating the cube." Taking $a = 1$, the construction calls for a segment of length b such that $b^3 = 2$, or $b = \sqrt[3]{2}$. Possible?

Problem 38. Use the information in Table 2.1 to show that the angle 60° cannot be trisected with straightedge and compass.

Problems 7 and 8 informed us that if a segment of one unit in length is given, then segments of any length $\frac{n}{m}$ and $\sqrt{n}$ with n and m positive integers can be constructed. However, it is a consequence of the theory mentioned above that the lengths required in (ii) and (iii) above cannot be constructed!

Architecture Inspired by Faith

The Roman emperor Constantine recognized the Christian faith in the fouth century and it became Rome's state religion soon thereafter. In refocusing attention from this world to the next, Christianity offered biblical narratives and supernatural explanations to guide the course of human action. Neither the inquisitive mindset with which Greek philosophy and mathematics analyzed the world nor the confident spirit with which Roman engineering shaped it received much promotion. When Constantine moved the imperial capital from Rome to Byzantium (soon thereafter renamed Constantinople), the Roman Empire and its Christian faith split into western and eastern parts. The split between Christian east and west, initially cultural and political, became a deep division in the eleventh century with the formal separation of the Christianity into a Catholic and an Orthodox faith. In the seventh century, the Islamic faith sprang into existence in the Arabian peninsula and spread quickly. In time, the territories around the Mediterranean Sea and to its north and east would give rise to three broad clusters of cultures: An Eastern Byzantine, a Western Catholic, and an Islamic civilization. Each was governed by theocracies and shaped by religion. Art and architecture developed principally to give visual expression to faith and to build houses of worship. This chapter studies some of the incredible edifices that these faiths inspired. To set the stage, we describe in very broad and brief strokes some of the characteristics of each of these civilizations.

The eastern part of the Roman Empire became the Byzantine Empire. It retained much that was Roman in government, law, and administration. The principal cultural ingredients were Greek, but blended with them were influences from Syria, Egypt, Persia, and lands to the east. Christianity was the dominating, unifying, and organizing force. The emperor derived his authority from God, surrounded himself with elaborate ceremonial, and ruled with absolute power. The Byzantine Empire would endure for some eleven centuries and serve as a bridge between the ancient and modern worlds. By the early sixth century, Byzantine artists and craftsmen fused Greek, Roman, Christian, and eastern elements to develop a richly ornamented art and architecture. Inspired more by Roman forms than classical Greek orders, Byzantine architects preferred arches, vaults, and domes to columns, friezes, and pediments. Their churches featured hemispherical domes on top of square or octagonal arrangements of arches. They surrounded the central dome with supporting configurations of vaults, half-domes, and arcades. Byzantine builders relied heavily on brick and mortar, but they also made extensive use of marble. Byzantine artists decorated the surfaces of the interiors with murals and delicate mosaics of brightly colored stone and glass on backgrounds of blues and golds. Following established traditions, their icons depicted religious motifs in stylized forms. Jewelers set gems into columns and walls, metalworkers added silver and gold, woodworkers carved screens and railings, and weavers hung tapestries, laid rugs, and draped embroidery and silk over altars. Never before had an art so rich in color and ornamentation decorated the curving surfaces of vaults and domes. The great church of the Hagia Sophia in Constantinople was the most brilliant example of this architectural form. The Byzantine Empire was under constant pressure from Islamic expansion. After the eighth century it consisted only of today's Turkey and the Balkans (including Greece) and when Constantinople fell to Islamic forces in the middle of the fifteenth century, the Byzantine Empire ceased to be.

The Islamic religion emerged in the 620s from Medina and Mecca on the Arabian peninsula. Driven by the simple theology *There is no god but Allah and Muhammad is His Prophet*, the armies of Allah conquered what is today the Middle East (Saudi Arabia, Egypt, Palestine, Syria, Iraq, Iran, and Afghanistan) in less than 30 years. From the eighth century forward, several Islamic dominions encompassed today's Middle East, reached around the eastern Mediterranean, across northern Africa, and controlled most of the Spanish peninsula. Initially, the Islamic empire was governed from Damascus (in today's Syria), but soon Baghdad (in today's Iraq) was established as the new capital. Baghdad, not far from the location of the ancient city of Babylon, quickly became a thriving center of commerce, culture, and learning. By the ninth century, its population of 800,000 was larger than that of Constantinople. Emissaries were sent to acquire the philosophical, mathematical, and scientific texts of the Greeks (and those from Persia and India). The new Chinese technology of making paper from plant fibers provided a durable and cheap alternative to fragile papyrus and expensive parchment. The learned texts were translated into Arabic and copied by hand. Libraries holding thousands of volumes attracted circles of scholars who studied and discussed them. Between the ninth and fourteenth centuries, Islamic scientists, mathematicians, and geographers engaged these works and extended their range. The reputation of Islamic scholarship grew and its fame spread. Islamic civilizations had laid claim to the legacy of Greek learning and inherited its treasure. At first, the nomadic Arabs had little need for permanent buildings. Muhammad himself had no use for architecture, observing that the "most unprofitable thing that eats up the wealth of the believer is building." But soon Islamic rulers began to gather artists, masons, stucco workers, wood carvers, mosaic and tile makers from all over their territory. Guided by Byzantine, Syrian, and Persian architects, these craftsmen built splendid buildings, especially mosques, with marble columns, stone arches, powerful vaults, and rising domes. Wood, metal, brick, stucco, terra cotta, and tile became media for a rich abstract art of glass mosaics, glazed tiles, and ornamented friezes.

Beginning in the sixth century, migrating and invading Germanic, Nordic, and Asiatic peoples came into the region of Western Rome and brought an end to its organizational order. Over the next several centuries, a process of conquest, interaction, adaptation, and assimilation integrated these peoples with local populations. Monasteries were established and monks spread the Christian faith to the territories east of the Rhine and north of the Danube. By the end of the year 1100, new population groups with their own ethnic, cultural, and linguistic characteristics had emerged. They formed a mosaic of many kingdoms and duchies that covered the area of today's British Isles, Germany, Italy, France, most of the Spanish peninsula, Scandinavia in the north, and Hungary and Poland in the east. Many of these kingdoms and duchies were loosely confederated as the Holy Roman Empire (the name intended to invoke both the centrality of Christianity and the power of ancient Rome). They were united enough to launch a sequence of crusades designed to recapture the sacred sites of the Holy Land from Islam. Women and men gained inspiration, direction, and a sense of purpose from their Christian faith. The local church was the visual manifestation of this spirit and the center of the life of the village or town. It was home for worship, meeting place for the community, and school for the children. From the eleventh to the fourteenth centuries, many large cathedrals rose. They received pilgrims who came to venerate relics of saints and to stop en route to the great religious shrines. The Roman basilica, an administrative building with a rectangular floor plan, a central aisle framed by interior colonnades, and a tiered pitched roof, evolved into the Romanesque cathedral with its Roman arches, small windows, and massive walls. Because they were vulnerable to fires, timber ceilings and roof structures gave way to masonry vaults. Master masons and master builders responded to the challenge of supporting the heavy weight of these vaults with systems of columns, piers, and arches. In time, they constructed gravity-defying skeletons of stone to brace and buttress the vaults and soaring spires of the cathedrals of the Gothic age. The portals, windows, statues, and walls of these monumental churches told (and tell) the story of the Bible and the history of the faith in delicately sculpted stone and magnificently colored panels of glass. In soaring beyond what would seem to be earthbound limits, these cathedrals stood (and stand) as magnificent architectural symbols of the transcendence of the realm of God and spirit over the world of man and matter.

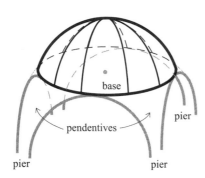

Figure 3.1. The dome of the Hagia Sophia in the abstract

The Hagia Sophia

A masterpiece of Byzantine architecture, the Hagia Sophia (in Greek, *hagia* = holy, *sophia* = wisdom) is one of the great buildings of the world. It was constructed in an incredibly short time between 532 and 537 during the reign of the Byzantine emperor Justinian. Its two architects were mathematicians and scientists, skilled in geometry and engineering. All of their talents were needed for the execution of the unprecedented design of this monumental church. The central part of the structure consists of four large semicircular arches arranged in a square and topped by a dome in the shape of a section of a hemisphere. Figure 3.1 depicts the essence of the design. The four curved triangular structures created by the circular base of the dome and the four arches are called pendentives. The massive supporting columns are known as piers. The dome of the Hagia Sophia is about 105 feet in diameter at its base and rises 180 feet above the floor at its highest point. One pair of the large arches under the dome open into half-domes and in turn into recesses that together provide the church with a continuous clear space of 250 feet in length. The circular arcade of 40 windows around the base of the dome gives the impression that the dome is floating above the soaring space that it creates. The interior surfaces were covered with marble, murals, and golden mosaics. The delicate use of color in the composition of the mosaic of Plate 6 tells us how sophisticated and splendid this art form would become. The interior, with its domes, arches, and vaults, had a celestial quality. The light that poured in from windows at many levels to touch the interior surfaces enriched their artistry and splendor. The elaborate religious ceremonies that this space hosted were officiated by the Greek Orthodox clergy in the sanctuary around the altar. The participation by the members of the imperial court near the entrance reflected the Byzantine duality of church and empire. The experience that any witness to these services would have had must have been spectacular. At around the year 1000, these witnesses included emissaries sent by Prince Vladimir of Kiev to the centers of the four great religions of this region of the world: Islam, Judaism, and Latin and Byzantine Christianity. Their glowing description—of their visit to the Hagia Sophia—surely influenced the prince when he decided that his Russian state would embrace Byzantine Christianity:

> We knew not whether we were in heaven or on earth. For on earth there is no such splendor or such beauty, and we are at a loss how to describe it. We know only that God dwells here among men, and their service is fairer than the ceremonies of other nations. For we cannot forget that beauty.

After the city fell to the Ottoman Turks in 1453, the Hagia Sophia was converted into a mosque and its wonderful mosaics were plastered over.

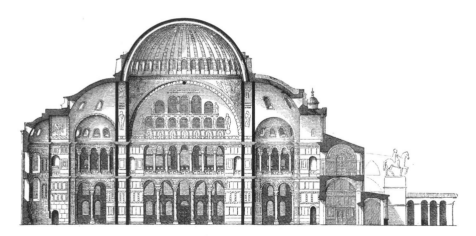

Figure 3.2. Section of the Hagia Sophia. From Wilhelm Lübke and Max Semrau, *Grundriß der Kunstgeschichte*. M. Auflage. Paul Neff Verlag, Esslinger, 1908

However, Plate 7 tells us that the interior of the Hagia Sophia was still a spectacular space 400 years thereafter.

Let's have a look at the basic structural aspects of the Hagia Sophia. Figure 3.2 shows a cross section of the church through the dome, half-domes, and recesses. The shell of the dome is made of brick and mortar and is about $2\frac{1}{2}$ feet thick. Its inner and outer surfaces are sections of spheres that have the same center. Their circular cross sections and the common center are highlighted respectively in black (for the outer circle and the center) and white (for the inner circle). Forty ribs radiate down from the top of the dome, not unlike the ribs of an umbrella. Descending between the 40 windows, they support the dome and anchor it to its circular base. The basic structural challenges facing the builders of the Hagia Sophia were the same as those faced by the Roman architects of the Pantheon four centuries earlier. The fact that brick and mortar have insufficient tensile strength meant that the hoop stress on the shell generated by the downward push of the weight of the dome needed to be controlled by a strong supporting structure at the dome's base. As we saw in Chapter 2, "The Pantheon," in the Roman Pantheon this structure is the massive, symmetric, closed cylinder from which its dome rises. It is apparent from the brief description already given that the design of the Hagia Sophia has a geometry that is more complex than that of a closed cylinder. Its dome rests on four large arches and the pendentives between them. Two of the arches open into half-domes to form the long interior space of the church. The other two arches, as both Plate 7 and Figure 3.2 show, are closed off by walls that are perforated by rows of windows and arcades. So unlike that of the dome of the Pantheon, the support structure of the Hagia Sophia is asymmetric. This is problematic because it means that the dome of the Hagia Sophia was (and is) supported unevenly around its base.

Let's pause to consider the forces that the shell of the dome above the row of 40 windows generates. Figure 3.3 provides a detail of the cross section of

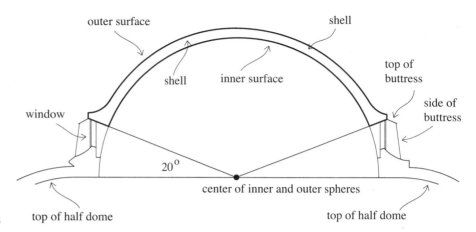

outer surface

shell

shell   inner surface

top of
buttress

side of
buttress

window

20°

center of inner and outer spheres

Figure 3.3   top of half dome

top of half dome

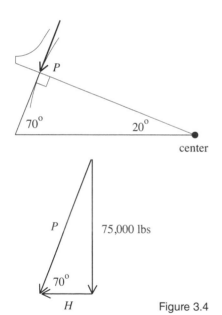

Figure 3.4

the dome that is abstracted from Figure 3.2. The location of the windows, the ribs between them, as well as the supporting buttresses are shown. The two circular arcs are the cross sections of the inner and outer surfaces of the shell. The information about the dimensions and building materials of the dome that follows is taken from recent studies. The two rays emanating from the common center of the two circles make an angle of about 20° with the horizontal. The inner and outer spherical surfaces of the shell have radii of $r = 50$ feet and $R = 52.5$ feet, respectively. The difference of 2.5 feet is the thickness of the shell. The average weight per cubic foot of the brick and mortar of the shell is about 110 pounds. Chapter 7, the section "Volumes of Spherical Domes," applies basic calculus to derive the estimate of 27,600 cubic feet for the volume of the shell of the dome above the circular gallery of windows. This implies that the weight of that part of the shell is approximately 27,600 ft^3 × 110 lb/ft^3 ≈ 3,000,000 pounds. Averaging this weight over the 40 supporting ribs, we get a load of about 75,000 pounds per rib. This means that if P is the slanting push by a rib, then the vertical component of P has a magnitude of about 75,000 pounds. It follows from Figure 3.4 that $\sin 70° = \frac{75,000}{P}$, so that therefore, $P \approx \frac{75,000}{\sin 70°} \approx 80,000$ pounds. The horizontal component H of the push P satisfies $\tan 70° = \frac{75,000}{H}$. Therefore $H \approx \frac{75,000}{\tan 70°} \approx 27,000$ pounds. This is an estimate of the force with which a typical rib pushes outward against the base of the dome.

The architects of the Hagia Sophia were aware of the challenge that the outward thrust of the dome would present (although not in numerical terms) and they took measures to contain it. A rectangular roof structure that features four corners of heavy masonry above the pendentives braces the dome at its base. It can be seen in Figure 3.5 below the circular array of windows. This structure and the four main arches are carried by the four stone piers already mentioned. They rise from foundations of solid rock. These piers are highlighted in black in Figure 3.2. The outward thrust of the dome is

Figure 3.5. Today's Hagia Sophia from the south. The four spires are minarets added after the conversion of the church to a mosque. Photo by Terry Donofrio

contained in two ways. In the long open direction, it is channelled downward and absorbed by the supporting half-domes and the sloping structure beyond them (as shown in Figures 3.2 and 3.5). This is similar in principle to the way a Roman arch transfers its loads downward. In the perpendicular direction the outward thrust is contained by two great external arches under each side of the rectangular roof. One of these arches can be seen in Figure 3.5.

The Hagia Sophia has had a difficult history. The stresses that the building is under makes it particularly vulnerable to the earthquakes that are common in both Greece and Turkey. An earthquake led to a partial collapse of the dome only 20 years after its completion. By 563 the dome had been completely rebuilt. This is the dome discussed above. It is still in place today. The 40 buttresses that brace the 40 ribs between the windows of the dome were added at that time. They are visible in Figure 3.5. Additional earthquakes in the tenth and fourteenth centuries did major damage to the dome and extensive repairs were required each time. These repairs also responded to basic structural problems that had arisen over time. They included the correction of deformations of the main piers. The forces that the dome and the two great exterior arches generate in the direction of the two arches are absorbed by the half-domes and the structures behind them. But the two great arches were deflected outward by the thrust of the dome and huge buttresses were added on the sides of the arches to stabilize them. Two of these buttresses can be seen in Figure 3.5. The structural elements that were added to contain the thrust of the dome do not intrude on the interior space of the church, but they do take a toll on its outward

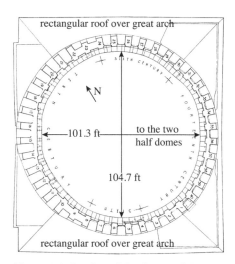

rectangular roof over great arch

rectangular roof over great arch

Figure 3.6. Section of the Dome of the Hagia Sophia at its base. From William Emerson and Robert L. van Nice, "Haghia Sophia, Istanbul: Preliminary Report of a Recent Examination of the Structure," *American Journal of Archaeology*, vol. 47, no. 4 (Oct.–Dec. 1943), p. 424

appearance. The moundlike exterior of the Hagia Sophia lacks the rising elegance of its interior.

As a consequence of the action of the stresses, the damage to the structure, and the extensive repairs that addressed them, the base of the dome is no longer a circle, but an oval. Figure 3.6 shows the horizontal cross section of the dome at its base. Notice that the diameter in the open direction of the two half-domes is about $3\frac{1}{2}$ feet shorter than the diameter between the two great exterior arches. This is consistent with both the stabilizing role played by the two half-domes and the outward deflection that the great arches experienced. Figure 3.6 also provides the positions of the ribs—numbered from 1 to 40—between the windows at the base of the dome and indicates when the various sections of the dome were repaired.

By the middle of the nineteenth century, the great building was once again in need of large-scale repairs. The sultan of the time called upon the Fossati brothers, a pair of Swiss architects, to carry them out. To better contain the outward forces of the dome of the Hagia Sophia, the Fossatis placed an iron chain around its base. This was a strategy that had already been used a century earlier to brace the dome of St. Peter's in Rome. Plate 7 is a lithographic plate from a set of 25 plates fashioned by one of the brothers to record the results of the reconstruction. Today, 1500 years after it was built, the Hagia Sophia—converted to a museum in 1935—is still a grand structure.

It has been said that the architects of the Hagia Sophia made use of mathematics in its design and its execution. While geometry clearly played a role, simple geometric considerations cannot give much information about the stability of a massive building. This was realized much later by Galileo, who observed that geometry alone can never ensure structural success. There is no evidence to suggest that applied mathematics was advanced enough at the time the Hagia Sophia was built to provide even the most elementary analysis of the loads that the structure would have to bear. There seems little doubt that the architects of the Hagia Sophia relied—directly or indirectly—on Roman vault designs and methods of construction rather than theoretical analyses.

Splendors of Islam

Islamic architecture finds its most prominent and distinctive expression in the mosque. A mosque is place of prayer. Its central feature is the mihrab, an ornamented prayer niche that points in the direction of Mecca. Great mosques often open to a large surrounding rectangular courtyard with a fountain that Muslims use for ritual washing. Every mosque has at least one tower, or minaret, from which the call to prayer is issued. Important mosques have several minarets. As the Islamic faith forbids representations of animals, humans, and divine beings, mosques are decorated instead with

Figure 3.7. Dome of the Rock, Jerusalem, 688–692. Photo by I. van der Wolf

intricate, colorful geometric patterns (called arabesques) and calligraphic elements (in Arabic script).

One of the most important early Islamic shrines is the Dome of the Rock in Jerusalem depicted in Figure 3.7. Built between 687 and 691, it is positioned over the rock from which Muhammad is said to have ascended to paradise. It is a place where pilgrims come to honor the prophet. The building has the shape of an octagon. A dome of wooden construction and diameter of 67 feet rises from its center on a cylindrical masonry drum. It is constructed with an inner and an outer shell, each supported by 32 converging timber ribs. The inner surface of the dome is painted with geometric and calligraphic designs. The exterior is covered with boards that are finished with lead and gold leaf. Unlike the dome of the Hagia Sophia, this wooden dome did not lead to structural complications. It is light and sits on its drum like a lid on a pot.

The plan of the Dome of the Rock is laid out in Figure 3.8. Start with a circle and embed a regular octagon into it. Figure 3.8a adds four diameters to the circle (as dashed lines) and the two squares *ABCD* and *EFGH* that they determine. The two squares intersect at eight points. These points determine the locations of the eight piers labeled 1 through 8 in Figure 3.8b. These piers, along with the columns between them (their locations are not shown in the figure), support the octagonal roof. The same eight points are the endpoints of two pairs of parallel segments. The four points of intersection of these segments determine the inner circle that sets the location of the drum of the dome. These points of intersection also provide the position

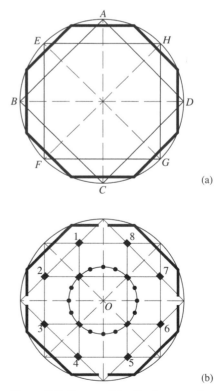

(a)

(b)

Figure 3.8. Dome of the Rock, geometry

Figure 3.9. Dome of the Rock, interior.
© Damon Lynch

of four piers. The four piers and the four sets of three columns between them carry the drum and the dome. Figure 3.8b shows the circle (centered at *O*) as well as the locations of the four piers and twelve columns. Figure 3.9 depicts the interior of the Dome of the Rock including the piers, columns, entrances, and the rock of the prophet. Notice that the columned interior is influenced by earlier Byzantine traditions. But the magnificent execution of the Dome of the Rock was a symbol of the ascending power of Islam.

Soon an Islamic culture began to flower in the cities of Córdoba, Granada, Seville, and Toledo of Arabic Spain or al-Andalus (from which today's region Andalusia in southern Spain derives its name). During a time when the people of Europe north of the Pyrenees suffered through the Dark Ages, the inhabitants of al-Andalus enjoyed gardens, fountains, running water, and lit streets. Córdoba was the capital and cultural epicenter. It was a city of philosophers, poets, and scholars. Thousands of students learned philosophy, law, science, mathematics, and geography in its university. Islamic engineers and craftsmen rebuilt the great Roman stone bridge with its 16 arches and span of about 800 feet over the city's river, the Guadalquivir (Arabic for "the great river"). The bridge still spans the river today. Córdoba's great mosque still stands as well. Built in several stages from 786 to 988, it is a large rectangle of about 590 by 420 feet in plan. Its interior space—now with a Catholic church spliced into its center—is astonishing. Plates 8 and 9 and Figure 3.10 show some of its remarkable artistry. Notice the double arches, the horseshoe arches, the three-leafed arches, the pointed arches, and the splendid detail with which they are constructed. Consider the dome in Figure 3.10. Each pointed arch at the base of the dome is part of a configuration of arches that transfers the loads of the dome down to the vertical wall below. Such

Figure 3.10. The dome rising over the Mihrab of the Great Mosque of Córdoba. Photo by Richard Semik

a structure is called a squinch. This structural device, common in Islamic architecture, provides an attractive alternative to the pendentive.

Late in the twelfth century a magnificent minaret was built for the mosque of Seville. After Seville returned to Christian control, the mosque was converted to a church. Early in the fifteenth century this earthquake-damaged structure was torn down and during the next 200 years a huge new cathedral was built on the site. It exceeded the Hagia Sophia in size and was the largest Christian church at the time of its completion. Fortunately, the original minaret survived. The 230-foot-high minaret was extended with the addition of a story for bells and a spire. It became the bell tower for the new cathedral. A weathervane at the top of the spire gives the tower its name, Giralda. The word derives from the Spanish *gira* meaning "that which turns." Ironically, the turning weathervane has the form of a woman that represents the unshakeable virtue of Faith. Plate 10 depicts the bell tower. It is still a prominent landmark of the city of Seville today. The delicate lacelike patterns in stucco and brick as well as the arched structures and balconies are wonderful examples of Islamic design.

By the end of the eleventh century the Arabic Islamic empire began to decline and a Turkish dynasty gained control of large parts of Islamic territory and in particular of Persia (a region that includes today's Iran). However, the building of magnificent mosques continued. An impressive example is the Friday Mosque in Isfahan. Its brick domes date from the beginning of the construction in the eleventh century. Figure 3.11 provides a study of the brickwork that gives shape to the pointed arches and squinches that undergird two octagonal domes. Figure 3.12 shows a circular dome also supported by squinches. Notice the ribs embedded within its shell. They are arranged in a double pentagonal design. They extend down to the base and play a supportive structural role. Figures 3.11 and 3.12 both afford a look at the techniques of construction of the architectural features beneath the gilded finishes of the interior of Córdoba's great mosque.

Expanding along with its territory was the genius of Islam to borrow advances from other cultures, to absorb them, and to infuse them with its own

Figure 3.11. Interior structure of the Friday Mosque, Isfahan, Iran. Photo by seier+seier

Figure 3.12. A circular dome of the Friday Mosque, Isfahan. Photo by seier+seier

spirit. Islamic architecture is a striking example. The architects of Islam made use of the structural forms (arches, arcades, domes) that they encountered in the territories that they conquered, but they transformed them with their own creative energy. What resulted was much more than imitation and adaptation, but new designs and structures. Aspects of Islamic architecture—its decorative elements, pointed arches, and ribbed structures—began to appear in central Europe more than a century later.

Romanesque Architecture

While Eastern Christianity preferred a centralized concept for its churches, Western Christianity turned to the basilica (from royal house, in Greek *basilikos* = royal), a large, colonnaded, rectangular hall with a sloping roof that the Romans used for transacting business and legal matters. The rectangular inner space of a basilica was usually divided into three or more sections by parallel colonnades. The Roman magistrates sat at the opposite end of the entrance in a vaulted recess, often on a raised platform. When the basilica became the model for early Christian churches, the central longitudinal space became the nave (in Latin, *navis* = ship) and the vaulted recess became the apse (in Greek, *apsis* = arch or vault). The apse was the place for the main altar. The two-tiered sloping roof structure created space for the placement of windows, the clerestory windows (clerestory comes from the Latin for clear story), high on the walls of the nave. The original St. Peter's church, depicted in Figure 3.13, illustrates this architectural form. Built in Rome in the fourth century, it was razed in the sixteenth century to make way for the new St. Peter's, today's Roman landmark and powerful symbol of Catholic Christianity.

Figure 3.13 tells us that the roof of Old St. Peter's was supported by configurations of wooden beams. The study that follows focuses on a simplified version of the triangular component at the top. The horizontal and vertical segments in the middle of the triangle are omitted, and the remaining three beams are assumed to be rigid and attached to each other by pins. Figure 3.14 illustrates this element (today referred to as a simple truss). We will assume that the entire gravitational load L on the triangle (in pounds or tons, for instance) acts at C and that the configuration is stable. The triangular structure transmits the load L downward along the two slanting beams to the points A and B. At these points it is supported by the two vertical walls. We'll assume that the structure is symmetric, that the loads at A and B are equal, and that the angles at A and B with the horizontal are both α. We'll let P be the magnitude of the slanting downward push by each of the two beams.

Figure 3.15a separates the push P at A into its horizontal and vertical components. The force diagram of Figure 3.15b tells us how the structure

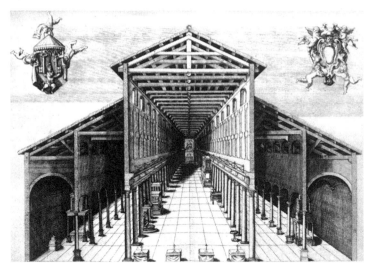

Figure 3.13. Old St. Peter's Basilica in Rome, built during 320–337. Sketch by Jacopo Grimaldi, c. 1590. From G. B. Costaguti, *Architettura della Basilica di S. Pietro in Vatican*, Rome, 1684

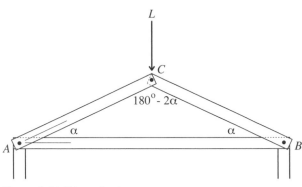

Figure 3.14. Triangular truss

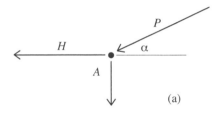

(a)

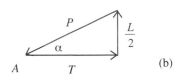

(b)

Figure 3.15. Two force diagrams

responds at A. The vertical wall pushes upward with a force of $\frac{L}{2}$ to counteract the vertical component of P and the horizontal beam, called a tie-beam, pulls inward with a force of $T = H$ to counteract the horizontal component of P. Notice that the two slanting beams are being compressed and that the tie-beam is under tension. Figure 3.15b tells us that

$$\sin\alpha = \frac{L/2}{P} \quad \text{and} \quad \tan\alpha = \frac{L/2}{T}.$$

After rearranging these two equations,

$$P = \frac{L}{2\sin\alpha} \quad \text{and} \quad T = \frac{L}{2\tan\alpha}.$$

It follows, as expected, that if both L and α arc known, then the push P and the pull T can be determined. By the symmetry of things, a study of the forces at the point B gives the same conclusions. The above equations can also be derived by analyzing the forces on the structure at the point C. (See Problem 7.)

Suppose that $\alpha = 25°$ (this is close to the corresponding angle in Figure 3.13) and that $L = 10{,}000$ pounds. Then

$$P = \frac{10{,}000}{2\sin 25°} \approx 11{,}800 \text{ pounds} \quad \text{and} \quad T = \frac{10{,}000}{2\tan 25°} \approx 10{,}700 \text{ pounds}.$$

Observe that the tension $T \approx 10{,}700$ pounds in the horizontal tie-beam is larger than the vertical load of $L = 10{,}000$ pounds. This is considerable, but timber resists both tension and compression well. The strength of timber (in both tension and compression) can range from 2100 pounds per square inch (psi) for soft woods to 5700 psi for hard woods. Specifics depend on the particular internal structure of the wood, its moisture content, and the direction of the grain.

The Germanic invasions of the central European realm were over by the seventh century. A recovery from the Dark Ages began and in subsequent centuries the region became Christianized. By around the year 1000 churches were being built in larger numbers. A modified version of the Roman basilica plan begins to emerge. The basic design is expanded near the apse with the addition of a rectangular section perpendicular to the nave. This extension, called a transept, gives the plan the shape of a cross. The space at the intersection of the nave and transept is called the crossing. The transept creates new spatial separations in the interior of the church. The chancel is the area between the apse and the transept. It includes the main altar for the clergy and the choir, the space for singers and musicians. The nave on the other side of the transept is the area for those who come to worship.

One style in which the basilica plan was executed came to be called Romanesque. There are variations, but common elements include the sloping roof structure with the clerestory windows and the liberal use of semicircular Roman arches. Massive masonry vaulting rose over the nave, aisles, and transept (often covered by wood framed roofs like those of Old St. Peter's depicted in Figure 3.13). Semicircular barrel vaults as well as groin vaults—obtained by intersecting two semicircular barrel vaults at right angles—were common. The weight of these vaults presented structural challenges. Heavy columns, piers, and walls were introduced to counter the large thrusts that they generated. Along with the ever-present Roman semicircular arches, these massive masonry elements and the small windows that perforated the walls became characteristic features of Romanesque architecture.

A Kaiserdom, or imperial cathedral, was erected in the city of Speyer in Germany in the middle of the eleventh century in the basilica plan. Work to enlarge the entrance area with a second transept was completed in the early twelfth century. The crypt, a burial chamber below the nave, is the resting place of several Holy Roman emperors and German kings. With a length of 444 feet it is the largest Romanesque church in existence. In spite of damage from wars and fires and the successive and extensive restorations that followed, the cathedral has retained the overall form and dimensions of the original structure. On the exterior the building is anchored by two pairs of tall towers that frame the chancel on the one end and the transept at the entrance on the other. Over each of the two crossings is an octagonal dome. The dome and the two towers near the apse are seen in the foreground of Plate 11. The cathedral has a colonnaded gallery that goes around the entire exterior just below the roofline. A similar gallery adorns the two domes below their base. The nave, towers, and domes are all roofed with pale green, weathered copper that contrasts with the pinkish red of the building stone.

On the inside, two arcades with heavy semicircular arches supported by massive piers separate the nave from the two aisles. The high groin vault over the nave is segmented by semicircular arches that reach across the nave. These arches are supported by columns that flow down along the wall of the nave to merge with the massive piers of the arcades.

A number of splendid Romanesque churches were erected in France at stopping points along pilgrimage routes. The Basilica Sainte Madeleine in Vézelay, Burgundy, is one the finest. The present church was built in the first part of the twelfth century (but it was restored several times over the centuries). The interior of the basilica in Vézelay is very similar in structure to that of the Cathedral of Speyer (but it is smaller and its vaults are lower). It is depicted in Plate 12. The ceiling of its nave consists of a sequence of groin vaults at regular intervals, all perpendicular to the long vault over the nave. Notice how these groin vaults create the spaces for the clerestory windows. The groin vaults are separated from each other by striped semicircular arches that merge into vertical columns that in turn join the piers of the arcades along the nave. These arches and columns partition the nave into a series of segments called bays. The original apse was destroyed by fire soon after its completion and was rebuilt in the Gothic style with its characteristic rows of large windows.

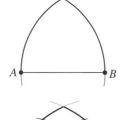

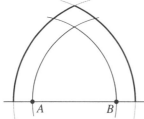

Figure 3.16. The Gothic arch

Soaring Gothic

By the beginning of the twelfth century, the Romanesque form was gradually giving way to the Gothic style. The word "Gothic" was first used in the Italian Renaissance as a negative term for all art and architecture of the Middle Ages, suggesting that it was of the quality of the work of the barbarian Goths. Today the term Gothic Age refers to the period of art and architecture immediately following the Romanesque. It is regarded to be an era of outstanding artistic achievement.

The most easily recognized feature of the Gothic style is the pointed or Gothic arch. Figure 3.16 shows one common design. Simply take a segment *AB* and draw two circular arcs of radius equal to the length of the segment, one centered at *A* and the other at *B*. To draw the outer perimeter of the arch, keep the two centers but increase the radii of the arcs. A look at Figure 3.17 tells us that as a structural device, the Gothic arch has advantages and disadvantages when compared to the semicircular arch. To span the same space, the Gothic arch needs to be higher. However, when it comes to transferring loads, the Gothic arch transfers them with a smaller horizontal component and a larger vertical component. This means that the outward thrusts generated by Gothic arches are less and therefore more easily contained. The fact that the vertical components are larger means that the materials from which

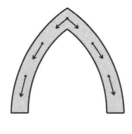

Figure 3.17. A comparison of forces

it is constructed are under greater compression. In this regard, however, we know that stone, masonry, and concrete stand up to compression very well.

We'll now take a look in quantitative terms at the outward thrust that a Gothic arch generates. The diagram of Figure 3.18a is the simplified model of a Gothic arch that we will consider. Our study will assume that the total load L that the arch supports acts at the very top and that this load includes the weight of the slanting elements of the arch. The angle that these elements make with the horizontal is α in each case. The slanting parts of the arch transfer the load downward. Can we compute, or at least estimate, the horizontal component H of the thrust that is generated? We will answer an equivalent problem instead. If we were to reinforce the arch by adding a horizontal beam as shown in Figure 3.18b, what would the pull T by this beam have to be to make the arch stable? This is the horizontal force H that we are looking for. The question about the pull T has transformed the problem into the context of Figures 3.14 and 3.15 of the preceding section. Applying the analysis undertaken there tells us that

$$H = T = \frac{L}{2\tan\alpha}.$$

Suppose, for example, that $L = 5000$ pounds. Let's take α successively equal to 30°, 45°, 60°, and finally 75°, to see how H depends on the steepness of the arch. Because $\tan 30° = 0.58$, $\tan 45° = 1.00$, $\tan 60° = 1.73$, and $\tan 75° = 3.73$, we find that the corresponding horizontal forces H are

$$4{,}330, \quad 2{,}500, \quad 1433, \quad \text{and} \quad 670$$

pounds, respectively. As Figure 3.17 already suggested in qualitative terms, the horizontal thrust H decreases dramatically as the steepness of the arch increases.

Our analysis also points to the important role played by the materials used in the components of a structure. If an arch is firmly braced as shown in Figure 3.18b, then it may be able to contain on its own the outward thrust produced by the loads that it supports. An arch made of a strong material such as high-grade steel or reinforced concrete may be able to respond in the same way. However, an arch made of stone, masonry, and ordinary concrete offers less internal resistance. Such an arch will be able to support massive loads only if the horizontal thrusts that the loads generate are counteracted by components of the structure acting against the sides of the arch.

The most important structural element of Gothic architecture is the ribbed vault. To understand the Gothic ribbed vault, return first to the ceiling of the basilica of Vézelay in Plate 12. Recall that the nave is segmented into bays, and that the ceiling of each bay is a groin vault obtained by the intersection of two cylindrical barrel vaults. Each of these groin vaults is bounded by two striped semicircular arches. Have a careful look at Plate

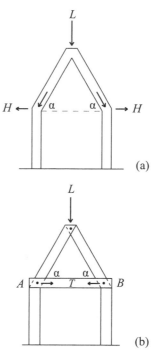

Figure 3.18. Horizontal thrust generated by an arch

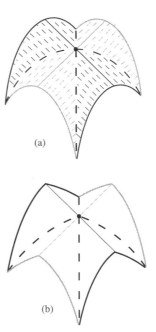

Figure 3.20. From the groin vault to the Gothic ribbed vault

Figure 3.19. Groin vault at Vézelay. Photo by Vassil

12 and notice that the two cylindrical surfaces of each groin vault give rise to two intersecting circular arcs. They are the dashed curves highlighted in Figure 3.19. Figure 3.20a is an abstract and reoriented version of the groin vault of Figure 3.19. The two barrel vaults, one coded in black and the other in gray, intersect perpendicularly. The two highlighted dashed lines represent the circular arcs at which they intersect. (The fact that one of the arcs is drawn as a straight line is a consequence of the chosen orientation.) Now modify Figure 3.20a as follows. Replace each of the two cylindrical surfaces by a curved surface that has pointed Gothic arcs as vertical cross sections. With this change, the semicircular arcs at the boundaries of the two barrel vaults depicted in Figure 3.20a are transformed into the two pairs of Gothic arcs depicted in Figure 3.20b. The semicircles that form the boundary of Figure 3.20a are transformed into Gothic arcs as well. What has happened so far is only a small geometric change. But now comes the Gothic architect's golden idea: Build the two crossing Gothic *arcs*—at this point they are only geometric curves—into structurally relevant, load-bearing, Gothic *arches*. In combination with the Gothic arches at the boundaries they form the interlocking grid of ribs shown in Figure 3.20b. This grid is the structural skeleton of the Gothic ribbed vault.

The ribbed vault was critical to the development of Gothic architecture. It gave builders new possibilities of design and construction and came to

Figure 3.21. Ribbed vaults over several bays in the ceiling of a Gothic cathedral. Photo by Magali Ferare

Figure 3.22. Sets of three converging ribs along with supporting columns. Photo by BjörnT

dominate medieval architecture. In a ribbed vault, the ribs are the primary structural members. The spaces between them are structurally less relevant and can be filled in with thinner masonry materials. So the ribbed vault is lighter and hence structurally more resilient than the more massive barrel and groin vaults. It was also easier to construct and it made a number of architectural innovations possible. Building a vault requires centering (the temporary wooden structure that we have already encountered in Chapter 2 in the context of the Roman arch) that supports the masonry until the shell of the vault is completed and the mortar has set. In the construction of a groin vault, the ceiling of an entire bay must be supported in this way. The construction of a ribbed vault is simpler. Builders laid two intersecting diagonal arches across the bay and supported them with light centering high on the nave walls. As part of an interlocking grid of ribbed arches this divided the vault into smaller triangular cells. One such arrangement is shown in Figure 3.21. After the ribs had set, these triangular cells were filled in with masonry or concrete. Again, no extensive centering was required.

The configuration of the ribs in the ceiling of the nave and transept determine the rest of the structural scheme of a Gothic church. The points of convergence of the ribs determine the placement of the supporting vertical columns that run down the sides of the nave. See Figure 3.22. These

Figure 3.23. A sequence of flying buttresses bracing corresponding sets of ribs and columns. Photo by Harmonia Amanda

columns flow into thicker piers toward the bottom. The final piece of the structural puzzle has to do with the outward forces that the vaulted ceiling and especially the ribs generate. See Figures 3.14 and 3.18 and the discussions that explain them. Since these thrusts are concentrated at the points where the ribs converge and along the columns that support them, they can be countered by flying buttresses. Figure 3.23 shows how these half-arches push against the wall of the nave from the outside along the vertical pressure lines determined by the ribs of the vaults and the supporting columns. The fact that the walls between the columns are of lesser relevance to the structural integrity of a Gothic church means that they can be punctured to provide large areas for windows. Large, finely crafted, stained glass windows (often magically illuminated by the light from the outside) are characteristic features of Gothic architecture. The delicate masonry work that frames the glass panels of these windows is known as tracery. The rose window depicted in Figure 3.24 provides an artistic example.

The structural scheme of a Gothic cathedral starts from the ribbed vault and flows logically downward and outward. The ribbed vault enabled Gothic master architects to build churches with soaring interiors that were higher, more elegant, and more delicate than earlier, sturdier, and more massive Romanesque churches.

A number of splendid Gothic cathedrals were built in France, England, Germany, and Austria. Regarded by architectural historians to be one of the

Figure 3.24. A large rose window. Photo by Harmonia Amanda

finest is the Cathedral of Notre Dame in the town of Chartres, an important center of pilgrimage about 50 miles from Paris. Construction of the cathedral began in the middle of the twelfth century, but a fire late in the century destroyed much of what had been built (and much of the town). The cathedral was rebuilt within 25 years (a short time for Gothic cathedrals) and was finished by the year 1220. The ribbed vaults, supporting columns, flying buttresses, and rose window of Figures 3.21, 3.22, 3.23, and 3.24 are those of the Cathedral of Chartres. Plate 13 shows a comprehensive view. Flying buttresses run along the sides of the nave and around the apse. The pattern is interrupted by the transept, braced not by flying buttresses but by small towers. The spires at the entrance act as piers against the pressure generated by the segments of the vaults near the entrance. The two spires are very different. One is 349 feet high and dates from the twelfth century. It is in plain early Gothic style. The other is 377 feet tall and was built in the sixteenth century. It is in French high Gothic style. The cathedral is famous for its stained glass windows, many in beautiful blue hues dating from the thirteenth century. Plate 14 depicts the rose window of Figure 3.24 viewed from the interior. In the same way that colored tiles with intricate designs are central to the artistry of Islam and delicate golden icons and mosaics are the hallmark of Byzantine art, stained glass windows in splendid hues exemplify the art of the Gothic Age.

We have seen that in its essence, Gothic architecture is geometry executed in stone. The upcoming discussion will tell us how very literally true this was.

From the Annals of a Building Council

Milan, capital of the northern Italian state of Lombardy, emerged from a period political instability to a position of power in the latter part of the fourteenth century. It annexed large areas of territory from neighboring Italian states and began to amass wealth. A new cathedral was planned to signal a new status for the city. The Gothic style that had started in France had spread to England, Germany, and Austria, where it found expression in a number of huge and soaring cathedrals. The new church of Milan was envisioned to rival the largest and most impressive of these in design and size. The duke of Milan established a Building Council to oversee the construction. What happened next is revealed in the *Annali della Fabbrica del Duomo di Milano.* Ordinarily, records of medieval construction provided little beyond tracking supplies, financing, and employment. But the *Annals* of the Building Council provide rare and valuable insights into planning efforts, building processes, and medieval approaches to questions of design and structure.

The members of the Building Council were inexperienced, but very confident in their judgment nonetheless. The council decided that the church

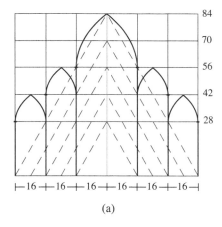

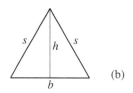

Figure 3.25

should be built in the basilica plan with two aisles on each side of the nave and it approved the design of the vaults and their supporting piers. The council's decisions are recorded in the *Annals*. Figure 3.25a summarizes many of them. It provides the configuration and shape of the vaults, the locations of the central axes of the piers, and the heights of the piers. The *Annals* list the dimensions in Milanese braccia. (A Milanese braccio corresponds to about 1.95 feet.) Figure 3.25a tells us that the council's design is based on a triangular scheme, included in dashed lines in the diagram. For instance, Figure 3.25a informs us that the width of the interior of the church was to be $96 \times 1.95 \approx 187$ feet, that the main vault over the nave was to rise to a height of $84 \times 1.95 \approx 164$ feet above the floor, and that the piers supporting it would be $56 \times 1.95 \approx 109$ feet high. The five triangles (with base at the bottom of the figure) that determine the design are isosceles and similar (because corresponding sides rise at the same rate of $\frac{28}{16}$). These triangles are close to being equilateral. Consider, for instance, the triangle with base $b = 32$ and height $h = 28$. Figure 3.25b tells us that its side s has length $s = \sqrt{16^2 + 28^2} = \sqrt{1040} \approx 32.25 \approx b$. This computation also tells us that the cross sections of the proposed vaults are close to the design specified in Figure 3.16.

Ground was broken in 1386. By the early 1390s, the foundations were finished and the piers were started. But as the piers grew, problems with the foundations surfaced. The council decided to stop construction and to review its plans for the church. A master Gothic architect from Germany was brought in as a consultant. The proposed heights of the piers, dimensions of the nave and aisles, the heights of the vaults, as well as the configuration of the buttressing structures came under rigorous review. The German expert voiced strong concerns about the structural soundness of what had already been built, especially the piers and the exterior buttressing. He recommended that the heights of the vaults be brought into line with those of Gothic cathedrals north of the Alps. He argued that the square rather than the triangle is the geometry that should be followed, and in particular that the height of the main vaults should be raised from 84 to 96 braccia to match the 96-braccia width of the interior. The Building Council convened a great forum to discuss the recommendations of the German expert. The *Annals* summarizes the discussions and outcomes in a question and answer format. There are eleven Q&As in all. The questions inform us about the issues that the German expert raised. The answers provide the council's final word on the matter. The most important exchanges include the following:

Q: Whether the portions of the rear as well as the sides and interior—namely, both the crossing and the other, lesser, piers—have sufficient strength?

A: It was considered, replied, and stated upon their soul and conscience, that in aforesaid the strength, both of the whole and separate, is sufficient to support even more.

Q: Whether work on the exterior piers or buttresses is to proceed as it was begun or be improved in any way?

A: It was said that this work was pleasing to them, that there is nothing to be altered, and that, on the contrary, work is to proceed.

Q: Whether this church, not counting within the measurement the tower which is to be built, ought to rise according to the square or the triangle?

A: It was stated that it should rise up to a triangle or to the triangular figure, and not farther.

The record shows that all of the German master builder's proposals are rejected. His concerns about the strength of the piers are responded to by an appeal to "soul and conscience." There was "nothing to be altered" in the exterior buttressing. The church should "rise up to a triangle" and the square—the idea that the height of the vault over the nave should match the width of the interior—is dismissed. In fact, the council responds to several questions about the heights of the piers by *lowering* the interior structural elements of the earlier design. The height of 28 braccia for the piers supporting the vaults over the outer aisles is retained, but all the vaults and interior piers are lowered. The revised heights for the piers and vaults are provided by Figure 3.26a. The dashed triangle at the center of the diagram provides the heights of the vaults over the inner aisles as well as the heights of the piers that support the vaults over the nave. This dashed triangle has a height of $52 - 28 = 24$ braccia and a base of $4 \cdot 16 = 64$ braccia. Because $24 = 8 \cdot 3$, $32 = 8 \cdot 4$, and

$$\sqrt{24^2 + 32^2} = \sqrt{8^2 \cdot 3^2 + 8^2 \cdot 4^2} = 8 \cdot \sqrt{3^2 + 4^2} = 8 \cdot 5,$$

it follows that this triangle is put together from two 3-4-5 right triangles as shown in Figure 3.26b. The larger dashed triangle toward the top of Figure 3.26a determines the height of the vault over the nave. It is easy to check that it too is a composite of two 3-4-5 right triangles. In lowering the heights of the vaults and piers in the interior of the church, the council abandoned the geometry of equilateral triangles of Figure 3.25a and replaced it by the geometry of 3-4-5 right triangles of Figure 3.26a. This concept prevailed and the heights of the piers and vaults listed in Figure 3.26a are close to those of the cathedral as it was built.

The rulings issued by the Building Council reveal a vision for a church with a profile that was lower and broader than the soaring profiles of earlier French, English, and German Gothic churches. The council's thinking was that if the nave is low enough and the inner aisles high enough, then the vaults of the nave could be adequately buttressed by the walls and vaults of the aisles on each side. These in turn would be supported by the structures of the outer aisles so that there would be no or little need of flying buttresses.

The council dismissed the German master builder and soon thereafter another. During the next several years the piers were completed, but final

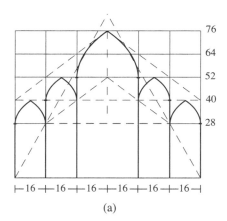

(a)

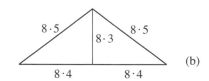

(b)

Figure 3.26

decisions about the vaults were postponed. Construction proceeded in spite of the fact that fundamental questions about the design of the church remained unresolved. Dissatisfied with the earlier German input, the council now called in French experts for advice. Two of the three French experts sensed the futility of their involvement and quickly departed. The third remained to raise alarm about the structure and to present extensive criticisms. He regarded the support for the proposed tower over the crossing inadequate. He thought the structures of the apse, in particular the piers, their foundations, and the buttressing too weak to be able to sustain the thrusts from the tower and the vaults. He was adamant that these structures needed to have greater strength. He also believed the buttresses around the exterior perimeter of the church to be too weak. Proposing to double their widths, he recommended that the ratio of the thickness of exterior buttresses to that of the piers be 3 to 1 rather than $1\frac{1}{2}$ to 1, as the council had specified. He also raised an aesthetic point about the heights of the piers along the nave in relation to the heights of the capitals that topped them. The proposed height for these capitals exceeded what the conventions of the time called for. (These piers and capitals are studied at the top of Figure 3.28.)

History would repeat itself. The *Annals* record that the Building Council rejects the concerns raised by the French master builder. The council responds to each of the points that he brought forward. Both experts and "formal oaths" had confirmed the soundness of the foundations of the church (that reached 14 braccia into the ground). The piers had a solid stone core and were reinforced with spikes of leaded iron. For additional support, they would be joined to each other with large iron binding rods above their capitals. The council asserts that Milanese stone and marble used in the buttressing is twice as strong as French stone, that the buttressing is correctly conceived, and that it could carry even greater weights. It concludes that no additional buttressing would be required for any part of the church. To drive home its defense of the existing design, the council turns to the vaults and asserts—rather incredibly—that they had been specified "to have pointed arches made according to the type suggested by many other good and expert engineers, who say concerning this, that pointed arches do not exert a thrust on the buttresses."

The council simply knew best and rejected the repeated warnings voiced by the German and French Gothic master builders about the soundness of the structure. Construction proceeded. The main altar of the cathedral was consecrated in 1418, but the pattern of delays would persist. In the 1480s new concerns arose about the proposed vault and tower over the crossing. True to form, expert advice was ignored again (even that of Leonardo DaVinci, as we will see in Chapter 5). Given the great cost of the construction, major delays were also brought on by the changing political and financial fortunes of the state of Lombardy. The cathedral would not be completed until 1572. The studies of Figures 3.27 and 3.28 confirm that the final design combines the

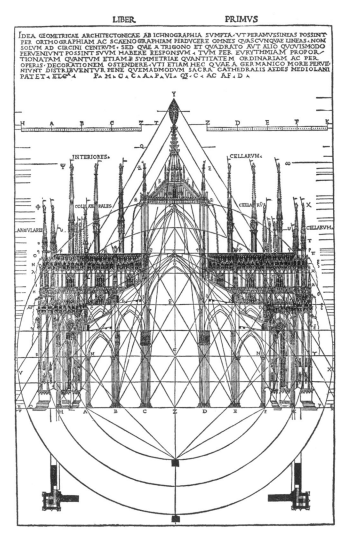

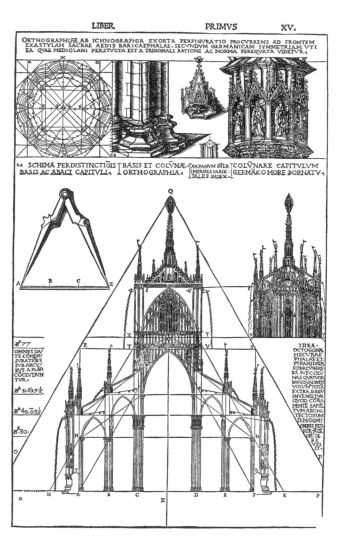

Figure 3.27. Section of the Duomo, Milan, including piers, vaulting over nave and aisles, vaults of crossing tower, and transept. From Cesare Cesariano's Italian edition of Vitruvius Pollio's *De Architectura*, Como, 1521. Marquand Library of Art and Archaeology, Princeton University, Barr Farre Collection

Figure 3.28. Section of the Duomo, Milan, with details about structural elements, in Cesare Cesariano's Italian edition of Vitruvius Pollio's *De Architectura*, Como, 1521. Marquand Library of Art and Archaeology, Princeton University Library, Barr Farre Collection

geometries of both equilateral and 3-4-5 right triangles. In addition to the vaults and piers, these studies include designs for the transept and the tower over the crossing. The small spires that rise over the roof play a structural role. The weight of these pinnacles adds rigidity to the columns and piers that they top. Figures 3.29, 3.30, and 3.31 depict Il Duomo di Milano as it looks today. The delicate and dense forest of marble ornamentation on the piers, buttresses, walls, and windows of the exterior is the work of the nineteenth century.

Figure 3.29. Some of the main piers of the nave with capitals and iron binding rods, and Gothic vaulting of the Milan Cathedral. Photo by Giovanni Dall'Orto

It is not surprising that the construction of the great stone monuments of the Gothic Age was led by master masons. The master masons oversaw the quarrying of the stone and supervised the cutting and shaping of the stone blocks. They also directed the hoists and cranes that raised the heavy stone blocks and masonry materials to place them in position at the site. (Discussion 3.3 pursues these matters further.) The account of the *Annals of the Building Council of the Cathedral of Milan* informs us about the *art* of building, and in particular about the importance of the experience and skill of the master mason and the relevance of the strength of stone and masonry. But the *Annals* also tell us that the art of building must be guided by precise scientific principles, or *scientia*. This had little to do with what we call structural engineering today. There was a general sense of the relevance of loads and thrusts and there were rules of thumb about the relative thicknesses of structural members. However, at no point in the controversy between the Gothic builders and the Building Council was either side able to provide a sound argument for the reliability of any component of the structure of the Milan Cathedral. Even rough estimates of thrusts such as those illustrated in the previous section on Romanesque architecture were completely beyond their

Figure 3.30. The triangular Gothic facade with its pinnacled buttresses of the Milan Cathedral. Photo by Skarkkai

conceptual horizon. So what was the *scientia* that guided the Gothic architect? The only science available at the time: geometry. In fact, geometry was more than a science, it was a divine activity. God's compass in Plate 15 is a symbol of His act of Creation. God created the universe following geometric and harmonic principles. He was the great geometer and the medieval master builder followed his lead. Simple geometry and related numerical relationships were used to determine the dimensions of relevant structural members including the widths, heights, and spacing of piers, walls, and vaults. Geometry not only informed dimensions, it also determined structure. For a Gothic building to be sound its design had to conform to an integrated geometric and numerical scheme. The configuration of the relevant components of the structure needed to be consistent with the geometric scheme of the whole. The Gothic concept of structure relied on geometric and numerical relationships and not on considerations of loads, thrusts, and stresses.

The Milan Cathedral documents what has just been described. Figures 3.25a, 3.26a, 3.27, and 3.28 show the scheme of triangles that determines the sizes and positions of the important structural components and the way they are interconnected. Each of the main piers along the nave is 4 braccia wide. The 4-braccia width determines the 2-braccia height of its base. The relationship between the base and the capital of a main pier is 1 to 4. So its capital

Figure 3.31. A view of the exterior, including the apse, transept, and the tower over the crossing of the Milan Cathedral. Photo by Fabio Alessandro Locati

is 8 braccia high. The 4-braccia width of a main pier has a 1 to 3 relationship to the 12-braccia distance between adjacent main piers. The distance between two piers, center to center, is 16 braccia. Twice this is the width of 32 braccia of the nave. An appeal to the square determines that 32 braccia is also the length of the shaft of a main pier (the pier without its capital). The 40-braccia height of the vaults over the outer aisles is the same as that of a main pier. So we see that the dimensions of the essential components of the structure are provided by a sequence of numerical ratios that arise from the geometric interconnectedness of the scheme as a whole. At the end, it is the web of numerical connections that attests to the stability of the structure.

What if two different geometric standards are proposed? How would one choice prevail over the other? Such a matter is resolved by drawing on information provided by traditional building practice and the expertise of the masons. This is how the decision about the vertical geometry—whether equilateral triangles or 3-4-5 right triangles—would have been made. This is also how the disagreement between the Building Council and the French master builder over the appropriate ratio of the thicknesses of the piers to that of the buttressing—whether 1 to 3 or 1 to $1\frac{1}{2}$—would have been resolved. It was in this way that a structure evolved and emerged as a compromise between practical knowhow, or *ars*, and geometric form, or *scientia*.

Let's return once more to the contentious disputes between the Gothic experts and the Building Council. Which side did history declare the winner?

While the concerns of the Gothic masters probably had merit, the decisive argument favors the council. The fact is that the cathedral, built according to the council's design, still stands. The poor foundations, weak piers, inadequate buttresses, and the vaults with their "pointed arches that do not exert a thrust" have survived for five centuries. The disproportionate capitals of the main piers with their decorative niches and statues of saints—see Figures 3.28 and 3.29—became a distinctive feature of the cathedral. However, the flying buttresses that the council wanted to avoid were added later in the nineteenth century. The construction of the cathedral—given its novel design—had been a grand experiment. The account in the *Annals* informs us that its success depended more on luck than the skill and know-how of its builders and experts on the council. It is remarkable that Gothic master masons (like Egyptian, Greek, and Roman builders before them) were able to execute the astonishing structures of their age, in spite of their very limited means of construction and their lack of understanding of loads and thrusts.

The Magic of Venice and Pisa

By the tenth century the city-states of Venice and Pisa—both strategically placed on the Mediterranean—were developing a flourishing trade that connected the European realm with both the Byzantine Empire and the Islamic world. The merchants of these cities shipped and traded timber, fur, wool, metals, grain, cloth, spices, and silks and became wealthy and influential. Ideas flowed back and forth along the trade routes as well and it is not surprising that the architecture of these cities was to become influenced by Byzantine and Islamic forms.

The two most prominent and significant examples of Venetian architecture are the church of San Marco and the Palace of the Doges, the residence of the Venetian dukes. (Doge is Venetian Italian for duke as derived from the Latin *dux*, meaning leader.) The church of San Marco was completely rebuilt in the years from 1063 to 1089 in the Byzantine style. The new church was crowned by five domes, a larger central dome surrounded by four others in a cross-shaped configuration. The domes were made of brick and supported by massive arches, vaults, and piers. With a low profile and circular row of windows at the base, each of them was a smaller version of the dome of the Hagia Sophia. Brick and terra cotta were the primary building materials of the church. Since this included the decorative elements, the church would have made a plain, if not stark, impression. This began to change dramatically in the thirteenth century. During a crusade led by the Republic of Venice, Constantinople was sacked and much of artistic value was taken. With the spoils of the crusade the interior of San Marco was transformed. Figure 3.32 gives a sense of the profusion of golden mosaics, splendid inlaid marble, and rich ornamentation that the solid brick surfaces gave way to.

Figure 3.32. Canaletto, San Marco: the interior, c. 1755. Oil on canvas. The Royal Collection © Her Majesty Queen Elizabeth II

The outside of San Marco was transformed as well. Figure 3.33 depicts the new facade from a vantage point on the piazza in front of the church. Islamic ogee arches, arches that first curve inward and then outward as they rise, were added to the top of the facade. The largest of them soars high over the central entrance and frames four striking bronze horses taken from Constantinople. Each of the other four ogee arches caps a semicircular panel of rich mosaics. Plate 16 provides a glimpse of the opulence of the facade. Columns and rectangular panels were added to give decorative detail to the five tall indented arching structures of the lower facade.

The change that was most striking, however, involved the five low Byzantine domes of the church. Each of them was capped by a lofty, bulging, outer shell. These shells are supported by configurations of wooden beams that are anchored on the original domes. They are roofed with lead sheets

Figure 3.33. St. Mark's Place. From Thomas Roscoe, *The Tourist in Italy*, illustrated from drawings by Samuel Prout Esq., London, 1831

and topped by onion-shaped lanterns. The curving geometry of these domes gave the church an exotic new profile. Figure 3.33 and Plate 16 depict these new domes as they exist today. Figure 3.34 lets us look below the surface. It shows a cross section of three of the original domes (highlighted in black) as well as the timber structures that support their outer shells. The idea to enhance these domes by providing them with spectacular new profiles may have come from the Dome of the Rock depicted in Figure 3.7. It might also have been inspired by reports about one of the historic mosques of Islam, the Great Mosque of Damascus. An Arab visitor from Islamic Spain describes the structure of the central dome of this mosque as follows in 1184:

> Then we hastened on to the entrance to the interior of the Dome, passing . . . over the planking of great wood beams which go all around the inner and smaller dome, which is inside the Outer Leaden Dome . . . and there are here two arched windows, through which you look down into the Mosque below. . . . This dome is round like a sphere, and its structure is made of planks strengthened with stout ribs of wood, bound with bands of iron. The ribs curve over the dome and meet at the summit in a round circle of wood. The inner dome, which is that seen from the interior of the Mosque, is inlaid with wooden panels. They are all gilt in the most beautiful manner, and ornamented with colour and carving. The Great Leaden Dome covers this inner dome that has just been described. It is also strengthened by wooden ribs bound with iron bands. The number of these ribs is forty-eight, and between each rib is a space of four spans. The ribs

Figure 3.34. San Marco, north–south cross section. From L. Cicognara, A. Diedo, and G. Selva, *Le fabbriche ei monumenti conspicue di Venezia* vol. 1, Venice, 1838. Marquand Library of Art and Archaeology, Princeton University Library. Presented by Dr. Allan Marquant

converge above, and unite in a centre-piece of wood. The Great Dome rests on a circular base.

Given the flourishing commercial activity between Damascus and Venice during this period, there is little doubt that such reports reached Venice.

Increasingly powerful rulers and increasingly wealthy merchants began to build splendid palaces and residences. The Palace of the Doges was constructed from 1309 to 1424. The palace served as the residence of the doge and as the political and legal center of Venice. Its design also combines diverse influences. The ogee arches of the upper arcade of the facade have an Islamic character and are similar to the three leafed arches of the Great Mosque of Córdoba. Refer to Plate 9 and Figure 3.10. The pattern of diamonds formed by inlaid stone in alternating pink and white adds life to the large flat surfaces of the facade and has an Islamic quality. The delicate tracery of the window in Figure 3.36 recalls that of the windows of the Cathedral of Chartres in Figure 3.24. The carved figures on the left and right of the window are also typically Gothic. (The winged lion below the window is the symbol of the Venetian Republic.) Finally, the arcaded openness of the palazzo anticipates the architecture of the Renaissance. (The architecture of the Renaissance will be taken up in Chapter 5.)

The routes that carried commerce back and forth across the Mediterranean also brought Byzantine and Islamic influences to the city of Pisa. The Cathedral of Pisa was planned in 1063 and largely completed by 1118. Laid out in the basic basilica plan, it is one of the best examples of Romanesque architecture in Italy. Byzantine influences are evident in its interior.

Figure 3.35. The facade of the Palace of the Doges, Venice. Photo by Benjamin Sattin

A Byzantine mosaic depicting Christ the Savior is in a dominant position behind the main altar. It is similar to that of Plate 6 but of lesser artistic value. In addition, the cathedral has two tiers of arcades on each side of the nave that are reminiscent of a similar feature in the Hagia Sophia. One of the two upper tiers is shown in Figure 3.37. The figure also shows that the rectangular space over the crossing is vaulted by a combination of supporting walls, arches, and squinches at the four corners. The squinches recall earlier Islamic precedents. The elongated octagonal drum and dome that these structures support are depicted in Figure 3.38. The arcade of pointed arches around the base of the exterior of the dome is of later construction.

During a time when the rest of Europe built its monuments in the soaring but also stark Gothic style, artists and builders in Venice and Pisa combined Byzantine and Islamic ideas with Romanesque and Gothic forms to create wonderful new architectures.

We'll conclude with two footnotes that update the stories of Venice and Pisa. Venice sits on an island that is surrounded by a lagoon. This setting adds to its magic, but threatens its existence. The subsoil of the island has always been unstable, so that the city needed to be built on a forest of vertically embedded wooden piles. Parts of this substructure are now failing. Regular floods make the problem worse. Efforts to save Venice have been under way for years. The most recent is the construction of a large system of flood gates. Pisa faced a problem of a smaller scale. The cathedral's bell tower began to lean soon after its construction began in 1175. The architects had failed to adequately secure the tower's foundations within the soft, wet soil.

Figure 3.36. Details on the facade of the Palace of the Doges, Venice. Photo by Deror Avi

Figure 3.37. The arches and squinches that support the vault and dome of the Cathedral of Pisa. Photo by JoJan

Construction continued, but the higher floors were built to compensate for the lean. As a result, when it was finished in 1350 at a height of 185 feet, the tower curved upward slightly from the direction of the lean toward the vertical (so that it has the curvature of a banana). In spite of the corrections, the lean slowly increased over the centuries and the tower came to be known as the "leaning tower of Pisa." When the lean had reached about 5.5° from the vertical in 1990, the tower was closed and corrective measures were taken. The lowest level of the tower was reinforced with steel and 600 tons of lead was attached to its base on the side opposite to the lean. This intervention seems to have stabilized the tower.

Problems and Discussions

The first set of problems deals with loads and thrusts in both qualitative and quantitative terms. The three discussions that follow deal with topics related to concerns of the chapter. [In several of the problems you will be supplied

with numerical data. You will need to decide how many of the digits are reliable so that you can then round off your final answers accordingly.]

Problem 1. The computation of the loads that the dome of the Hagia Sophia generates was based on an estimate of the volume of the shell of the dome and the density of the materials used in its construction. What difficulty does this computational strategy run into if it is applied to the dome of the Pantheon of Rome?

Problem 2. After reviewing the phenomenon of hoop stress (in Chapter 2, "The Pantheon") and the design and geometry of the dome of the Hagia Sophia, assess its structural soundness.

Figure 3.39 depicts the vertical cross section of a shell of a dome much like that of the Hagia Sophia above its gallery of 40 windows. The inner and outer boundaries of the shell lie on concentric spheres. In Figure 3.39a, C is the common center of the two spheres, r is the radius of the inner sphere, and θ is the angle that determines the extent of the shell. The horizontal circular base of the dome along with its center are shown Figure 3.39b. The circular base has radius b and its distance from the top of the inside of the shell is a. The shell is assumed to have a rib structure like that of the Hagia Sophia and the vectors labeled by P denote the push of two opposing pairs of ribs against the base of the shell.

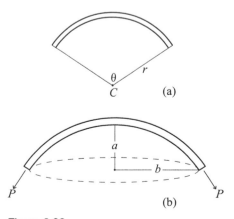

Figure 3.39

Problem 3. Recall that for the dome of the Hagia Sophia $r = 50$ feet and $\theta = 140°$. Conclude that $b = 47$ feet and $a = 33$ feet.

An earthquake caused the partial collapse of the original dome of the Hagia Sophia soon after its construction had been completed and the dome was rebuilt. Not much seems to be known about the original dome other than the fact that it was lower and flatter than the rebuilt dome. However, under the assumption that the basic structure of the original dome was the same as that of the rebuilt dome, it is possible to draw a number of speculative conclusions about it. These are developed in Problems 4 and 5 below. The assumptions made about the original dome are that Figure 3.39 depicts its essential structure, that the size of its circular base was the same as that of the rebuilt dome, and it had a rib structure with 40 ribs. In view of the results of Problem 3, we will take the radius of the circular base of the original dome to be $b = 47$ feet and assume that the distance from this circular base to the top of the inside of the shell was $a = 23$ feet, 10 feet less than that of the rebuilt dome. The fact that the original dome was flatter means that the inner radius r of Figure 3.39 must have been larger.

Problem 4. Given the assumptions that have been made, show that for the original dome of the Hagia Sophia $r = 60$ feet and $\theta = 104°$ (both approximately). [Hint: Use the Pythagorean Theorem to find r. Then notice that $\sin \frac{\theta}{2} = \frac{b}{r}$.]

The fact that the shell of the dome of the Hagia Sophia is $2\frac{1}{2}$ feet thick made it possible to derive the estimate of 27,600 cubic feet for its volume. This volume computation is carried out in the section "Volumes of Spherical Domes" of Chapter 7. Assume that the original shell also had a thickness of $2\frac{1}{2}$ feet. The results of Problem 4 in combination with a similar volume computation provide the estimate of 23,300 cubic feet for the volume of the original shell.

Problem 5. Assume that the masonry of the original shell weighed the same 110 pounds per cubic foot as that of the rebuilt shell. Conclude that the original shell weighed approximately 2,560,000 pounds. Refer to the section "The Hagia Sophia" and derive the estimates $P \approx 81,000$ pounds for the push of one rib against the base of the original dome and $H \approx 50,000$ pounds for the horizontal component of P. Compare these estimates with those for the rebuilt dome and discuss the differences.

It is important to note that the basic underlying assumption of the study above—as well as the one that preceded it in the section "The Hagia Sophia"— is that the opposing pair of ribs shown in Figure 3.39b and the loads that the ribs carry are modeled by the simple truss depicted in Figure 3.14 and analyzed in the section "Romanesque Architecture."

Problem 6. Suppose the truss in Figure 3.14 is an abstract model of a rib segment of the vault of a Gothic cathedral. What structural feature of the cathedral plays the role of the tie-beam connecting A and B? Assume that the load L is 10,000 pounds. Take $\alpha = 25°$, $\alpha = 50°$, and $\alpha = 75°$, and compute in each case the horizontal component of the outward thrust that the slanting beam generates at A (or B).

Problem 7. Go to Figure 3.14 of the text and consider the forces acting at the point C. Use the force diagram of Figure 3.40 to compute the upward push P at C as well as the horizontal component H of this push in terms of the load L and the angle α.

Problem 8. Suppose that the load L on the arch depicted in Figure 3.41 consists only of the combined weight of the two slanting members. Let α be the indicated angle, let d be half the span of the arch, let h be the height, and let l be the length of the slanting elements of the arch. Let w be the weight per unit length of these elements. Explain why $L \approx 2w\sqrt{d^2 + h^2}$. Let H be the horizontal thrust that the load generates and show that $H \approx wd\sqrt{1 + \frac{d^2}{h^2}}$. Now let w and d be fixed and discuss what happens to both L and H as h varies.

Problem 9. Discuss the outward forces generated by a spire of the Cathedral of Chartres in light of the conclusion of Problem 8.

Problem 10. Look up the history of the French Gothic Cathedral of Beauvais and write a paragraph that explains the nature of and reasons for the structural failures.

Discussion 3.1. The Mathematics of Symmetry. The Alhambra palace in Granada in Andalusia, Spain, built in the thirteenth and fourteenth centuries, is a dazzling example of Islamic architecture. The walls and floors of the Alhambra are adorned with intricate geometric designs. A few of them are depicted in Plate 17. Most anyone looking at them would agree, instinctively and intuitively, that each of the designs is symmetric in its own particular way. But can this intuitive notion of "symmetric" be formulated mathematically? The key to the answer is the concept of a mathematical *group*.

Let n be a positive integer and let $1, 2, \ldots, n-1, n$ be the list of the first n consecutive integers. Any rearrangement of these n numbers in a different order is a permutation of $1, 2, \ldots, n-1, n$. The "rearrangement" that leaves the original list unchanged is included and is called the identity permutation. We will write permutations numerically in the form $\left(\begin{smallmatrix} 1 & 2 & \cdots & n \\ \_ & \_ & \cdots & \_ \end{smallmatrix}\right)$ with the rearranged list of the n numbers appearing in the second row. This way of writing a permutation emphasizes the changes that have been made. Since any one of n integers can be placed under the 1, and any one of the remaining $n-1$ under the 2, and so on, it follows that there are a total of $n(n-1)$ $(n-2)\cdots 2 \cdot 1$ permutations. This product is written as $n!$ and referred to as n factorial. Any two permutations P and Q can be combined, or multiplied, to

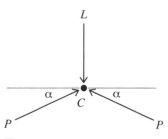

Figure 3.40

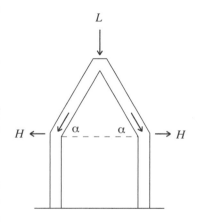

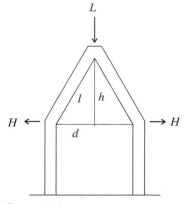

Figure 3.41

form the permutation PQ by first applying P and then Q. The identity permutation I satisfies $PI = IP = P$ for any permutation P. Since the rearrangement that determines a permutation P can be reversed, there is a permutation Q such that $PQ = QP = I$. This Q is the inverse permutation of P. It is also a fact that for any three permutations P, Q, and R, the product of P times the product QR is equal to the product PQ times R, or in symbolic form, $P(QR) = (PQ)R$. Any set with a multiplication that satisfies the identity, inverse, and multiplicative properties just described is known as a group. The number of different elements in a group is called the order of the group. The particular example just described is the permutation group of n elements. Notice that its order is $n!$. Group theory is an important discipline that is central in modern mathematics. The aspects of group theory relevant to the discussion below were developed by the French mathematician (also political activist and revolutionary) Évariste Galois (1811–1832), who understood their role in the problem of the solvability of polynomial equations. (The young genius died in a duel soon after he made his discoveries.)

Let's have a detailed look at the case $n = 4$, where there are $4! = 24$ permutations. They are provided by the following 24 rearrangements of the listing 1, 2, 3, 4 :

1, 2, 3, 4	1, 2, 4, 3	1, 3, 2, 4	1, 3, 4, 2	1, 4, 2, 3	1, 4, 3, 2
2, 1, 3, 4	2, 1, 4, 3	2, 3, 1, 4	2, 3, 4, 1	2, 4, 1, 3	2, 4, 3, 1
3, 1, 2, 4	3, 1, 4, 2	3, 2, 1, 4	3, 2, 4, 1	3, 4, 1, 2	3, 4, 2, 1
4, 1, 2, 3	4, 1, 3, 2	4, 2, 1, 3	4, 2, 3, 1	4, 3, 1, 2	4, 3, 2, 1

The corresponding permutations are $I = \left(\begin{smallmatrix} 1 & 2 & 3 & 4 \\ 1 & 2 & 3 & 4 \end{smallmatrix}\right)$, $\left(\begin{smallmatrix} 1 & 2 & 3 & 4 \\ 1 & 2 & 4 & 3 \end{smallmatrix}\right)$, $\left(\begin{smallmatrix} 1 & 2 & 3 & 4 \\ 1 & 3 & 2 & 4 \end{smallmatrix}\right)$, and so on, with $\left(\begin{smallmatrix} 1 & 2 & 3 & 4 \\ 4 & 3 & 2 & 1 \end{smallmatrix}\right)$ being the last. Take the product PQ of $P = \left(\begin{smallmatrix} 1 & 2 & 3 & 4 \\ 1 & 3 & 2 & 4 \end{smallmatrix}\right)$ and $Q = \left(\begin{smallmatrix} 1 & 2 & 3 & 4 \\ 4 & 3 & 2 & 1 \end{smallmatrix}\right)$. Because P is applied first and then Q, this product sends $1 \to 1 \to 4$, $2 \to 3 \to 2$, $3 \to 2 \to 3$, and $4 \to 4 \to 1$. Therefore, $PQ = \left(\begin{smallmatrix} 1 & 2 & 3 & 4 \\ 4 & 2 & 3 & 1 \end{smallmatrix}\right)$.

Consider the polygons depicted in Figure 3.42. With the exception of the initial isosceles triangle 3.42a and the diamond shape 3.42c, they are all regular polygons. The numerical labels for their vertices are fixed (but the numbers are not part of the design). Each of the polygons is regarded to be a rigid frame. Start with the isosceles triangle of Figure 3.42a. Reflect, or flip, the triangle around the vertical axis shown and return it to the plane. This move interchanges the position of the vertices 1 and 2, but the triangle looks exactly the same after this move. In fact, it looks as if it had not been moved at all. Any move of a polygon of Figure 3.42 that repositions the polygon in such a way that one cannot tell whether it has been moved or not is a symmetry transformation, or briefly a symmetry, of the polygon. Since any symmetry will move the vertices around, it is a permutation of the numbers that label them. When a symmetry is considered, the focus will be on the end result, namely on the permutation of the vertices that it produces. In particular, two symmetries are the same if they produce the same permutation of

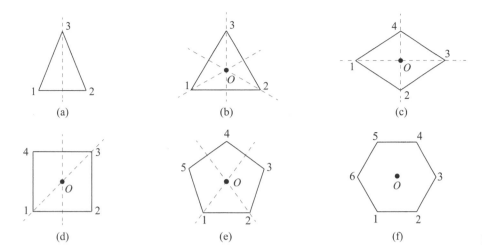

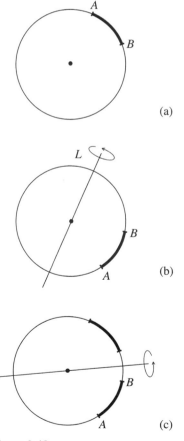

(a)

(b)

(c)

Figure 3.42

the vertices. So which permutations of the vertices are symmetries? Certainly the identity is. Since our polygons are regarded to be rigid, the only way they can be moved is by a rotation around a point, a reflection or flip around an axis, a translation, namely a lateral shift in the plane with no rotation, or a combination of such moves. Since a symmetry repositions the polygon on top of itself, it follows that it can be brought about by a rotation around the center of the polygon if there is no reflection involved, and if there is, by a combination of a reflection around a line through the center followed by a rotation around the center. Consider the circle of Figure 3.43a along with the arc AB. Figure 3.43b shows the circle flipped around the axis L and then rotated to a new position. Figure 3.43c shows that this combination of a flip followed by a rotation can be accomplished by just a flip alone around a different axis. Applying this to our discussion of the polygons of Figure 3.42, we see that any symmetry of a polygon is either a rotation around the center or a reflection around a line through the center. It follows that the symmetries are precisely those permutations that such rotations and reflections determine. Observe that the product of any two symmetries (first apply one, then the other) of any of the polygons is a symmetry. Because the inverse of a rotation is a rotation and a reflection is its own inverse, the inverse of any symmetry is a symmetry. It follows that the symmetries of any of the polygons form a group, the symmetry group of the polygon. It is a subgroup of the permutation group of n elements, where $n = 3, 4, 5,$ or 6 depending on the polygon under consideration. The isosceles triangle of Figure 3.42a has two symmetries, the identity and the reflection already discussed. As permutations of the vertices 1, 2, 3 they are $I = \left(\begin{smallmatrix} 1 & 2 & 3 \\ 1 & 2 & 3 \end{smallmatrix}\right)$ and $\left(\begin{smallmatrix} 1 & 2 & 3 \\ 2 & 1 & 3 \end{smallmatrix}\right)$. Because the reflection is the only symmetry (other than the identity), this isosceles triangle only has bilateral symmetry. Bilateral symmetry is common in architecture. Notice its presence in the structures depicted in Figures 2.12, 2.42, 3.7, 3.13, 3.29, and 3.30 of the text.

Figure 3.43

Turn next to the equilateral triangle of Figure 3.42b. It has the two symmetries already identified for the isosceles triangle. But it has additional symmetries. For example, a counterclockwise rotation by 120° around its center O returns the triangle to the same position. It determines the permutation $1 \to 2$, $2 \to 3$, $3 \to 1$ of the vertices. Notice that the permutation $1 \to 3$, $3 \to 2$, $2 \to 1$ determined by the clockwise rotation through 120° can also be achieved by the counterclockwise rotation through 240°. It is not difficult to see that this is so more generally. Let θ with $0° \le \theta \le 360°$ be any angle. If a permutation of the vertices of a polygon is the result of a clockwise rotation through θ, then the same permutation is achieved by a counterclockwise rotation through $360° - \theta$. For this reason, we'll consider counterclockwise rotations only.

Problem 11. Consider the six permutations of the vertices 1, 2, 3 of the equilateral triangle:

$$\begin{pmatrix} 1 & 2 & 3 \\ 1 & 2 & 3 \end{pmatrix} \begin{pmatrix} 1 & 2 & 3 \\ 2 & 3 & 1 \end{pmatrix} \begin{pmatrix} 1 & 2 & 3 \\ 3 & 1 & 2 \end{pmatrix} \begin{pmatrix} 1 & 2 & 3 \\ 2 & 1 & 3 \end{pmatrix} \begin{pmatrix} 1 & 2 & 3 \\ 3 & 2 & 1 \end{pmatrix} \begin{pmatrix} 1 & 2 & 3 \\ 1 & 3 & 2 \end{pmatrix}$$

Check that all of them are symmetries. For each permutation, identify the angle if it is a rotation, or the axis if it is a reflection. Let P be the reflection around the axis through vertex 1 and Q the reflection around the axis through vertex 2. Compute the product PQ. Which one of the six symmetries from the list is the result?

Let's consider the square of Figure 3.42d and list its symmetries. The rotations are the counterclockwise rotations through 0° (the identity), 90°, 180°, and 270°. There are also four reflections. They are those around the following four axes through O: the horizontal, the vertical, the one determined by vertices 1 and 3, and the one determined by vertices 2 and 4. So the symmetry group of the square consists of four rotations and four reflections. So only 8 of the $4! = 24$ permutations of the numbers 1, 2, 3, 4 provide symmetries of the square.

Problem 12. Write each of the eight symmetries of the square in numerical notation. Describe how the symmetries

$$P = \begin{pmatrix} 1 & 2 & 3 & 4 \\ 1 & 4 & 3 & 2 \end{pmatrix} \quad \text{and} \quad Q = \begin{pmatrix} 1 & 2 & 3 & 4 \\ 4 & 3 & 2 & 1 \end{pmatrix}$$

as well as their products PQ and QP move the square.

Problem 13. List all the symmetries of the diamond of Figure 3.42c. How many rotations and how many reflections are there? Compare the symmetry group of the diamond with that of the square.

Problem 14. The regular pentagon of Figure 3.42e has 10 symmetries, 5 rotations and 5 reflections. What are the angles of the five rotations and what are the axes of the five reflections? The axes of two of the reflections are shown in the figure. Express these two reflections in numerical notation.

Problem 15. Consider the regular hexagon of Figure 3.42f. It has 12 symmetries, 6 rotations and 6 reflections. What is the angle of the rotation $\left(\begin{smallmatrix} 1 & 2 & 3 & 4 & 5 & 6 \\ 2 & 3 & 4 & 5 & 6 & 1 \end{smallmatrix}\right)$? What are the angles of the other five rotations? What is the axis of the reflection $\left(\begin{smallmatrix} 1 & 2 & 3 & 4 & 5 & 6 \\ 4 & 3 & 2 & 1 & 6 & 5 \end{smallmatrix}\right)$? Specify the axes of the other five reflections.

Problem 16. Draw hexagons with symmetry groups of orders 1, 2, and 4. Then draw hexagons with symmetry groups of orders 3 and 6. [Hint: For orders 2 and 4 put two trapezoids together. For orders 3 and 6 combine two equilateral triangles.]

Consider a polygon. If it has n vertices it is an n-gon. An n-gon is regular if its vertices can be placed on a circle and spaced evenly around it. The center of the circle is also the center of the regular n-gon.

Let a regular n-gon be given. Let R be the counterclockwise rotation around the center O through the angle $\left(\frac{360}{n}\right)^\circ$. Then the rotations

$$R, R^2 = RR, R^3 = RRR, \ldots, R^n = I$$

are all symmetries of the n-gon. Now let L be any line through O such that each of the two points of intersection of L with the n-gon is either a vertex or the midpoint of an edge. There are exactly n such lines and the n reflections that they determine are symmetries of the n-gon.

Problem 17. Explore the assertions of the paragraph above for the examples of regular polygons of Figure 3.42 and then more generally. Show that every symmetry of a regular n-gon is equal to one of the rotations or reflections that is described.

The discussion above applies not only to polygons but to any design. The extent of the symmetry of a design is assessed by the study of its symmetry transformations or symmetries. The "moves," or transformations, of the design that do not change its appearance are the symmetries of the design. Let's start with the design on the left side of Plate 17. It consists of the repetition of the flowerlike unit with border in lighter color. Notice the bilateral symmetry of the design. Observe also that the appearance of the design is unchanged by translation (an upward or downward shift by one unit, for instance). In the examination of the other four designs of Plate 17 we will restrict our attention—as with the earlier polygons—to rotations and reflections and the symmetry groups that they determine.

Figure 3.44 is a design from the collection of Plate 17. Consider the part of the design that falls inside the smaller of the two circles. Observe that if the 16 tips of the star formation at the center are connected in a consecutive

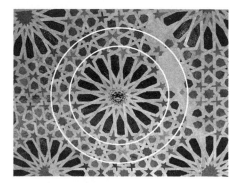

Figure 3.44. Photo by Jebulon

way, a regular 16-gon results. Notice that the symmetries of the design are precisely the symmetries of the regular 16-gon. (You'll need to ignore the arrangement of birds at the very center.) It follows that the design has 16 rotations, 16 reflections, and a symmetry group of order 32. Consider the larger design that falls inside the larger circle (complete the design in your mind) and analyze it in the same way.

Problem 18. Study the remaining three designs in the collection of Plate 17. In each case, describe the relevant regular polygon, the rotations and reflections, and determine the order of the symmetry group.

Problem 19. Have a look at the rosette window of the Cathedral of Chartres depicted in Figure 3.24. Discuss what the symmetries of the window are by making reference to its rotations and reflections. What is the order of the symmetry group?

The picture of the rosette window of the Cathedral of Chartres raises a basic question that will be taken up at the end of Chapter 5. Why does this circular window appear to be elliptical in the figure?

Problem 20. What emerges from our mathematical discussion of symmetry is the suggestion that the larger the group of symmetries of a figure is, the more symmetric it is. Does this conform to your conception of symmetry? For instance, do you think that the circle (what is the order of its symmetry group?) is more symmetric than the hexagon?

Discussion 3.2. Norman Architecture. By the end of the eleventh century, a duke of Normandy (today a region on the English channel in northern France) ruled over much of France and England. He would become known as William the Conqueror. A Norman style of architecture developed in the lands the Normans controlled. The Cathedral St. Etienne in Caen (where William is buried) and the cathedral in Durham are two prominent examples. The construction of St. Etienne was started in 1066 and completed in the thirteenth century. Work on the Durham Cathedral began in 1093 and was largely completed within 40 years. The semicircular arches and massive walls that flank the naves of the two cathedrals confirm that both are in the Romanesque tradition. Those of the Durham Cathedral are depicted in Figure 3.45. However, the vaults over the nave are ribbed vaults, rather than the expected groin vaults. Historians of architecture regard this important structural innovation, characteristic of later Gothic churches, to have originated with these two cathedrals. The ribbed dome of the Mosque of Córdoba (see Figure 3.10) was constructed earlier, but it is differently configured and on a much smaller scale. The exteriors of the walls of the nave of the two cathedrals are reinforced by vertical structures at precisely the

Figure 3.45. Photo by Oliver Bonjoch

locations where the ribs converge. This feature later evolved into the flying buttress of Gothic architecture.

Discussion 3.3. Medieval Building Practices. The structures discussed in this chapter, including the great Romanesque and Gothic cathedrals, were built by hand. Wood and stone were cut and shaped by simple tools, lifted with simple hoists powered by men, donkeys, or oxen, and put into position on flimsy scaffolds. Stone was worked with axe, hammer, and chisel. Mortar was mixed and worked in wooden bins and carried to the walling masons on portable troughs. Timber was cut, split with a wedge and hammer, prepared, and given a smooth finish with axes. A large variety of different one-handed and two-handed saws were used. Drills with bits made of shell and turned by wooden braces made holes for pins, nails, and metal clamps. Planes, chisels, hammers, and mallets completed the carpenter's tool kit. With a master mason in charge (the title architect came into use only later), a team of professional quarrymen, stone masons, bricklayers, carpenters, blacksmiths, plasterers, tilers, painters, glaziers, and sculptors, assisted by apprentices and unskilled laborers, would raise a structure. Master craftsmen taught apprentices all aspects of the craft on site. After they were certified as journeymen, they received a wage for their work (the word journey is derived from the French word *journée*, meaning a day's work and also the compensation received for one day's labor). Such construction teams traveled from project to project, and ideas about design and building methods traveled with them. However, the knowledge of the master masons was carefully held and secretly kept. Only the most experienced and accomplished journeymen rose to the level of master in their craft and only the most capable master craftsmen were entrusted to direct building projects and given the title master builder. The written documentation about medieval building practices and methods is thin. Not even the archives and libraries of Europe's Catholic monasteries contain much. These records mention building contracts and sizes of buildings, but beyond comments that these should be built "according to the traditional model," there are few details about the construction. An early record that survived are the notebooks of Villard de Honnecourt from the period 1225–1250. These notebooks cover topics ranging from geometric problems, to designs of vaults and roof trusses, to matters of masonry, carpentry, and ornamentation. There are sketches of interior and exterior elevations and plans (representations of vertical and horizontal sections) as well as details that specify the position of the ribs of a vault and the thickness of walls. Even in later years, insights such as those recorded in the *Annals of the Building Council of the Cathedral of Milan* (during 1392 and 1400–1401) are rare. However, the pictorial record provided by people who saw the work being done is rich. Figures 3.46 and 3.47 depict two examples of many. They give insight into the activities at a construction site and the equipment that

Figure 3.46. Life of the Offas, pen-and-ink drawing, late fourteenth century, British Museum, London, Cotton Nero D I, fol. 23. From Günther Bending, Medieval Building Techniques, Tempus, 2004. Marquand Library of Art and Archaeology, Princeton University Library

Figure 3.47. Illumination, mid-thirteenth century, Old Testament miniatures. The Pierpont Morgan Library, New York. MS M638, fol. 3. Photo by David A. Loggie, 1990

was used. Figure 3.46 also shows a master builder receiving instructions from a royal sponsor.

The Middle Ages were an age of royal courts, lords, nobles, knights, vassals, peasants, and serfs. It was a time of constant warfare. The construction teams that built the soaring cathedrals also built the powerful castles and walled towns that provided protection. The fortified French town of Carcassonne built from 800 to 1300 is a wonderfully preserved example.

Transmission of Mathematics and Transition in Architecture

Fundamental changes occurred in the territories around the Mediterranean Sea and to its north and east during the period from the twelfth to the fifteenth centuries. The energy of the Islamic Dominions was diminishing. A Christian Reconquest ended the Islamic occupation of Spain in the west, and Mongol invasions ended the influence of the Seljuk Turkish dynasty in the east. The Byzantine Empire, its territories under constant pressure from Islamic expansion, was stagnating and no longer creative. A rising Ottoman Turkish dynasty finally ended its existence in the fifteenth century. However, for Catholic Christian Europe, this was a dynamic period of progress and growth. It was a time when this region began to develop its current identity and its present institutions. Improved methods of agriculture freed people to move into towns and cities. Manufacturing and commercial activities grew. Hospitals were built and banking developed. Merchants and manufacturers organized themselves into associations known as guilds. Guilds organized the professions, set prices and wages, and promoted and regulated trade. Trade routes developed through the Black and Mediterranean Seas, along the Atlantic coast, across the North and Baltic Seas, along the Rhone, Seine, Danube, Rhine, Dnieper, Volga, and other great rivers, and through Alpine mountain passes. They connected important European cities such as Venice, Genoa, Pisa, Florence, Milan, Vienna, Augsburg, Lisbon, Bordeaux, Paris, London, Hamburg, Danzig, Riga, and Kiev with Baghdad, Damascus, Cairo, and Constantinople and delivered timber, wool, cotton, silk, sugar, salt, spices, wheat, dried fruit, wine, fish, and furs. Christian Europe grew in population and wealth.

An important concurrent development was the flow into Christian Europe of classical Greek and Roman tracts and treatises about philosophy, law, rhetoric, science, and mathematics. These had been preserved in Islamic libraries and engaged by Islamic, Jewish, and Byzantine scholars. They were translated into Latin and copied many times over in centers established in Toledo, Spain, in Palermo, Sicily, and in other cities after they returned to Christian control. These manuscripts were widely circulated among scholars. Universities, often sponsored by the Church, were founded in the twelfth and thirteenth centuries in major European centers such as Paris, Oxford, Bologna, and Padova. Thousands of students pursued law or medicine, or prepared for careers in government or the Church. Universities became centers for intellectual discourse, debate, and dispute, in which Greek reason contended with Catholic faith. Aristotle's views about the nature of man and his soul, the natural sciences, and the universe challenged established explanations. How are truths derived by reason related to truths provided by scripture and sacred authority? Is the universe subject to laws that the human mind can understand and analyze? Or does the absolute freedom of God to do as he wills preclude the existence and hence the comprehension of such an order? Both sides of these issues had powerful proponents that included the greatest thinkers of the time. It is significant that these debates were carried out, not between religious and secular adversaries, but within the intellectual circles of the Catholic Church. After intense and protracted disagreements, Thomas Aquinas, professor of philosophy and theology in Paris and priest of the Dominican order, produced a synthesis of faith and reason. He affirmed—within a universe sustained by God—the presence of patterns and laws that the human mind could grasp.

Beginning in the twelfth century, European scholars took up Greek science and mathematics by delving into transcriptions of the works of Euclid, Archimedes, Apollonius, and Ptolemy. They learned what the Greeks knew about numbers and geometry from Euclid's *Elements*. They studied the curves obtained by cutting a cone with a plane from the *Conic Sections* of Apollonius and the works of Archimedes. The ellipse and the parabola are the most relevant examples. European scholars engaged the Greeks' remarkable mathematical theory of the movements of the planets in Islamic expositions of Claudius Ptolemy's *Almagest* and they saw the incredible maps of the known world in Ptolemy's *Geographike*. At around the same time, European scholars acquired knowledge of the Hindu-Arabic number system from the work of Islamic mathematicians, and European traders saw how Islamic merchants put this system to use. It would take time for all this to be absorbed and for the Europeans to make their own contributions. In the sixteenth century, major advances were made in algebra. Most significantly, in the early part of the seventeenth century, Galileo discovered that the trajectories of thrown objects are parabolic arcs and, after Copernicus had correctly put the Sun at the center of the solar system, Kepler observed that the orbits of the planets are ellipses. Toward the middle of the seventeenth century, Descartes and Fermat combined Greek geometry with the Hindu-Arabic number system into a single mathematical structure. This coordinate geometry quantified Euclidean geometry in the plane and in space, and in the process fused algebra and geometry.

In the same way that European scholars made significant progress in science and mathematics by engaging the works of classical thinkers, European builders were inspired by the forms and structures of classical Greece and Rome. In the fifteenth and sixteenth centuries, they took Greek and Roman columns, arches, pediments, and porticos, refined their design, and used them as elements in a new harmonious, ambitious, and glorious architecture. This is the story of the architecture of the Renaissance that will be taken up in Chapter 5. One of the creative geniuses who developed this new art of building was Filippo Brunelleschi from Florence. He had taken up the study of the structures of ancient Rome and learned about the design and method of construction of the vaults and domes of its baths, basilicas, amphitheaters, and temples. At the beginning of the fifteenth century, the Cathedral of Florence, with its distinctly Gothic interior, was still without the dome that was to cover the gaping octagonal hole over its crossing. This dome would be of unprecedented size and generate unprecedented loads. Brunelleschi planned the dome and supervised its construction. He countered the dome's large loads and thrusts by incorporating Roman structural features in its design. The Cathedral of Florence signals the beginning of the new art of building and marks the transition of architecture from the Gothic Age to the Renaissance.

From the time structures were first built, architects have drawn lines, circles, and circular arcs on flat stretches of ground, parchment, and paper to compose their designs, They have executed their designs by realizing sections of planes, cylinders, and spheres as walls, columns, vaults, and domes. Mathematicians study lines, circles, and other curves in the coordinate plane, and planes, spheres, and other shapes in coordinate space, as idealized mathematical abstractions. That the mathematical study of curves and shapes should inform the curves that architects draw and the shapes that they execute should not be surprising. The purpose of this chapter is to tell the story of coordinate geometry starting with Greek contributions and incorporating Islamic advances, to use coordinate geometry to study lines, circles, planes, and spheres, and, finally, to illustrate how this study clarifies the rising shape of the octagonal dome of the Cathedral of Florence. This is only a very modest application of coordinate geometry to architecture. The last sections of Chapter 5 provide a richer application to the study of perspective, and several sections of Chapters 6 and 7 illustrate how the application of coordinate geometry (with calculus added) provides information critical for the understanding of architectural structures.

Remarkable Curves and Remarkable Maps

The translations from which Europe learned the mathematics of Greece included studies of the parabola and the ellipse. Along with the hyperbola, these are the conic sections, the curves that can be obtained by intersecting or cutting a double cone with a plane. (The word *section* comes from the Latin word for cut.) These curves (see Figure 4.1) were known to Euclid, but

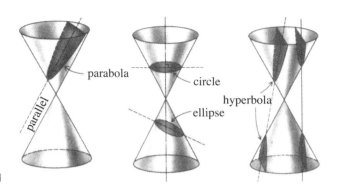

Figure 4.1

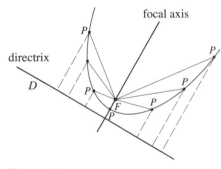

Figure 4.2

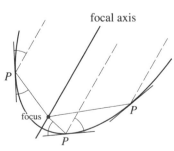

Figure 4.3

it was the Greek Apollonius (about 262–190 B.C.) who analyzed them thoroughly and extensively. We will have a look at his analysis of the parabola and ellipse (but omit his analysis of the hyperbola).

In this section, a plane—a perfectly flat mathematical plane—is given. So that distances can be measured, say between two points in the plane, a unit of length, say the inch, foot, or meter, is provided as well. For two points P and Q in the plane, PQ or QP will label the segment that connects them. If PQ appears in an equation, it will be understood to be the length of the segment PQ or, equivalently, the distance between P and Q.

Let's start with the parabola. Fix both a line D and a point F in the plane with F not on D. The parabola determined by D and F is the collection of all points P in the plane such that the distance from P to F is equal to the (perpendicular) distance from P to D. The line D is called the directrix of the parabola and the point F is called the focus of the parabola. The line through the focus perpendicular to the directrix is the focal axis of the parabola. Figure 4.2 illustrates what has been described. It shows the focus, directrix, and focal axis and locates several points P on the parabola. In each case, the length of the segment from P to F is equal to the length of the dashed segment from P to D. We'll recall—without proofs—just two basic propositions about the parabola from the work of Apollonius.

Proposition P1. Let P be any point on a parabola and consider the tangent line at P. The angle at P that the tangent makes with the line from P to the focus is equal to the angle at P that the tangent makes with the line from P parallel to the focal axis. (Figure 4.3 illustrates this statement for three different choices of the point P.)

Proposition P2. Label the point of intersection of a parabola with its focal axis by O and consider the line through O that is perpendicular to the focal axis. Let A and C be any two points on this line with C different from O. Let B and D be the points on the parabola with the property that AB and

CD are both parallel to the focal axis. (Figure 4.4 shows what has been de-scribed.) Then

$$\frac{AB}{CD} = \frac{OA^2}{OC^2}.$$

We continue with a mathematical discovery of the legendary Archimedes (287–212 B.C.). A parabolic section is a region obtained by taking a parabola and cutting it in some way. Figure 4.5 shows a parabolic section given by the cut *AB*. Take that point *C* on the parabola with the property that the tangent line at *C* is parallel to the cut and consider the triangle *ABC*. Com-bining fundamental properties about the parabola from Apollonius (that include Proposition P2) with an argument that conforms to high standards of mathematical precision and rigor, Archimedes verified that the area of a parabolic section is equal to exactly $\frac{4}{3}$ of the area of the inscribed triangle. His proof is a spectacular example of early calculus.

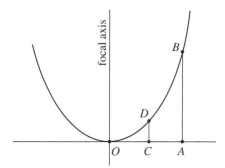

Figure 4.4

We turn to the ellipse next. Fix any two points F_1 and F_2 in the plane and a constant *k* that is greater than the distance between F_1 and F_2. Consider the collection of all points *P* such that the distances from *P* to F_1 and *P* to F_2 add to *k*, or, put another way, such that $PF_1 + PF_2 = k$. This collection of points is the ellipse determined by the points F_1 and F_2 and the constant *k*. In the context of Figure 4.6, notice that *P* is on the ellipse precisely if the lengths of the solid segment from *P* and the dashed segment from *P* add up to *k*. The points F_1 and F_2 are the focal points of the ellipse. The line through the focal points is the focal axis of the ellipse. The midpoint *C* of the segment F_1F_2 is the center of the ellipse. Any segment from a point on the ellipse, through the center, to the point on the ellipse on the other side is a diameter of the el-lipse. One-half the length of the diameter through the two focal points is the semimajor axis of the ellipse, and one-half the length of the perpendicular diameter is the semiminor axis. Consider a circle with center *C* and radius *r*. Why is it an ellipse with focal points $C = F_1 = F_2$ and $k = 2r$?

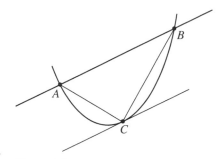

Figure 4.5

Two basic propositions about the ellipse—again without proofs—and again from the work of Apollonius follow next.

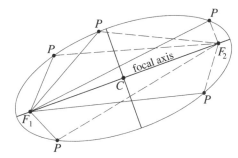

Figure 4.6

Proposition E1. Let *P* be any point on an ellipse and consider the tangent line to the ellipse at *P*. Then the angle at *P* that the tangent makes with the line from *P* to one focus is equal to the angle at *P* that the tangent makes with the line from *P* to the other focus. (Figure 4.7a illustrates this statement for four different choices of *P*.)

Suppose that the ellipse is a circle. Then the two focal points are the same point (the center of the circle), and the solid and dashed segments of Figure 4.7a are the same radius of the circle. So Figure 4.7b and Proposition E1 in-form us that the angle that the radius makes with the tangent is equal to 90°.

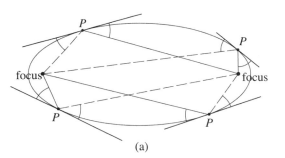
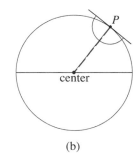

Figure 4.7 (a) (b)

We can conclude for any circle and any radius of the circle, that the tangent to the circle at the point where the radius meets the circle is perpendicular to the radius. We used this fact in the study of the ovals of the Colosseum in Chapter 2.

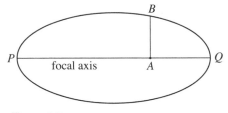

Figure 4.8

Proposition E2. Let P and Q be the points of intersection of the ellipse and its focal axis. Let A be any point on the focal axis between P and Q and let B be a point both on the ellipse and on the perpendicular to the focal axis through A. (Note that there are two possibilities for B.) Figure 4.8 shows the points in question. Then the ratio $\frac{PA \cdot AQ}{AB^2}$ is the same, no matter where the point A is chosen between P and Q.

Let's have another look at Propositions P2 and E2. Return to Figure 4.4. Choose C so that $OC = 1$ and let $CD = c$. Now let $OA = x$ and $AB = y$. Because the point A can be anywhere on the horizontal through O, both x and y are variables. By an application of Proposition P2, $\frac{y}{c} = \frac{x^2}{1^2}$ and therefore $y = cx^2$. We next turn to Proposition E2. Let C be the center of the ellipse of Figure 4.8 and let a be the length of PC. Consider the perpendicular to PQ at C and let b be the length of the segment from C to the ellipse. Refer to Figure 4.9. With $A = C$, we get that $\frac{PA \cdot AQ}{AB^2} = \frac{a \cdot a}{b^2} = \frac{a^2}{b^2}$. Now let A be any point on PQ. Let $x = CA$ and let $y = AB$. As A can be anywhere on PQ, both x and y are variables. If A is to the right of C, then $PA = a + x$, $AQ = a - x$, and we get that $\frac{PA \cdot AQ}{AB^2} = \frac{(a+x)(a-x)}{y^2}$. Check that this equality also holds if A is to the left of C. Proposition E2 tells us that the ratio $\frac{PA \cdot AQ}{AB^2}$ is the same for any A, and therefore that $\frac{(a+x)(a-x)}{y^2} = \frac{a^2}{b^2}$. After a little algebra, $a^2 - x^2 = a^2 \cdot \frac{y^2}{b^2}$. So $1 - \frac{x^2}{a^2} = \frac{y^2}{b^2}$, and hence $\frac{x^2}{a^2} + \frac{y^2}{b^2} = 1$.

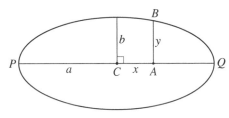

Figure 4.9

In the case of the parabola, notice that the equation $y = cx^2$ relates the distance x from the point B to a fixed vertical axis to the distance y from B to a fixed horizontal axis. A similar thing is true for the ellipse. The equation $\frac{x^2}{a^2} + \frac{y^2}{b^2} = 1$ relates the distance x from the point B to a fixed vertical axis to the distance y from B to a fixed horizontal axis. If you have seen an xy coordinate system in action, you will realize that Apollonius had such a system in his grasp! Note, however, that he did so only for the special cases that he considered: the parabola and the ellipse (and the hyperbola).

Claudius Ptolemy, astronomer and mathematician, studied the skies in the second century A.D. and devised an elaborate scheme of circles that described how the Sun, Moon, and planets move from the vantage point of a fixed Earth. Ptolemy developed this sophisticated mathematical model of the solar system in his treatise *Almagest* (the name is an Arabic derivative of its Greek title *Megiste Syntaxis*, *Megiste* = greatest, *Syntaxis* = system or collection). It would be the accepted theory of planetary motion until Galileo confirmed the Sun-centered explanation and Kepler introduced his elliptical orbits in the seventeenth century. The very same Claudius Ptolemy was also a master mapmaker. His work *Geographike Syntaxis* laid the foundation of the science of cartography. It reached western Europe from Constantinople as a Greek manuscript and was translated into Latin early in the fifteenth century. Remarkable maps were reconstructed by medieval cartographers from the precise positional information that the text supplied. Plate 18 depicts an early printed version of Ptolemy's map of the world. Important areas of Europe around the Mediterranean Sea are easily recognized. North Africa as well as the Arabian peninsula are also easily made out. India and the Far East on the other hand are off target. Central and southern Africa is unknown territory. Most relevant to our current discussion are the intersecting sets of vertical lines of latitude and horizontal lines of longitude of the map. Ptolemy devised this carefully spaced grid to organize his map and to provide precise positions of its key features. He used his grid like a coordinate system. The coordinates of the 8000 locations that the *Geographike* contains enabled the reconstruction of the maps. The strip of numbers at the lower boundary of the map tells us that Ptolemy's lines of latitude divide the known part of the globe into 180 degrees. Ptolemy knew about earlier Greek estimates of the circumference of the Earth and had a sense that he had mapped about half the globe (of course, he had no knowledge whatever about the missing half). Christopher Columbus used Ptolemy's map to make the case to Queen Isabella and King Ferdinand of Spain that he could reach Asia by sailing west. The fact that Ptolemy's map underestimated the size of the globe, and hence the distance that needed to be traveled, might have contributed to the success of Columbus's argument.

We have seen that both Apollonius and Ptolemy had an understanding of important special instances of a coordinate system. But how far were they from devising an abstract coordinate system in the plane with which any curve could be examined with numerical precision? The fact is that they lacked a critical element.

A Line of Numbers

The contributions of the Greeks to mathematics are nothing short of astounding. They axiomatized geometry (today's Euclidean geometry) by

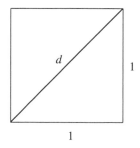

Figure 4.10

presenting it as a mathematical structure that starts with a few central definitions and statements and then sets out everything else in a cohesive and logical way. They studied the conic sections and developed all the basic properties of the ellipse, parabola, and hyperbola. Finally, they invented methods that anticipated modern calculus. But in spite of all their ingenuity, they ran into a problem with their arithmetic. The Pythagoreans knew, as the Babylonians before them, that the lengths of the sides of some figures were measurable "on the nose" with a given unit of length. The right triangle with sides of length 3, 4, and 5 is a simple example. But take what is perhaps the simplest shape of all, the square with sides of length 1, and consider its diagonal. See Figure 4.10. Today we know that this length can be expressed as a number d with a decimal expansion. But how did the Greeks regard this number? The Pythagorean Theorem told them that d needed to satisfy $d^2 = 1^2 + 1^2 = 2$. But now what? The Pythagoreans held as deeply rooted the belief that numbers explain everything. For them, numbers were *rational numbers*, that is, numbers expressible as fractions $\frac{n}{m}$ for integers n and m. Therefore, it was obvious for the Pythagoreans to ask: for which integers n and m is $d = \frac{n}{m}$? In time, the Pythagoreans came to realize that such an equality cannot exist and hence that d *cannot* be equal to one of their numbers. The realization that the diagonal of a square, so simply constructed, was of a length that their system of numbers could not capture was a problem for the Pythagoreans. It threw the Pythagorean fraternity into turmoil and probably contributed to its disappearance.

The proof of the fact that an equality of the form $d = \frac{n}{m}$ cannot hold is contained in the work of Aristotle (and it is this proof that is commonly reproduced today). We will establish this fact with a more recent geometric argument. The strategy will be to *assume* that d is equal to $\frac{n}{m}$ for some positive integers n and m, and to show that this leads to a contradiction. Because $2 = d^2 = \frac{n^2}{m^2}$, it follows that $n^2 = 2m^2$. So *there is* a square of side length n that is equal in area to two smaller squares of side length m. If this square is not the smallest square *with side length a positive integer* that splits up in this way into two smaller identical squares *with side length a positive integer*, select a smaller one. If this smaller one is not the smallest, select a smaller one yet. Because this process has to stop, there must be a smallest square with integer side that can be split up as two smaller identical squares with integer sides. Let's say that this smallest one is depicted in Figure 4.11 along with the two squares

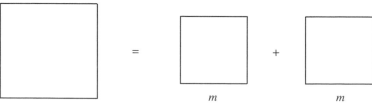

Figure 4.11

that are together equal to it in area. Now slide the two smaller squares of Figure 4.11 on top of the larger square as shown in Figure 4.12. This move creates three squares, one in the center and two identical ones at the corners. That the side lengths of these squares are positive integers follows from the fact that n and m are positive integers. Let the area of the larger square be A and that of the two smaller ones B. After studying Figure 4.12, conclude that $n^2 = 2m^2 + 2B - A$, and therefore (because $n^2 = 2m^2$) that $A = B + B$. The fact that the area A is smaller than n^2 contradicts the fact that Figure 4.11 depicts the smallest square that can be split into two. This is the contradiction we were after. Therefore, an equality of the form $d = \frac{n}{m}$ where n and m are integers is impossible. So d cannot be a rational number. The Greeks did develop a theory of such incommensurable magnitudes, later called irrational numbers, but they were not incorporated into their number system. Neither their numerical notation nor their arithmetic extended to them. So at the end of the day, the number system of the Greeks was not large enough to capture their geometry.

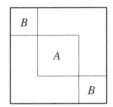

Figure 4.12

What was needed was a number system with which all lengths could be recorded. Only such a system could establish a link between geometry on the one hand and arithmetic and algebra on the other. Such a system needed to come with a notation for numbers so devised that efficient procedures for addition, multiplication, and division could be developed. Both Greek and Roman approaches were completely inadequate on all counts. For example, consider the Roman numbers

I, II, III, . . . , IX, X, . . . , XXXIV, . . . , LXXVI, . . . , XCIII, . . . , CCCLIX, . . . , MMMDCCXV, . . .

If you were asked to multiply and then divide the last two Roman numbers of this list within the Roman scheme, you would find this cumbersome in the first case and exasperating in the second.

Mathematicians from the Islamic world began to make important progress toward the development of a new number system. By the ninth century the scholar al-Khwarizmi working in Baghdad had adopted the ten symbols 1, 2, 3, 4, 5, 6, 7, 8, 9, and 0 as well as the strategy for writing larger numbers from Indian mathematics. In this scheme, the Roman numerals above are written 1, 2, 3, . . . , 9, 10, . . . , 34, . . . 76, . . . , 93, . . . , 359, . . . , 3715, ... This *Hindu-Arabic* number system is *positional*. The symbol 3 means a different thing in each of the numbers 93, 34, 359, and 3715, namely three, thirty, three hundred, and three thousand, in accordance with its position in each of these numbers. A treatise of al-Khwarizmi pointed to the importance of 0 by instructing his readers "when nothing remains, put down a small circle so that the place be not empty, but the circle must occupy it." The Latin title *Algoritmi de Numero Indorum* of this work tells us that our word *algorithm* is derived from the name of its author. Another of al-Khwarizmi's treatises deals with the problem of solving equations. Its Arabic title *Hisab al-jabr* . . . (referring to the procedure of rearranging and combining terms) informs us that our word *algebra* also has Arabic origins. Islamic scholars developed effective and sequential procedures, or algorithms, for carrying out addition, subtraction, multiplication, and division for their numbers. Their number system also enabled them to express lengths numerically, including the length of the diagonal of the 1 by 1 square (the number d of our discussion) as well as the circumference of the circle with diameter 1 (the number now written as π).

It took time for the Islamic system of numbers to reach western Europe and gain acceptance. Even a pope was involved in the promotion of this effort. In the eleventh century, when Pope Sylvester II was the young monk Gerbert, his abbot sent him from France to Islamic Spain to study mathematics. Great institutions of learning, such as University of Córdoba, were making advanced education available to thousands of students. Gerbert not only learned the new number system but also absorbed the questioning and probing spirit of Islamic scholarship. This is the spirit that would lead to the establishment of the first European universities about 100 years later.

Another promoter of Islamic mathematics in western Europe was Leonardo of Pisa (about 1175–1250), known today as Fibonacci. (This name, conferred by a mathematician of the eighteenth century, is a contraction of the Latin *filius Bonacci*, meaning "son of Bonacci" or possibly "son of good nature.") Leonardo's father was a trade official who facilitated the commercial dealings of the merchants of Pisa in what is today the north African country of Algeria. Young Leonardo joined him there and became acquainted with the mathematics of Euclid, Apollonius, and Archimedes. Taught methods of accounting by Islamic scholars, he learned how to calculate "by a marvelous method through the nine figures" together with the figure 0 "called zephirum in Arabic." On his return he published the *Liber Abaci* in 1202. This historic book (contrary to the suggestion of the title, it has nothing to do with the abacus) introduced western Europe to Islamic arithmetic and algebra, as well as the practice of using letters instead of numbers to generalize and abbreviate algebraic equations. It provided western Europe with the first thorough exposition of the Hindu-Arabic numerals and the methods of calculating with them.

The acceptance of the new system—not yet equipped with the decimal point—was slow. As late as the year 1299, the merchants of Florence were forbidden to use it in bookkeeping. They were told instead either to use Roman numerals or to write out numbers in words. The historical record informs us that Hindu-Arabic numbers appeared on gravestones in German states in 1371. They were imprinted on coins of Switzerland in 1424, of Austria in 1484, of France in 1485, of German states in 1489, of Scotland in 1539, and of England in 1551. We saw them on the historic map of Ptolemy printed in a German state in 1482 (depicted in Plate 18). In an architectural context they were used in 1487 by Leonardo da Vinci within his design of a vault and in 1521 in an etching of an elevation. (Both involve the Cathedral of Milan. Refer to Figures 3.28 and 5.24.)

Decimal fractions began to gain use in Europe after they were developed in a 1585 publication by the Flemish mathematician Simon Stevin (1548–1620). *De Theinde*, or *The Tenth*, is a 29-page booklet that presents an elementary and thorough account of them. It was written for the benefit of "stargazers, survey-ors, carpet-makers, wine-gaugers, mint-masters and all kind of merchants." The French version, *La disme*, appeared in the same year. An English trans-lation, *Dime: The Art of Tenths, or Decimal Arithmetic*, was published in 1608 in London. (This translation inspired Thomas Jefferson to propose a decimal currency for the United States and possibly the name dime for the 10-cent coin.) A little later, John Napier (1550–1617), a Scotsman and one of the in-ventors of logarithms, put the "dot on the *i*" by adding the decimal point. What is now our modern number system was notationally complete.

With the decimal or base ten number system (and a given unit of length) it is possible—and this is the very important point—to express *any length* in terms of a number. Whether the length is rational, that is, of the form $\frac{n}{m}$ for integers n and m, or irrational, such as the number d in Figure 4.10, does not matter. For instance, the lengths $5\frac{1}{4}$ and $71\frac{2}{3}$ are expressed as 5.25 and 71.6666 . . . , meaning that $5\frac{1}{4} = 5 + \frac{2}{10} + \frac{5}{10^2}$ and $71\frac{2}{3} = 71 + \frac{6}{10} + \frac{6}{10^2} + \frac{6}{10^3} + \frac{6}{10^4}$ . . . . In the same way, the lengths $d = \sqrt{2}, \sqrt{3}, \sqrt{5}, \sqrt{6}, \sqrt{7}, \ldots$, that arise in the evolving spiral of Figure 4.13 by successive applications of the Py-thagorean theorem, as well as those given by other square roots, cube roots, and higher roots, can all be written with this system. For example, $\sqrt{2} = 1.414213562 \ldots$, $\sqrt{3} = 1.732050808 \ldots$, $\sqrt{5} = 2.236067977 \ldots$, the golden ratio $\phi = \frac{1+\sqrt{5}}{2} = 1.618033989 \ldots$, and $\pi = 3.141592654 \ldots$. In these five examples the full infinite progression of numbers is needed to achieve equality. Stopping at any point gives only an approximation. For example, $\sqrt{2} \approx 1.41421$ and $\sqrt{2} \approx 1.41421356$ are approximations. (Discussion 4.2 later in this chapter considers these matters.)

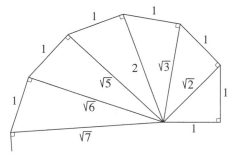

Figure 4.13

The collection of all numbers, positive and negative, that are expressible in this decimal notation is called the set of real numbers. Take a unit of length and a straight line that extends infinitely in both directions. Fix a point on the line and label it with the number 0. Using distance as mea-sure, mark off the numbers 1, 2, 3, 4, . . . , on the right of 0 and −1, −2, −3, −4, . . . , on the left of 0. In the same way, every real number corresponds to a point on the line and, in turn, every point on the line corresponds to a real number. Positive numbers are on the right of 0 and negative numbers on the left. Figure 4.14 illustrates what has been described. It also shows how to posi-tion $\sqrt{2} = 1.4142. \ldots$ A line with this numerical structure is a number line.

Observe that the distance between the points −5 and 3 on a number line is 8. This is also equal to the absolute value of −5 − 3 or, equivalently, to the

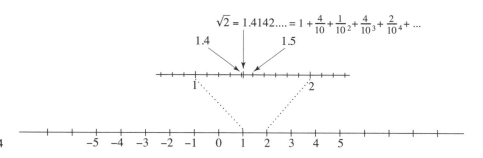

Figure 4.14

Figure 4.15

absolute value of $3 - (-5)$. In general, the distance between points a and b on a number line is equal to $a - b$ if $a \geq b$ and $b - a$ if $b \geq a$. Refer to Figure 4.15. In the first case, $a - b \geq 0$ and hence $a - b$ is equal to the absolute value $|a - b|$. In the second case, $a - b < 0$, so $b - a = -(a - b) = |a - b|$. Observe, therefore, that the distance between a and b is always equal to $|a - b| = |b - a|$. Now let $c = \frac{a+b}{2}$. The fact that

$$|a - c| = \left| a - \frac{a+b}{2} \right| = \left| \frac{2a - a - b}{2} \right| = \left| \frac{a - b}{2} \right| \quad \text{and}$$

$$|b - c| = \left| b - \frac{a+b}{2} \right| = \left| \frac{2b - a - b}{2} \right| = \left| \frac{b - a}{2} \right| = \left| \frac{a - b}{2} \right|$$

tells us that the distances from point c to both a and b are the same. Therefore, $c = \frac{a+b}{2}$ is the midpoint of the segment between a and b.

The real number system with its symbols, organization, and algorithms that Islamic mathematicians introduced and European mathematicians of the seventeenth century expanded into the coordinate plane and space (as we will see next) opened the way to mathematical constructions and computational strategies of enormous consequence. The coordinate plane (and its higher dimensional versions) would become the platforms from which mathematics built the advanced structures that made progress in modern science and engineering possible. This began toward the end of the seventeenth century, when Newton (1642–1727) and Leibniz (1646–1716) placed the foundations of calculus on top of the coordinate plane. Progress continued in the next two centuries and accelerated in the second part of the twentieth century with the invention of high-speed computers. Chapters 6 and 7 give a sense of these advances as well as their impact on architecture.

The Coordinate Plane

In the first part of the seventeenth century, two Frenchmen, working independently, abstracted what Apollonius and Ptolemy had done into a powerful tool for studying curves in the plane. The philosopher René Descartes

(1596–1650) and the lawyer Pierre de Fermat (1601–1665) took two perpendicular number lines, placed them into Euclid's plane, and aligned them so that the two 0s coincide. This setup is called the rectangular or Cartesian (for Descartes) coordinate system.

This section begins with a detailed description of a coordinate system. It then turns to the study of perfect abstract versions of the lines and circles that architects sketch on flat sheets of paper to compose their designs.

Place two number lines, both constructed with the same unit of length, into a plane in such a way that they intersect at their respective points 0 and are perpendicular to each other. A plane equipped with such a pair of number lines is a coordinate plane. We'll call one of the lines the x-axis (it is often placed horizontally) and the other line the y-axis (it is often placed vertically). (Needless to say, there are other ways to label these lines.) Now let P be any point in the plane. Draw a line through P parallel to the y-axis. This line intersects the x-axis and the point of intersection corresponds to a number, say a, on the axis. This number a is called the x-coordinate of P. In the same way, the line through P parallel to the x-axis intersects the y-axis at a number, say b, and this the y-coordinate of P. We say that P is the point with coordinates (a, b) and will often write (a, b) in place of P. By reversing the process just described, we can start with a pair of numbers, say (c, d), and arrive at a point, say R, that has coordinates (c, d). The dual relationship just described is illustrated in Figure 4.16. The point S gives rise to the coordinates $(3, -3.2)$ so that $S = (3, -3.2)$, and the pair of numbers 2 and 4$\frac{1}{2}$ determine the point $Q = (2, 4\frac{1}{2})$. The point $(0, 0)$ is called the origin and is

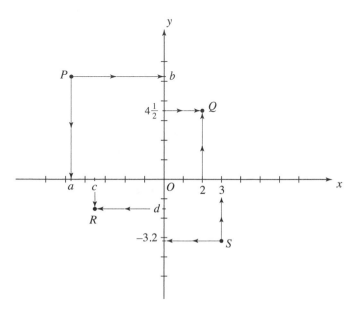

Figure 4.16

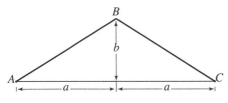

Figure 4.17

denoted by O. The notation 0 (zero) will be used for this point if the attention is on the x-axis or the y-axis.

Consider any equation with variables x and y. The graph of such an equation is the set of all points (a, b) in the plane with the property that the values $x = a$ and $y = b$ satisfy the equation. It is a simple matter to check that the point $(1, \frac{1}{2})$ is on the graph of $x^2 + 4y^2 - 2 = 0$ and that the point $(2, 1)$ is on the graph of $4x^3 - 27y^{\frac{1}{2}} + \frac{5}{y^2} = 10$. The graphs of the equations that we will consider (in this chapter and later) are often curves, such as lines, circles, parabolas, and ellipses. If the graph of a curve in the plane is the graph of an equation, we will refer to the equation as an equation of the curve.

Let's turn to the study of lines. We'll start with the roof of Old St. Peter's as it is depicted in Figure 3.13. Figure 4.17 shows the cross section of the roof as a combination of two slanting segments AB and BC. Focus on the segment AB. The ratio $\frac{b}{a}$ of the "rise" b over the "run" a is a measure of the steepness, pitch, or slope of the roof segment AB. This is the rate at which the segment AB rises. The larger b is relative to a, the steeper the roof is and the larger the slope $\frac{b}{a}$. What about the slope of BC? Its "rise over run" is the same $\frac{b}{a}$, but in proceeding from left to right along it, the roof falls. So we define the slope of BC to be $-\frac{b}{a}$. For instance, if $a = 15$ feet and $b = 9$ feet, then the slope of AB is $\frac{b}{a} = \frac{9}{15} = \frac{3}{5} = 0.6$, so that the roof segment AB rises at a rate of 0.6 foot for each horizontal foot. The slope of the segment BC is -0.6. This segment of the roof falls at a rate of 0.6 foot (or rises at a negative rate of -0.6) per horizontal foot.

Notice the importance of the selection of a preferred direction, namely *from left to right*. In our Western culture (but not in all cultures), this direction *is* preferred. When we write, we do so from left to right. The arrangement of the numbers on a horizontal number line—increasing from left to right—also reflects this preference.

Place an xy-coordinate system into the plane of Figure 4.17 and extend the roof segments AB and BC in both directions to get the lines L_{AB} and L_{BC}. Let $P_1 = (x_1, y_1)$ and $P_2 = (x_2, y_2)$ be two distinct points on L_{AB}. Refer to

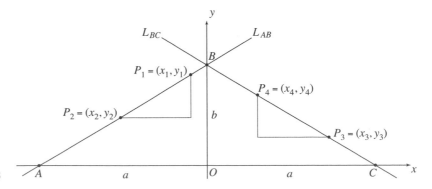

Figure 4.18

Figure 4.18 and observe that the triangle determined by P_1 and P_2 is similar to the triangle ABO. So the ratios of corresponding sides are equal and it follows that

$$\frac{y_1 - y_2}{x_1 - x_2} = \frac{b}{a}.$$

Now turn to the line L_{BC}, and let $P_3 = (x_3, y_3)$ and $P_4 = (x_4, y_4)$ be two distinct points on L_{BC}. By the similarity of triangles once more, $\frac{y_4 - y_3}{x_3 - x_4} = \frac{b}{a}$, or

$$\frac{y_3 - y_4}{x_3 - x_4} = -\frac{b}{a}.$$

Combining the two equalities just derived with earlier observations about the roof of Old St. Peter's tells us that the slope of a line L in the xy-plane should be defined as follows. Select any two distinct points $P_1 = (x_1, y_1)$ and $P_2 = (x_2, y_2)$ on L, and define the slope of L to be the ratio

$$\frac{y_1 - y_2}{x_1 - x_2}.$$

We have seen for the line L_{AB} that the ratio defining the slope is the same (namely $\frac{b}{a}$), no matter what pair of distinct points are selected. Similarly, for L_{BC} it is always $-\frac{b}{a}$. For the same reason, this is true for any line L. No matter what pair of distinct points on L are chosen, the ratio defining the slope is the same number. If L is vertical, then there is a problem. In this case $x_1 = x_2$, so that the ratio is undefined. So vertical lines *have no slope*. If L is horizontal, then $y_1 = y_2$, so that the slope of L is 0.

In Figure 4.19, L is a nonvertical line and $P_1 = (x_1, y_1)$ and $P_2 = (x_2, y_2)$ are two distinct points on L. Let the slope of L be m. So $m = \frac{y_1 - y_2}{x_1 - x_2} = \frac{y_2 - y_1}{x_2 - x_1}$. Now let $P = (x, y)$ be any point in the plane (not equal to P_1). Notice that $P = (x, y)$ is on L precisely if L and the segment $P_1 P$ are equally steep. So $P = (x, y)$ is on L precisely if the slope of the line through P_1 and P is equal to m. So the point $P = (x, y)$ is on L, precisely if $\frac{y - y_1}{x - x_1} = m$, or, in rewritten form, if $y - y_1 =$

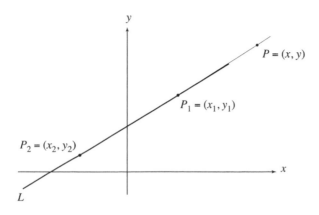

Figure 4.19

$m(x-x_1)$. This equation is also satisfied when $x=x_1$ and $y=y_1$ (as both sides are zero). It follows that a point (x, y) is on the line L precisely if it satisfies

$$y - y_1 = m(x - x_1).$$

An equation of the line L arranged in this way is said to be in *point-slope* form. Rewrite this point-slope equation of L as $y = mx + (y_1 - mx_1)$. With $b = y_1 - mx_1$, this becomes

$$y = mx + b.$$

Because the line crosses the y-axis at $(0, b)$, the number b is referred to as the y-intercept, and the equation $y = mx + b$ for L is in *slope-intercept* form. Putting $m = \frac{y_2 - y_1}{x_2 - x_1}$ into the point-slope form, we get the equation

$$y - y_1 = \left(\frac{y_2 - y_1}{x_2 - x_1}\right)(x - x_1)$$

of L. It is in *two-point* form.

It's time for some concrete examples. Consider the line L that the points $(-2, 3)$ and $(4, 1)$ determine. The equation $y - 1 = (\frac{1-3}{4-(-2)})(x - 4)$ of L is in two-point form. The slope of the line is $(\frac{1-3}{4-(-2)}) = -\frac{2}{6} = -\frac{1}{3}$. The equations $y - 1 = -\frac{1}{3}(x - 4)$ and $y - 3 = -\frac{1}{3}(x - (-2)) = -\frac{1}{3}(x + 2)$ of L are both in point-slope form. From this last equation, we get the slope-intercept equation $y = -\frac{1}{3}x + 2\frac{1}{3}$ of L. The line L with slope $\frac{5}{4}$ and the point $(7, -3)$ on it has equation $y + 3 = \frac{5}{4}(x - 7)$. It is in point-slope form. The rewritten version $y = \frac{5}{4}x - \frac{35}{4} - 3 = \frac{5}{4}x - 11\frac{3}{4}$ is the slope-intercept form of the equation of L.

Because vertical lines have no slope, their equations are expressed differently. For example, consider the vertical line through the point 3 on the x-axis. This line is the set of all the points of the form (x, y) with $x = 3$. So $x = 3$ is the equation of the line. In the same way, the vertical line through x_0 on the x-axis has equation $x = x_0$.

The equation of a line determines the line with complete precision. For example, the fact that the line L determined by the points $(-2, 3)$ and $(4, 1)$ has equation $y = -\frac{1}{3}x + 2\frac{1}{3}$ tells us that a point lies on L only if its coordinates satisfy this equation. In particular, the point $(0, 2\frac{1}{3})$ is on L. The point $(0, 2.3333)$ is very close, but it is not on the line.

Having investigated lines, we now turn to circles. The key to their study is the formula that specifies the distance between to points.

Consider any two points $P_1 = (x_1, y_1)$ and $P_2 = (x_2, y_2)$ in the plane and refer to Figure 4.20. By an observation made toward the end of the preceding section, the distance between the x-coordinates of the two points is $|x_1 - x_2|$ and the distance between their y-coordinates is $|y_1 - y_2|$. It follows that the segment P_1P_2 is the hypotenuse of the right triangle of Figure 4.21. Therefore, by the Pythagorean Theorem, $P_1P_2 = \sqrt{|x_1 - x_2|^2 + |y_1 - y_2|^2}$. This equation reflects an ongoing notational practice. When a label for a segment, such

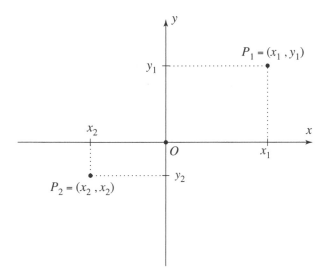

Figure 4.20

as P_1P_2, appears in a mathematical expression, it represents the length of the segment. Since $|x_1 - x_2|^2 = (x_1 - x_2)^2$ and $|y_1 - y_2|^2 = (y_1 - y_2)^2$, the distance between the points $P_1 = (x_1, y_1)$ and $P_2 = (x_2, y_2)$ is equal to

$$P_1P_2 = \sqrt{(x_1 - x_2)^2 + (y_1 - y_2)^2}.$$

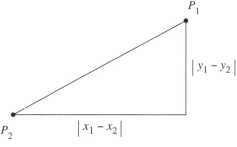

Figure 4.21

This is the distance formula for the distance between two points in the plane.

For example, the distance between $(1, -2)$ and $(5, 3)$ is $\sqrt{(1-5)^2 + (-2-3)^2} = \sqrt{4^2 + 5^2} = \sqrt{41}$. Because $\sqrt{(5-1)^2 + (3-(-2))^2}$ is also equal to $\sqrt{4^2 + 5^2} = \sqrt{41}$, the order in which the points are taken has, as expected, no effect on the result.

It is easy to check with the distance formula that the midpoint of the segment determined by the points $P_1 = (x_1, y_1)$ and $P_2 = (x_2, y_2)$ is the point

$$\left(\frac{x_1 + x_2}{2}, \frac{y_1 + y_2}{2} \right).$$

Consider the circle with center $C = (3, 2)$ and radius 4. Figure 4.22 tells us that a point $P = (x, y)$ in the plane is on this circle, precisely if the distance from P to the center $(3, 2)$ is equal to 4. By the distance formula, $P = (x, y)$ is on the circle exactly when $PC = \sqrt{(x-3)^2 + (y-2)^2} = 4$. Squaring both sides, we see that this is the same condition as

$$(x - 3)^2 + (y - 2)^2 = 16.$$

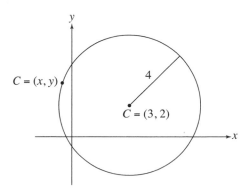

Figure 4.22

It follows that this is an equation of the circle with center $(3, 2)$ and radius 4. In the same way,

$$(x - h)^2 + (y - k)^2 = r^2$$

is an equation of the circle with center (h, k) and radius r. This is the standard equation of this circle. The circle with center the origin $O = (0, 0)$ and radius r has standard equation $x^2 + y^2 = r^2$.

The equation of a circle determines the circle with perfect precision. Consider the circle with radius 3 and center $(2, -5)$ and observe that its standard equation is $(x - 2)^2 + (y + 5)^2 = 9$. Check that $(2 + 2\sqrt{2}, -4)$ is a point on the circle. Why is the point $(6.8284, -4)$ very close to the circle, but not on it?

Coordinate System in Three Dimensions

The inventors of the coordinate plane, Descartes and Fermat, both realized that coordinate systems can also be constructed for the three-dimensional space in which we live and in which architects build. This section explains how this is done.

Start with a plane that has an xy-coordinate system and let O be the origin. Place a third number line (constructed with the same unit of length as the other two) through O perpendicular to the plane. We'll call it the z-axis. The setup just described is shown in Figure 4.23 and is called an xyz-coordinate system. In each case, the positive part of the axis is shown as a solid line with an arrow. The negative parts are dashed. The xy-, xz-, and yz-planes are the coordinate planes of the system. Take any point P in space. Push it in the direction parallel to the z-axis into the xy-plane and let (x_0, y_0) be the coordinates of this point. Let the distance from P to the xy-plane (or the negative of this

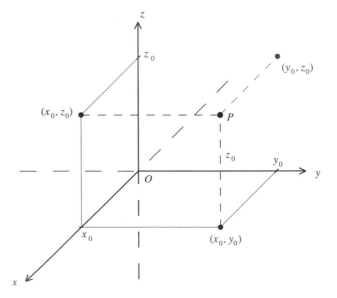

Figure 4.23

distance if P is below the xy-plane) be z_0. The numbers x_0, y_0, and z_0 are the x-, y-, and z-coordinates of P. Because x_0, y_0, and z_0 determine the location of P, we write $P = (x_0, y_0, z_0)$. Push P into the xz plane in the direction parallel to the y-axis and observe that (x_0, z_0) are the coordinates of the point obtained. Pushing P into the yz-plane parallel to the x-axis, we get the point (y_0, z_0). Figure 4.23 pictures what has been described. The process that started with the point P and provided its coordinates x_0, y_0, and z_0 can be reversed. For example, the three numbers $(2, -3, 4)$ determine a point in space as follows. Start with $(2, -3)$ in the xy-plane. From there, go up 4 units in the direction of the z-axis. In this way, the numbers $(2, -3, 4)$ give rise to a point. It follows from the way the point is determined that its x-, y-, and z-coordinates are 2, -3, and 4, respectively. In this way, any triple of numbers determines a point in space.

Using what is done in the coordinate plane as a model, we define the graph of an equation with variables x, y, and z to be the set of all points (x, y, z) in space whose coordinates satisfy the equation. The graphs of such equations are geometric shapes in space, usually curves and surfaces. If a curve or surface is the graph of an equation in x, y, and z, we say that the equation is an equation of the curve or surface.

For instance, let a, b, c, and d be constants and consider the equation

$$ax + by + cz = d.$$

What can we say about its graph? Let's start by looking at some specific cases. The equation $z = 0$ (obtained by taking $a = b = d = 0$ and $c = 1$) is satisfied by all points of the form $(x, y, 0)$ where x and y can be anything. But this is the xy-coordinate plane. What about $z = 3$? The points that satisfy it are all those of the form $(x, y, 3)$. This is the plane 3 units above and parallel to the xy plane. Can you describe the planes $x = 0$, $x = 7$, $y = 0$, and $y = -5$? Consider next the example $3x + 4y + 0z = 5$. Solving $3x + 4y = 5$ for y, we get $y = -\frac{3}{4}x + \frac{5}{4}$, the line in the xy plane with slope $-\frac{3}{4}$ and y-intercept $\frac{5}{4}$. The points that satisfy $4x + 5y + 0z = 5$ are those of the form (x, y, z), where (x, y) lies on the line $y = -\frac{3}{4}x + \frac{5}{4}$ in the xy-plane and z is free to have any value. So the graph of $4x + 5y + 0z = 5$ is the plane through the line $y = -\frac{3}{4}x + \frac{5}{4}$, perpendicular to the xy-plane.

Finally, let's look at the case where none of the constants a, b, c, or d is equal to 0. Again, the set of points satisfying $ax + by + cz = d$ is a plane. Which plane? Set $y = z = 0$ and solve for x to get that the point $(\frac{d}{a}, 0, 0)$ satisfies the equation. The points $(0, \frac{d}{b}, 0)$ and $(0, 0, \frac{d}{c})$ satisfy the equation as well. It turns out that the set of points satisfying $ax + by + cz = d$ is the plane that the triangle in Figure 4.24 determines. In the example of the figure, a, c, and d are positive and b is negative. The fact is that the graph of the equation $ax + by + cz = d$ is a plane, except when $a = b = c = 0$. What happens then?

Let's look at another example. Consider $x^2 + y^2 = r^2$ as an equation in the variables x, y, and z by writing it as $x^2 + y^2 + 0z = r^2$ and observe that it does not

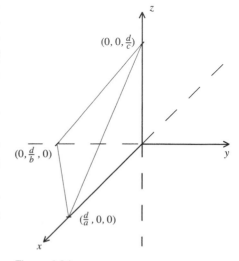

Figure 4.24

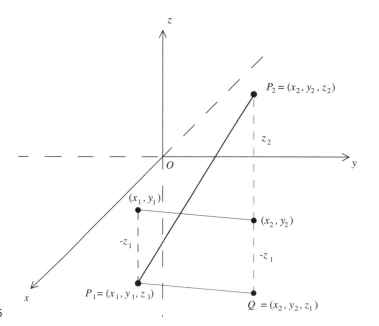

Figure 4.25

restrict z in any way. So z is free to be anything, and the graph consists of all points (x, y, z) with the property that $x^2 + y^2 = r^2$. This is an infinite cylinder with central axis the z-axis through the circle $x^2 + y^2 = r^2$ in the xy plane.

We turn next to the formula for the distance between two points in space. Let the points $P_1 = (x_1, y_1, z_1)$ and $P_2 = (x_2, y_2, z_2)$ be given. The points (x_1, y_1) and (x_2, y_2) in the xy-plane are obtained by pushing P_1 and P_2 into the xy-plane in the direction of the z-axis. Let Q be the point $Q = (x_2, y_2, z_1)$. Refer to Figure 4.25. Notice that the distance between P_1 and Q is the same as the distance between (x_1, y_1) and (x_2, y_2). By the distance formula in the plane this distance is

$$\sqrt{(x_1 - x_2)^2 + (y_1 - y_2)^2}.$$

The distance between P_2 and Q is $|z_2 - z_1|$. (In Figure 4.25, z_2 happens to be positive and z_1 negative, so that the distance is $z_2 - z_1$.) The Pythagorean Theorem applied to the triangle P_1QP_2 tells us that

$$(P_1P_2)^2 = \left(\sqrt{(x_1 - x_2)^2 + (y_1 - y_2)^2}\right)^2 + |z_2 - z_1|^2.$$

So $(P_1P_2)^2 = (x_1 - x_2)^2 + (y_1 - y_2)^2 + (z_1 - z_2)^2$ and therefore

$$P_1P_2 = \sqrt{(x_1 - x_2)^2 + (y_1 - y_2)^2 + (z_1 - z_2)^2}.$$

This is the distance formula for the distance between two points in space.

Suppose that a point $P = (x, y, z)$ is at a distance 4 from the point $(1, -2, 3)$. The distance formula tells us that $\sqrt{(x-1)^2 + (y-(-2))^2 + (z-3)^2} = 4$, and therefore

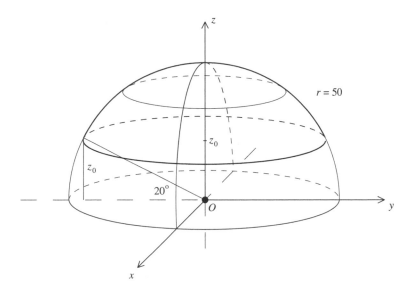

Figure 4.26

$$(x-1)^2 + (y+2)^2 + (z-3)^2 = 4^2.$$

It follows that this is an equation of the sphere with center $(1,-2,3)$ and radius 4. In the very same way,

$$(x-h)^2 + (y-k)^2 + (z-l)^2 = r^2$$

is an equation for the sphere with center (h,k,l) and radius r. This is the standard equation of the sphere with center (h,k,l) and radius r.

Let's return to Chapter 3 and the dome of the Hagia Sophia. Refer to Figure 3.3 and set up an xyz-coordinate system in a such way that $O=(0,0,0)$ is the common center of the spheres that determine the inner and outer surfaces of the shell and the z-axis is the central vertical axis of the dome. Recall that the radius of the inner surface of the dome is 50 feet, and consider the sphere $x^2 + y^2 + z^2 = 50^2$. Figure 4.26 depicts the upper half. Put in the 20° angle from Figure 3.3 and let z_0 be the height of the triangle that the angle determines. Note that $\sin 20° = \frac{z_0}{50}$, so $z_0 = 50 \sin 20° \approx 17$. Observe that the section of the sphere above the plane $z = z_0$ is a mathematical model of the inner surface of the dome above the row of 40 windows. To determine the circle at the base of this section of the sphere, plug z_0 into $x^2 + y^2 + z^2 = 50^2$ to get $x^2 + y^2 + z_0^2 = 50^2$ or $x^2 + y^2 = 50^2 - z_0^2$. Because $50 - z_0^2 \approx 50^2 - 17^2 = 2211$, this equation represents a circle with center $(0, 0, z_0)$ and radius $\sqrt{50 - z_0^2} \approx \sqrt{2211} \approx 47$ feet.

This mathematical model is probably a good approximation of the inner surface of the dome of the Hagia Sophia as it existed in the sixth century, but only a rough approximation of this surface as it is today. Recall from Chapter 3 that various reconstructions of the dome deformed its spherical geometry. In particular, its base is now an oval (refer to Figure 3.6).

This section and the preceding section both illustrate the dual relationship between algebra and geometry that a coordinate system provides. For a given algebraic equation (in two or three variables), the graph provides a visual sense of the relationship between the variables. In the other direction, when a curve or surface is given as the graph of an equation, the algebraic analysis of the equation can reveal exact numerical and geometric information about the curve or surface.

The Duomo of Florence

In 1296 the prosperous city of Florence began building a cathedral that was intended to surpass in grandeur those of its Tuscan rivals, Pisa and Siena. The cathedral Santa Maria del Fiore, "Our Lady of the Flower," was to have a rectangular nave with ribbed Gothic vaulting and three apses in the shape of half-octagons. The most distinctive aspect of its design called for a huge dome that was to soar over the crossing. During the first half of the fourteenth century work proceeded very slowly. But by the year 1418, the transept was finished, the two apses of the transept were vaulted with half-domes, and the massive octagonal drum, or tambour, from which the dome was to rise was in place. Refer to Figure 4.27. The drum was supported by an alternating combination of heavy Gothic arches and huge piers. It was made of a resilient sandstone with walls 15 feet thick. The diameter of its inner octagonal cross section was 145 feet and its upper edge soared about 180 feet above the cathedral's floor.

The time had come to vault the gaping octagonal opening with the dome. In 1418 the Opera del Duomo, the commission that oversaw the construction of the cathedral, announced a competition for the design and execution of the dome. Several structural aspects were already determined. The

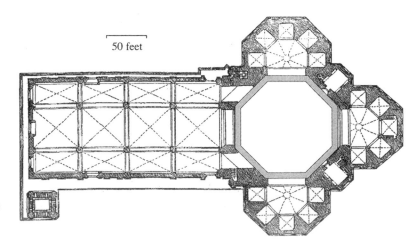

Figure 4.27. The plan of the Santa Maria del Fiore with its three apses and its octagonal drum

size of the drum determined the size of the base of the dome, and the shape of the drum meant that the dome would rise from its base with octagonal cross sections. However, the most daunting questions still remained to be addressed. Could the 145-foot space be spanned by a dome constructed with brick, stone, mortar, or concrete? Was it possible and feasible to build a traditional wooden centering structure that would reach 180 feet up from the ground to the drum and then rise from there to hold the dome up during its construction? To top it off, the Opera declared massive exterior buttresses to be unattractive and decreed that there could be no such structures. But how could the considerable hoop stresses expected from such a large masonry dome be contained without massive exterior buttressing? It is an understatement to say that the challenges that the construction of this dome presented were without precedent. How—or even if—such a dome could be built was not at all clear. The Opera had relied on the same "just build and figure out the next step when the time comes" strategy that had guided the Building Council in the construction of the Cathedral of Milan (in Chapter 3, "From the Annals of a Building Council").

Filippo Brunelleschi (1377–1446) responded to the Opera's announcement. He was a goldsmith by training, but he had spent an extended visit in Rome studying the brickwork, construction techniques, and methods of vaulting of the ancient Roman builders. Brunelleschi submitted a model to the Opera that was designed to convey both his concept and its practicality. He had used it as a test bed to validate his methods of construction. The use of scale models was common in the planning of large structures and goes back to antiquity. Made of wood, brick, and mortar, it was large enough to allow members of the commission to walk inside and inspect its interior. The central vertical cross sections of the dome were to be pointed Gothic arches. The steeper pitch of such a design would lessen the outward thrusts that the heavy dome would generate. Brunelleschi's design of the dome called for inner and outer masonry shells. The structurally powerful inner shell would support and be protected by a thinner outer shell that would give the dome a heightened profile. The most radical and controversial aspect of Brunelleschi's proposal was the idea that the dome could be built without the use of a centering structure that would reach up from the ground to support it until its construction was complete. The Opera was skeptical, but also impressed. In 1420, it appointed Brunelleschi as one of two *capomaestri*, or master builders, to lead the construction.

Work on the dome began in the same year. Records indicate that the number of craftsmen, including stonecutters, bricklayers, mortar specialists, and woodworkers, employed at any time was fewer than 100. This number did not include unskilled laborers. One of the first orders of business was the construction of hoists and cranes for moving and lifting building materials. Probably inspired by Vitruvius's descriptions of the machines that the Romans used to build their large structures, Brunelleschi invented a large hoist that made it possible, according to a later commentator, "for a single ox to raise a load so heavy that previously . . . [would have required] six pairs of oxen to move." The main hoist was ground-based and powered by yoked teams of oxen or horses. It consisted of a massive wooden frame and a novel system of wooden pulleys, cogwheels, gears, counterweights, and a rotating drum to spool or unspool the heavy rope. Loads of up to several thousand pounds could be raised. Cranes built with wooden masts, pivoting horizontal crossbeams, and adjustable counterweights operated much like modern construction cranes. Designed to sit high on working platforms supported by completed wall structures of the dome, they would move building materials to the points of construction. According to later accounts, Brunelleschi had a large area near the banks of Florence's Arno river cleared and leveled. By stretching and rotating ropes, his workmen could trace a full-scale plan of the dome and its components in the sand. Templates for the dome's stone ribs and other elements probably were made using such a huge diagram.

Progress on the dome depended on three critical operations: controlling the rising profile, raising working platforms, and laying the successive rings of bricks. By using Figure 4.28, let's first see in the abstract how the workers guided the upward-curving geometry of the dome. The key to its geometry is the inner edge of the octagonal drum. Take an xy-coordinate plane. Consider the unit circle $x^2 + y^2 = 1$ and put in the four points $S_1 = (1, 0)$, $S_3 = (0, 1)$, $S_5 = (-1, 0)$, and $S_7 = (0, -1)$. The four points S_2, S_4, S_6, and S_8 are the points of intersection of the circle and the two lines $y = x$ and $y = -x$. Because the angles formed by the axes and the lines $y = x$ and $y = -x$ are all 45°, it follows

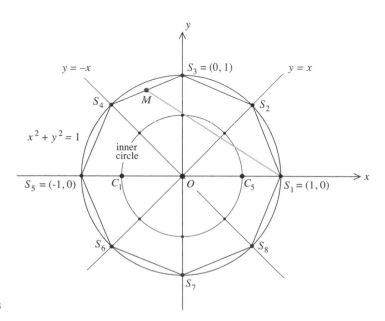

Figure 4.28

that these eight points are the vertices of a regular octagon. This octagon represents the inner edge of the drum. The point M is the midpoint of the segment joining S_3 and S_4. The gray segment represents a rope that is stretched from S_1 to M. The length of this rope is important. To compute it, check first that $S_4 = (-\frac{1}{\sqrt{2}}, \frac{1}{\sqrt{2}})$ and then use the midpoint formula from the coordinate plane section to get that

$$M = \left(\frac{-\frac{1}{\sqrt{2}}+0}{2}, \frac{\frac{1}{\sqrt{2}}+1}{2} \right) = \left(-\frac{1}{2\sqrt{2}}, \frac{\frac{1+\sqrt{2}}{\sqrt{2}}}{2} \right) = \left(-\frac{1}{2\sqrt{2}}, \frac{1+\sqrt{2}}{2\sqrt{2}} \right).$$

The distance formula from the same section now shows that the length of the rope is

$$S_1 M = \sqrt{\left(1 - \left(-\frac{1}{2\sqrt{2}} \right) \right)^2 + \left(0 - \frac{1+\sqrt{2}}{2\sqrt{2}} \right)^2} = \sqrt{\left(\frac{2\sqrt{2}+1}{2\sqrt{2}} \right)^2 + \left(\frac{1+\sqrt{2}}{2\sqrt{2}} \right)^2}$$

$$= \sqrt{\frac{(4 \cdot 2 + 4\sqrt{2}+1)+(1+2\sqrt{2}+2)}{(2\sqrt{2})^2}} = \frac{\sqrt{12+6\sqrt{2}}}{2\sqrt{2}} = \frac{\sqrt{6(2+\sqrt{2})}}{2\sqrt{2}} = \frac{\sqrt{6}}{2\sqrt{2}}\sqrt{2+\sqrt{2}}$$

$$= \frac{\sqrt{3}}{2}\sqrt{2+\sqrt{2}} \approx 1.600206\ldots \approx 1.6.$$

Keep the rope fixed at S_1 and rotate the point M over to get the point C_1 on the x-axis. Observe that

$$\frac{S_1 C_1}{S_1 S_5} = \frac{S_1 M}{S_1 S_5} \approx \frac{1.600}{2.000} = \frac{4\,(0.4)}{5\,(0.4)} = \frac{4}{5}.$$

So the point C_1 is very nearly four-fifths of the way from S_1 to S_5. Next place the rope at S_5 and stretch it toward S_1 to get the point C_5. As above, $\frac{S_5 C_5}{S_1 S_5} \approx \frac{4}{5}$ so that C_5 is nearly four-fifths of the way from S_5 to S_1.

Fix one end of the rope at C_1, stretch the other end to S_1, and swing this end of the rope vertically upward. Then place the rope at C_5 and swing it vertically upward from S_5. What is obtained in this way is an arch in a vertical plane that consists of two circular arcs that meet at a point. The points S_1 and S_5 are its springing points, C_1 and C_5 are the centers of the two circles, and T is the point at which they meet. This is illustrated in Figure 4.29a. Because the distances from S_1 to C_5 and S_5 to C_1 are one-fifth of the distance from S_1 to S_5, this arch is known as the pointed Gothic fifth or, in Italian, the *quinto acuto*. Next, do the very same thing with the segments $S_2 S_6$, $S_3 S_7$, and $S_4 S_8$, to get three more arches in the shape of the pointed Gothic fifth, each in its own vertical plane. The four arches rise from the springing points $S_1, S_2, \ldots, S_8$ on the inner octagon of the drum. This evenly spaced configuration of four arches determines the inner surface of the dome. This surface consists of the eight panels. Each panel is flat in the horizontal direction and rises from one of the eight sides of the octagon. The panels curve inward to meet at the point T. They are shown in Figure 4.29b. The central,

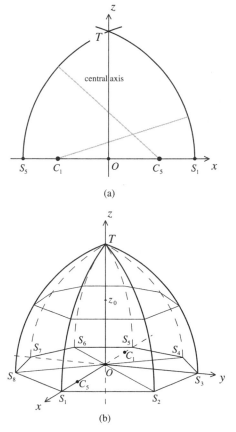

Figure 4.29

vertical z-axis is the line of intersection of the four vertical planes that the four pointed arches determine. The Gothic fifth of the inner shell also sets the geometry of its exterior. Unlike the flat domes of the Pantheon and the Hagia Sophia, the dome of the Santa Maria del Fiore would soar skyward.

Brunelleschi's workers constructed a platform that projected inward from the edge of the drum to the inner circle shown in Figure 4.28. The platform was supported from below by partial centering attached to the walls of the drum and from above by ropes attached to iron rings anchored in the brickwork. It was from this platform that workers stretched *corda da murare*, or "building strings," that determined the curving geometry of the eight rising panels of the inner surface of the dome. The inner edge of the drum is not a perfect octagon. Its eight sides vary from a longest of 56.5 feet to a shortest of 54.5 feet. As a consequence, there would also be variations in the shape of the eight curving panels.

The eight sections of the two shells of the dome soon began to grow, their shapes guided by those of the eight inner surfaces. For the first 20 or so feet the two shells are built with ashlar—big, cut, rectangular blocks of sandstone. From then on the shells are made with lighter brick. Both the stones and the bricks are bound together by mortar. The working crew was organized into eight teams so that the eight sections of the two shells could grow simultaneously, one octagonal ring at a time. During the construction of a ring the brick and mortar structure needed to be supported to prevent it from falling inward. However, a completed ring was stable on its own, compressed into place by the surrounding shell. The rising dome was stable after the completion of each ring of bricks in much the same way that the Pantheon is stable in spite of its oculus. Brunelleschi knew this when he proposed the idea that the dome could be built without an elaborate centering structure. Bricks of several different sizes and shapes were used, including rectangular, triangular, and dovetailed bricks, and bricks with flanges—several million of them in all. To provide the growing double shell with additional stability, the bricks are arranged in an interlocking herringbone pattern of horizontal rows and upward spirals. The brickwork at the inner surfaces of the two shells seen in Figure 4.30 gives a sense of the pattern.

Figure 4.30. Herringbone brickwork pattern in the two shells. Photo by Philip Holtzman and Rose Holtzman

Near the drum, the inner shell has a thickness of about 7.5 feet, the outer shell about 2.5 feet and the gap between them is about 4 feet. The thicknesses of the shells decrease and the gap between the shells increases as the dome rises. The two shells are connected to each other by a system of 24 masonry ribs, called spurs. These buttressing elements between the shells start near the base and continue to the top. There is one corner spur for each of the eight corners of the octagonal shell and two more intermediate spurs on each side. Figure 4.31 is a horizontal section of the dome that shows these structures. The spurs at the sides vary in thickness from about 6 feet at the lower elevations to about 1.5 feet higher up. Those at the corners are twice as thick. A main component of the structure of the dome is a system of powerful stone arches at nine different elevations, all in the upper two-thirds of the dome. They function to reinforce the outer shell near the corner spurs. They are thickest at the corners and taper outward in both directions to join the two intermediate spurs. Figure 4.31 cuts through a set of these arches.

The shells of the dome are reinforced by four pairs of surrounding sandstone chains. They are fashioned from 9-foot sections of rectangular sandstone blocks that are held together by iron clamps. In addition to the stone chains, there is a wooden chain made of chestnut beams strengthened by oak segments and attached to each other by iron pins. The vertical section of the dome in Figure 4.32 shows the drum and the two shells. The figure indicates the locations of the reinforcing chains and the positions of the nine sets of stone arches. It also provides the heights of some of the structural elements above the cathedral floor.

The construction of the dome was completed in 1436. The exteriors of the eight panels of the outer shell are covered with about 30,000 terra cotta tiles 15 inches wide and 20 inches high. (*Terra cotta*, Italian for "baked earth," is a brownish-red clay that is molded and made hard when subjected to high temperatures in a kiln.) Between the panels are arches made of segments of

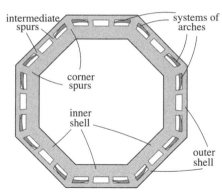

Figure 4.31. Horizontal section of the dome

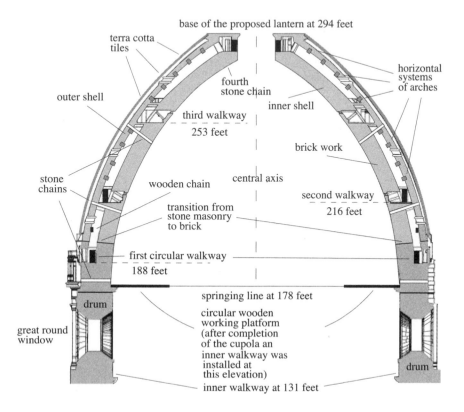

Figure 4.32. Vertical section of the dome through the central axis. Adapted from G. Fanelli and M. Fanelli, *Brunelleschi's Cupola: Past and Present of an Architectural Masterpiece*, pp. 180–81

Figure 4.33. Interior view of dome.
Photo by Chanclos

white marble that cover the exterior of the corner spurs. An octagonal open-ing of about 18.5 feet in diameter is built into the top. It is braced by a hori-zontal masonry ring structure that connects the inner and outer shells. This ring would later support a lantern. Made of white marble and capped with a conical roof, the lantern crowns the dome's exterior and provides light for the interior. See Plate 19 and Figure 4.33. After the lantern was topped with a sphere and a cross in 1471, the construction of the cathedral Santa Maria del Fiore was finally finished.

The church, 500 feet long, 125 feet wide, with a dome that rises to a height of 294 feet—376 feet if the lantern is included—was the largest church in the world. Today, St. Peter's in Rome (see Chapter 5, "Michelangelo's St. Peter's") and St. Paul's in London (refer to Chapter 6, "Evolving Structures") are domed structures of similar type that are slightly larger. Experts have estimated that the dome weighs about 67 million pounds (about 40 mil-lion for the inner shell, 19 million for the outer shell including the tiles and marble segments, 6 million for the spurs, and another 2 million pounds for

Figure 4.34. Photo by Gryffinder

the lantern). The fact that this is more than 20 times greater than the 3 million pounds of the dome of the Hagia Sophia (see Chapter 3, "The Hagia Sophia") confirms how monumental this dome is and how unprecedented it was at the time. The cathedral Santa Maria del Fiore is one of the most daring and impressive engineering feats ever undertaken.

Much of the decorative detail of both the outside and the inside of the church—see Plate 19 and Figure 4.33—was added later. The fresco *The Last Judgment* that unfolds on the 43,000-square-foot surface of the inside of the dome, depicted in Figure 4.33, was finished in the sixteenth century. It was in part inspired by the *The Last Judgment* that Michelangelo painted on the ceiling of the Sistine Chapel of the Vatican in Rome. The floor of the nave with its intricate geometry laid out in marble tiles was in place by the seventeenth century, and the elaborately decorated marble facade of the exterior dates from the nineteenth century.

The dome began to develop cracks even as it was being constructed. Over the years they have become more extensive. Four main cracks have been of particular concern. A detailed technical report issued in 1985 recognized them to be the result of the horizontal hoop stresses that the great weight of the dome generates. The masonry of the shells was not quite strong enough

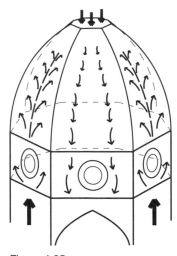

Figure 4.35

to resist them and Brunelleschi's system of stone and wooden chains has not sufficiently contained them. The fact that the dome is supported unevenly from below has added to the problem. Figure 4.33 shows the large pointed arches and the tops of the massive piers that alternate to support the drum. Not surprisingly, the drum is supported more stiffly by the four piers than by the four arches. In effect, then, the four sections of the dome over the arches pull down on the four sections of the dome over the piers. Figure 4.35 is a simplified diagram of the forces in question. The figure shows two of the piers, the arch between, and the sections of the dome above them. The upward push of the piers is represented by the two flows of upward vectors. The cluster of vectors pointing down represents the downward pull of the section over the arch. Figure 4.35 shows the outward deflection of the two upward flows of vectors that this downward pull effects. The tension that these deflected forces induce on the masonry combined with hoop stress have caused major cracks—indicated by the dashed curves in the figure—in each of the four sections of the dome supported by a pier. The cracks follow meridians, extend at least two-thirds of the way up the dome, and reach well below the great windows of the drum. They affect both the inner and outer shells and range in width from 1 to more than 3 inches. According to the structural engineers who have examined these cracks in recent years, the dome is not in danger of collapsing (unless an earthquake were to occur). Nonetheless, these cracks, other cracks, and additional structural problems are currently being monitored extensively by high-tech equipment and subjected to sophisticated mathematical analysis. It is the goal of this careful scrutiny to determine corrective measures that will secure the stability of this wonderful structure for centuries to come.

The Santa Maria del Fiore is a church in the Gothic tradition. Figure 4.33 shows that the massive arches that support the drum are pointed Gothic arches. Figure 4.34 informs us that the arches that frame the nave are also pointed and the vaults above the nave are ribbed vaults. The *quinto acuto* of the vertical sections of the dome are Gothic arches as well. However, the dome profited from Brunelleschi's studies of the vaults of Roman structures (for instance, their deployment of angled courses of bricks). These aspects and the confident spirit that pushed Brunelleschi to execute this extraordinary dome connect the Santa Maria del Fiore to the Renaissance.

Problems and Discussions

The problems deal with conic sections initially, and then turn to problems of Islamic origin. Thereafter, they involve coordinate systems in two and three dimensions, and the last few focus on the dome of the Cathedral of Florence. The section concludes with three discussions related to topics of the chapter.

Problem 1. To get a sense of the conic sections of the Greeks get a flashlight. Shine it against a flat wall. The wall is a physical plane that cuts the cone of light that the flashlight emits. If the flashlight is directed so that the tube housing the batteries is perpendicular to the wall, a circular area is illuminated. If the flashlight is good, then the circles will be clean and distinct. Tilt the flashlight away from the perpendicular position and an ellipse will appear. Tilt it so that the tube is parallel to the wall, and you will get a parabola. (What curves appear as the flashlight begins to be directed away from the wall?)

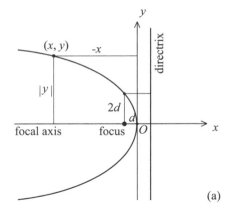

(a)

Problem 2. Consider the parabola in Figure 4.36a. Its focus is on the negative x-axis a distance d from the origin O. Why is the point $(-d, 2d)$ on the parabola? Let (x, y) be a point on the parabola. Use Proposition P2 to show that $y^2 = -4dx$. Now turn to the ellipse of Figure 4.36b. The numbers a and b are the semimajor and semiminor axes, respectively. Let (x, y) be a typical point on the ellipse and use Proposition E2 to show that $\frac{x^2}{b^2} + \frac{y^2}{a^2} = 1$.

Problem 3. The distance from the focus F of a parabola to its directrix is 3 units. The parabola is cut parallel to the directrix at a distance of 7 units from the directrix. Refer to Figure 4.37. Use the result of Archimedes to determine the area of the parabolic section. [Hint: First analyze the triangle SFS' to find the length of the cut.]

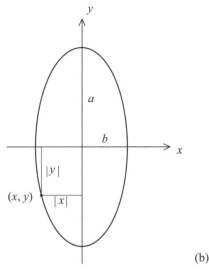

(b)

Figure 4.36

We have already met the influential Islamic mathematician al-Khwarizmi (around 780–850). Problems 4 to 7 come from his work *Hisab al-jabr....*

Problem 4. Solve the equation $(10 - x)^2 + x^2 + (10 - x) - x = 54$.

Problem 5. Solve $\frac{10-x}{x} + \frac{x}{10-x} = \frac{13}{6}$.

Problem 6. I multiply a third of a quantity plus a unit by a fourth of a quantity plus a unit, and it becomes 20 units. Determine the quantity.

Problem 7. I multiply a third of a quantity by a fourth of the quantity in such a way as to give the quantity itself plus 24 units. Determine the quantity.

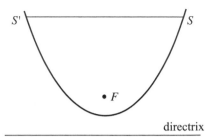

Figure 4.37

Abu Kamil (from around 850–930) was another great Islamic algebraist probably from Egypt. His *Book of Algebra* expanded the work of al-Khwarizmi. Problems 8 to 11 are taken from it.

Problem 8. Find a number such that if 7 is added to it and the sum multiplied by the root of 3 times the number, then the result is 10 times the number.

Problem 9. Find two numbers x and y such that $x + y = 10$ and $\frac{50}{x} \cdot \frac{40}{y} = 125$.

Problem 10. Find two numbers x and y such that $x + y = 10$ and $\frac{x}{y} + \frac{y}{x} = 4\frac{1}{4}$.

Problem 11. Let a and b be positive numbers with $a \geq b$. Show that $a + b - 2\sqrt{ab} > 0$. Then verify Abu Kamil's formula $\sqrt{a} \pm \sqrt{b} = \sqrt{a + b \pm 2\sqrt{ab}}$. Use it to show that $\sqrt{18} - \sqrt{8} = \sqrt{2}$.

The Persian mathematician and astronomer al-Kashi (1380–1429) worked in Islamic Samarkand (in today's Uzbekistan). Al-Kashi's text *The Key to Arithmetic* gives a description of decimal fractions and uses them to compute 2π to 16 decimal places (remarkably, this is equivalent to $\pi \approx 3.1415926535898732$). It also develops an algorithm for calculating nth roots. In addition, al-Kashi gave an explicit statement of the Law of Cosines (see Problem 15 of Chapter 2). In fact, in French mathematical circles the Law of Cosines is referred to as the Theorem of al-Kashi. Problems 12 and 13 are based on his work.

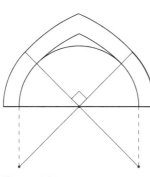

Figure 4.38

Problem 12. Al-Kashi provided a design for a pointed Islamic arch that is depicted in Figure 4.38. The construction starts with a semicircle. Study the figure and explain how it proceeds from there. [If you need a hint, study the Roman oval in Chapter 2, "The Colosseum."]

Problem 13. Start with al-Kashi's semicircle but modify the rest of his design to construct an Islamic ogee arch of the sort that appears on the facade of Saint Mark's Cathedral in Venice (as seen in Plate 16).

It took time for Islamic computational practices to be adopted in Christian Europe. The monk Gerbert (who later became Pope Sylvester II) as well as Leonardo of Pisa were influential in this regard. Leonardo's *Liber Abaci* drew heavily from the *Book of Algebra* of Abu Kamil.

Problem 14. Show that Gerbert's expression of the area of an equilateral triangle of side a as $\frac{a}{2}(a - \frac{a}{7})$ is equivalent to the approximation $\sqrt{3} \approx \frac{12}{7}$. How good is this approximation?

Problem 15. Here's a problem adapted from Fibonacci's *Liber Abaci*. Each of two people A and B has a number of coins. If A gives B nine coins, then A and B have the same number of coins, and if B gives A nine coins, then A has 10 times as many coins as B. How many coins do A and B have initially?

Problem 16. Find the British Isles, Saudi Arabia, India, and East Asia on Ptolemy's map in Plate 18 and comment on their shapes.

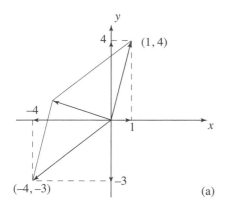

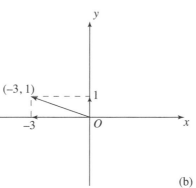

(a)　　　　　　　　　(b)　　Figure 4.39

Problem 17. Compute the distance between the points 8 and −5 on the number line and determine the midpoint between them.

The next block of problems studies coordinate geometry in the plane. The solutions of the first two require some of the basic properties of vectors described in Chapter 2, "Dealing with Forces."

Problem 18. Consider an *xy* coordinate system and the two vectors from the origin to the points $(1, 4)$ and $(−4, −3)$. Refer to Figure 4.39a. Compute the magnitudes of these vectors. The figure shows the decompositions of these vectors into their horizontal and vertical components. The resultant as obtained by the parallelogram law is also shown. Explain why this resultant is the vector from the origin to the point $(−3, 1)$ shown in Figure 4.39b.

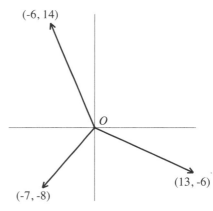

Figure 4.40

Problem 19. The vectors in Figure 4.40 all start at the origin of a coordinate system and end at the indicated points. Show that if they represent forces acting at the origin O, then these forces are in balance and produce no motion at O.

Problem 20. Go to Figure 4.17 of the text. Determine the slope of the roof AB if the angle BAC is 45°. Repeat this for the angle 30°, and then for 25°.

Problem 21. Explain what is going on in Figure 4.41. The two triangular regions have the same base of 13 units and the same height of 5 units and yet (as the subdivision of the areas shows) their areas are different!

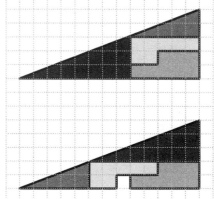

Figure 4.41

Problem 22. Consider the line in the *xy* plane that the two points $(5, 1)$ and $(−4, 7)$ determine. Sketch the line. Find equations for this line in two-point form, in point-slope form, and in slope-intercept form.

Problem 23. Sketch the line that has equation $y = \frac{1}{3}x - 4$.

Problem 24. Use the distance formula to check that the midpoint of the segment determined by the points $P_1 = (x_1, y_1)$ and $P_2 = (x_2, y_2)$ is the point

$$\left(\frac{x_1 + x_2}{2}, \frac{y_1 + y_2}{2} \right).$$

Problem 25. Place the two points $A = (-3, -7)$ and $B = (6, 8)$ in the coordinate plane. Find the length and the midpoint of the segment AB.

Problem 26. Consider the circle $x^2 + y^2 = 9$ in the xy plane. Translate (that is, shift) the circle 7 units down and 3 units to the right. What is the equation of the circle in this new position?

Problem 27. Consider the circle $(x - 3)^2 + (y + 2)^2 = 20$ and the line $y = x - 3$. Sketch both on an xy plane. Then find the points of intersection of the circle and the line.

Problem 28. Find the points of intersection of the circle with center (2, 3) and radius 5 and the line through the center with slope $\frac{1}{2}$. Find the coordinates "on the nose," then approximate them.

Problem 29. You are given a parabola with focus the point $(-4, 5)$ and directrix the line $y = -3$. Let $P = (x, y)$ be a point in the plane. Refer to Figure 4.42a and use the definition of the parabola and the distance formula to determine an equation of the parabola in the variables x and y.

Problem 30. You are given a parabola with focus the point $(-3, 4)$ and directrix the line $x = 5$. Use the definition of the parabola, the distance formula, and refer to Figure 4.42b to determine an equation of the parabola.

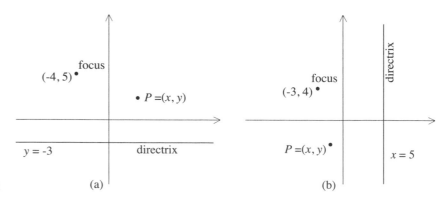

Figure 4.42   (a)   (b)

The problems that follow consider coordinate geometry in three-dimensional situations.

Problem 31. Consider the planes $3x - 4y - 2z = 1$ and $x - 2y + 3z = 4$. In each case, get a sense of the position of the plane by sketching the triangles that their points of intersection with the x-, y-, and z-axes determine.

The discussion about the geometry of the dome of the Hagia Sophia that concludes the "Coordinate System in Three Dimensions" section of this chapter requires an explanatory note. The equation $x^2 + y^2 = 50^2 - z_0^2$ does not involve the variable z, so that it is satisfied by any point (x, y, z) with (x, y) on the circle $x^2 + y^2 = 50^2 - z_0^2$ in the xy plane and z any value. It follows that the graph of $x^2 + y^2 = 50^2 - z_0^2$ in xyz space is the cylinder parallel to the z-axis, through the circle $x^2 + y^2 = 50^2 - z_0^2$ in the xy plane. The points (x, y, z) that lie on the circle through z_0 (this circle was the focus of the discussion) are therefore specified by two equations: $x^2 + y^2 = 50^2 - z_0^2$ and $z = z_0$.

Problem 32. The intersection of the sphere $x^2 + y^2 + z^2 = 8^2$ of radius 8 with the plane $y = -5$ is a circle. What is the radius of this circle? Find two equations with the property that a point is on this circle precisely if its coordinates satisfy both equations.

Problem 33. Identify the graph of the equation $x^2 + (y - 1)^2 + (z - 2)^2 = 9$.

We next turn to study the Santa Maria del Fiore of Florence, especially its dome.

Problem 34. Have a look at Figure 4.34. Comment on the structural relevance of the pairs of metal rods over the nave as well as those that connect the arches over the aisles. Would you expect these rods to be under compression or tension?

Problem 35. Refer to Figure 4.28. Determine the midpoint of the segment $S_7 S_8$. Show that the distance between this point and S_5 is the same as the distance between the points M and S_1 in the figure.

Problem 36. Figure 4.43 shows an abstract vertical section of the outer surface of the dome through a pair of opposing ribs in an xy coordinate plane. It is an arch in the shape of the Gothic fifth with a span of 10 units. What are the equations of the two circles on which the two arcs of the arch lie? The segment with y-coordinate h represents the base of the lantern of the dome. It is $3\frac{1}{3}$ units long. Show that $\frac{h}{10} \approx 0.65$. Is this consistent with the listed dimensions: an outer diameter of the drum of about 174 feet and a

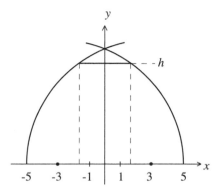

Figure 4.43

vertical distance from the base of the dome to the base of the lantern of about 116 feet?

Problem 37. Refer to the abstract version of the inner surface of the dome in Figure 4.29b and take the foot as the unit of length. Given that the inner diameter $S_1 S_5$ of the dome is 145 feet, show that $S_1 C_1 = 116$ feet, $C_1 S_5 = 29$ feet, and that $OC_1 = OS_5 - C_1 S_5 = 43.5$ feet. Show that the circle in the xz plane with center C_1 and through S_1 and T has equation $(x + 43.5)^2 + z^2 = 116^2$.

   i. Cut the inner surface of the dome with a horizontal plane at the elevation $z = z_0$ and show that the diameter of the octagonal cross section of the dome at that elevation is equal to $2\sqrt{116^2 - z_0^2} - 87$.
   ii. Show that this diameter is approximately 20 feet for $z_0 = 103$ and 12 feet for $z_0 = 105$. This conclusion is at variance with existing information asserting that the octagonal opening below the dome's lantern (refer to Figure 4.32) starts at an elevation of 105 feet above the top of the drum and has a diameter of about 18.5 feet. Provide a possible explanation for the discrepancy.

Problem 38. Refer to Figure 4.29b. Let c and d be the respective distances from C_1 to O and S_1 to O. Consider the circular arc $S_1 T$ and the circle on which the arc lies. This circle is the intersection of a sphere and a plane. Find an equation of the sphere and an equation for the plane. For a point (x, y, z) to be on the circle it needs to satisfy both equations. Do the same thing for the circular arc $S_2 T$.

Discussion 4.1. More Greek Geometry. Euclid's *Elements* discusses five remarkable solids, named after Plato. These Platonic solids are the tetrahedron, cube, octahedron, dodecahedron, and icosahedron. The tetrahedron has 4 sides (*tetra* is Greek for four), each an equilateral triangle; the cube has 6 sides, each a square; the octahedron has 8 sides (*octo* is Greek for eight), each an equilateral triangle; the dodecahedron has 12 sides (*dodeca* is Greek for twelve), each a regular pentagon; and, finally, the icosahedron has 20 sides (*icosi* is Greek for twenty), each an equilateral triangle. These five solid figures are the only three-dimensional convex shapes (think of convex as the requirement that all vertices can be placed on a circle) that have the property that all the sides are identical regular polygons. This is verified in the *Elements*, but the verification is not easy. However, it is not hard to show how these five Platonic solids can be "manufactured." Consider the arrangements of equilateral triangles, squares, and pentagons depicted in Figure 4.44. If you were to cut out each of them out you could: (a) Build the tetrahedron by taking the arrangement of four equilateral triangles and folding the three outer ones along the perforations so that the three points (all labeled

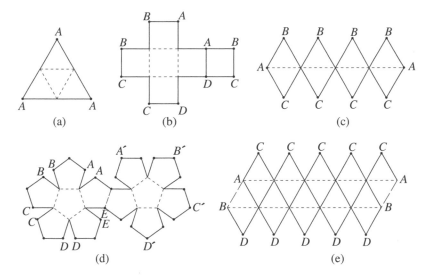

(a)   (b)   (c)

(d)   (e)   Figure 4.44

A) meet. (b) Form a cube by doing a similar thing with the configuration of six squares. Begin by folding along the perforations, so that each of the pairs of points *A*, *B*, *C*, and *D* meet. (c) Build an octahedron by taking the configuration of eight triangles, folding the perforated line into a square so that the points *A* meet, and then folding the triangles along the perforations so that all the points *B* meet and all the points *C* meet. (d) Form the clusters of six pentagons on the left into a pentagonal "cup" by folding along the perforations so that each of the pairs of points *A*, *B*, *C*, *D* and *E* meet. Form a second pentagonal cup by doing a similar thing with the six pentagons on the right. Then join the two cups so that *A'* and *A*, *B'* and *B*, and so on, meet. (e) Follow the instructions implicit in Figure 4.44e to put the icosahedron together. After the folds are made and the points are joined, the points *C* and *D* sit on top of a five-sided "hat." Because of the symmetry of the figure, such a pentagonal array emerges from each of the 12 vertices of the icosahedron. The dodecahedron and the icosahedron are shown in Figures 4.45d and 4.45e.

Plato thought these solids to be related to the structure of the universe. He associated one solid to each of the four classical elements of matter: the cube to earth, the octahedron to air, the icosahedron to water, and the tetrahedron to fire. The dodecahedron was what "god used for arranging the constellations on the whole heaven." More relevant than Plato's science is a connection between Platonic solids and architecture. When an icosahedron is built as a configuration of linked metal rods it becomes a very stable structure. Much more elaborate such structures can be designed by placing any number of points evenly on a sphere and connecting them with rods to form a symmetric mesh of triangles (on occasion including squares and other polygons). When covered with roofing such structures are called geodesic

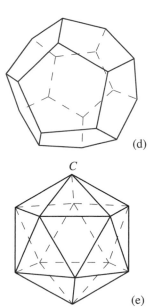

(d)

(e)

Figure 4.45

spheres or geodesic domes (when only a section of the sphere is considered). Such "omnitriangulated" surfaces are strong, stable, and also light in weight. They were built in the second half of the twentieth century and were used as greenhouses, auditoriums, weather observatories, and storage facilities. However, geodesic structures have disadvantages. These include the challenge of insulating and waterproofing the large number of edges and flat surfaces, and the fact that spaces enclosed by curving walls tend to be less practical than spaces enclosed by rectangular walls.

Problem 39. Study Figure 4.45e. Do you see a way to deform the icosahedron into a solid that has the property that all of its sides are the same equilateral triangle, but that is not convex?

Problem 40. Leonardo da Vinci studied the dodecahedron in his Notebooks. The diagram in Figure 4.46 is an example. What might da Vinci be attempting to do with this sequence of diagrams?

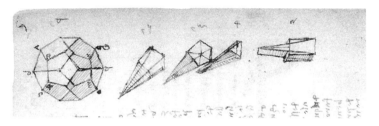

Figure 4.46. Sketches from p. 7 in volume 1 of the Forster Codex. © V & A Images/Victoria and Albert Museum, London

Maps have played a historically important role in the discovery, development, and control of the world's territories. We saw earlier that Ptolemy's map played such a role. Among the challenges that a mapmaker faces is a geometric challenge. The fact is that no part of a sphere can be flattened without deforming it. (For example, the rind of an orange wedge cannot be pushed against the flat surface of a table without tearing it.) This means that no matter what strategy is used to make a map of a region of the globe, distortion is inevitable. Take a look at a standard map of North America and find the distortions. The icosahedron has been used to make a map of the globe. Take a model of the globe that is a perfect sphere. Place an icosahedron into this sphere in such a way that all the vertices lie on the sphere. Take any point on the sphere and project it in the direction of the center of the sphere until it meets the icosahedron. In this way, any point on the globe corresponds to a point on the icosahedron and there is a representation of the globe on the icosahedron. To get a flat map of the world cut the icosahedron along some of its edges (in such a way that the cuts fall in ocean areas) and unfold.

Discussion 4.2. Rational and Irrational Numbers. An integer greater than 1 is a prime number if it can be divided only by 1 and itself. The numbers 2, 3, 5, 7, 11, 13, and 17 are examples. It is a fact that any integer greater than 1 can be written as a product of powers of prime numbers, and that this can be done in only one way (except for the order of the factors). For example, $324 = 2^2 \cdot 3^4$ and $4536 = 2^3 \cdot 3^4 \cdot 7$ and there is no other way of doing this (other than $324 = 3^4 \cdot 2^2$ or $4536 = 7 \cdot 2^3 \cdot 3^4$, for example). Notice that if a number is a square, then all primes in the product occur to an even power. For instance, $4536^2 = (2^3)^2 \cdot (3^4)^2 \cdot 7^2 = 2^6 \cdot 3^8 \cdot 7^2$. It is also the case that if all the primes in the product occur to an even power, then the number is a square. For example, $324 = 2^2 \cdot 3^4 = (2 \cdot 3^2)^2 = 18^2$.

It follows from what was just said that if m is a positive integer, then $\sqrt{m}$ is a rational number only if m is a square. How so? Certainly if $m = n^2$ for a positive number n, then $\sqrt{m} = n$ is rational. But what about the other way? If $\sqrt{m} = \frac{s}{t}$ for positive integers s and t, is it the case that m is a square? If $\sqrt{m} = \frac{s}{t}$, then $m \cdot t^2 = s^2$. If one of the primes in the factorization of m occurs to an odd power, then the same thing is true for $m \cdot t^2$. But this can't be because $m \cdot t^2 = s^2$ is a square. So all the primes in the factorization of m occur to an even power, and m is a square as asserted. It follows that with the exception of $\sqrt{4} = 2$, $\sqrt{9} = 3$, $\sqrt{16} = 4$, and so on, the numbers $\sqrt{2}$, $\sqrt{3}$, $\sqrt{5}$, $\sqrt{6}$, $\sqrt{7}$, $\sqrt{8}$, ... are all irrational. By the way, all of the above was known to the Greeks and can be found in Euclid's *Elements*.

Problem 41. Consider the numbers $\sqrt{n}$ for $n = 1, 2, \ldots, 100$. How many of these 100 numbers are rational and how many are irrational? Consider $\sqrt{n}$ for $n = 1, 2, \ldots, 1,000,000$. How many of these numbers are rational? [Hint: How many squares less than or equal to 1,000,000 are there?]

Problem 42. Numbers occurring as lengths in architectural plans and elevations are often irrational. For example, the diagonal of a 1 by n rectangle has length $\sqrt{1 + n^2}$. Show that any number of the form $\sqrt{1 + n^2}$ with n an integer is irrational.

Let r be any real number. Is there a pattern in its decimal expansion from which it is possible to tell whether r is rational or irrational? Suppose that in the decimal expansion of r there is a repeating block of numbers from some point on. For example, let's suppose that $r = 23.74865865865\ldots$ and that the block 865 keeps repeating. Then $100r = 2374.865865\ldots$ and $100,000r = 2374865.865865\ldots$. Notice that $100,000r - 100r = 2,374,865 - 2374 = 2,372,491$. So $99,900r = 2,372,491$. It follows that $r = \frac{2,372,491}{99,900}$ is a rational number. In the logically opposite direction, is it also true that the decimal expansion of any rational number has a block that keeps repeating? From the (longhand) division algorithm applied to, say t divided into s, we know that there is, at every step, a remainder that is less than t. So there are only a finite number of possibilities for such a remainder. If 0 is a remainder, then the

$$
\begin{array}{r}
8.5151 \ldots \\
132\,\overline{\smash{)}\,1124} \\
\underline{1056} \\
680 \\
\underline{660} \\
200 \\
\underline{132} \\
680 \\
\underline{20}
\end{array}
$$

Figure 4.47

division process stops and 0 is the repeating block. For instance, $\frac{581}{25} = 23.24$ $= 23.24000\ldots$. If 0 is not a remainder, the process keeps going, so that there must be a point when a remainder occurs again. But at that point the cycle repeats. Divide 132 into 1124, for instance. A look at Figure 4.47 tells us that the first remainder is 68, the next is 20, then 68 recurs, so there is a repetition after two steps. In other examples of the division process many steps may be needed before the same remainder reappears. But there will be a repeating cycle, so there is a repeating block in the decimal expansion. It follows that the rational numbers are precisely those real numbers that have a decimal expansion with a repeating block of numbers.

The decimal expansions of some irrational numbers are completely unpredictable. The number π is an example. The expansion $\pi = 3.1415926535898732\ldots$ has no pattern. The numbers that arise do so in a completely random way. One last point. The numbers obtained by cutting the expansion off, say 3.141, 3.14159, 3.141592653, and so on, are all rational (because $3.141 = \frac{3141}{1000}$, $3.14159 = \frac{314,159}{100,000}$, and $3.141592653 = \frac{3,141,592,653}{1,000,000,000}$, and so on). This illustrates the fact that any real number can be approximated by rational numbers to any desired degree of accuracy.

Discussion 4.3. Comparing the Stability of a Structure and Its Model. Consider a model made for the study of an architectural structure such as the one Brunelleschi made for the dome of the Cathedral of Florence. Let's assume that the materials that the model is built with are identical to those of the proposed structure. So the only difference between the two are the dimensions. The critical question is this: Does the stability of the model ensure that of the structure itself? Let's make things simple. Suppose that the proposed structure consists of a stone cylinder supported by two parallel horizontal wooden beams that rest on a rigid foundation at their ends. The cylinder has a diameter of 1 foot and a weight of 400 pounds. The two wooden beams that support the cylinder are $2\frac{1}{2}$ feet long with 1 inch by 1 inch square cross sections. See Figure 4.48a. Let's say that all the dimensions of the model are one-twelfth of those specified for the structure.

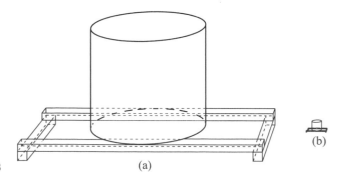

Figure 4.48   (a)   (b)

Refer to Figure 4.48b. Because 1 foot has 12 inches, 1 cubic foot has $12 \times 12 \times 12 = 1728$ cubic inches. It follows that the cylinder of the model has a diameter of 1 inch and weighs $\frac{400}{1728} \approx 0.23$ pound. The wooden beams of the model are each $2\frac{1}{2}$ inches long and have a cross section of $\frac{1}{12}$ inch square. Such a model is readily built with toothpicks and quarters. A stack of 20 quarters weighs slightly more than 0.25 pound and is easily supported by two toothpicks. But even if you're not a gambler, you'd probably bet that there is no way two ordinary $2\frac{1}{2}$ foot long 1- by 1-inch wooden beams would support 400 pounds in the way described above.

Problem 43. Build the quarter/toothpick model that was described. Then devise and test some version of the cylindrical structure to settle the bet.

It is a fact that the structural soundness of a model does not imply that of the building it represents. The basic reason is that weight is proportional to volume, and volume varies with the cube of its linear dimensions. Therefore, an increase in the dimensions has a disproportionate effect on the loads that the structure is subjected to. It is reasonable to assume that Brunelleschi was aware of the unreliability of his model as a predictor of the stability of his dome. However, history credits Galileo with this insight. He described it in his *Discourses about two New Sciences* of 1638.

5

The Renaissance:
Architecture and the Human Spirit

During the fourteenth and fifteenth centuries, five major political powers took control of the Italian peninsula: the three major city-states of Florence, Venice, and Milan, the States of the Vatican with Rome as center, and the kingdom of Naples. In a time of constant warfare and shifting alliances and borders, the fortunes of smaller city-states, such as Siena, Genoa, and Pisa, rose and fell. In spite of this, manufacturing increased in towns and cities, their markets grew, and commercial activity expanded. Soon a class of merchants and bankers shared wealth, lands, and influence with nobles and aristocrats. As commerce grew, so did the need for legal and contractual arrangements. Merchants needed to know how to read, write, and compute. Bankers needed to provide bookkeeping methods, extend flexible credit, and assess collateral. Lawyers had to negotiate partnerships and trade agreements. As business and bureaucratic practices became more complex, there was a growing need for competent functionaries. The importance and scope of commercial, financial, and legal activity created a demand for an education with a more practical and less theological focus, an education that provided professional skills, broad competence, and worldly attitudes.

A humanistic program of studies, or *studia humanitatis*, took shape to provide it. This included the reading of the ancient authors and the study of grammar, law, rhetoric, history, and moral philosophy. Members of the merchant and ruling classes looked for counsel not only in the teachings of the Church, but also in the practical knowledge of Rome and the secular philosophy of Greece. In the Middle Ages, God was believed to be the creative source of everything and humans lived to serve him. As the citizens of the Italian states learned from the wisdom of antiquity and responded to the challenges that faced them, they discovered *their own* God-given creative energy and *their own* capacity to reason, think, act, imagine, and build. This realization was the dawn of the Renaissance (meaning rebirth in French): an age of splendid achievement in literature, painting, sculpture, and architecture. Two powerful paintings capture the driving forces of this epoch.

The fresco *School of Athens* by Raphael (1483–1520), painter of portraits and frescos and one of the master artists of the time, depicts the great figures of classicism and of the Renaissance in conversation in the same scene.

In Plate 20, under the arch at the center, we see a white-bearded Plato in conversation with Aristotle. Plato the idealist is on the left pointing skyward. He is portrayed as Leonardo da Vinci (1452–1519), the multifaceted genius who painted the *Mona Lisa* and the *Last Supper* and who captured with his graphic artistry the anatomy of nature and the workings of machines of his own design. Aristotle, the man of good sense, is at Plato's right holding out a moderating hand. Below them, sprawled on the steps, is Diogenes, a Greek philosopher who made a virtue of extreme poverty. Leaning on a large white marble slab in the foreground is the Greek philosopher Heraclitus, depicted as a young, absorbed Michelangelo. Michelangelo (1475–1564), with his sculptures *David, Moses*, and the *Pietà*, his frescos on the ceiling of the Sistine Chapel of the Vatican, and the splendid buildings that he executed, would become the most versatile of the master artists of the Renaissance. To the left of Michelangelo we see Pythagoras reading from a large book. In the lower right corner, a geometer (probably Euclid, possibly Archimedes), depicted as the great Renaissance architect Bramante (1444–1514), bends down to demonstrate a geometric construction. Standing close by with his back turned is Claudius Ptolemy holding a globe of the Earth. Raphael has placed himself in the same group wearing a black cap. The building surrounding them is most likely Bramante's vision for the new St. Peter's basilica that was begun in Rome at that time. This was an appropriate setting for this meeting of minds, given that institutions of the Catholic Church had hosted the intellectual confrontation between the Christian religion and Greek philosophy and sanctioned their synthesis within Catholic theology.

Michelangelo's fresco *Creation of Adam*, shown in Figure 5.1, depicts a powerful God creating man in his image. Representing a human being infused with a divine spark, it is a poetic expression of the nobility of humanity. The

Figure 5.1. Michelangelo's *Creation of Adam*, a scene from the Sistine Chapel, 1508–1512, Vatican

paintings of Raphael and Michelangelo are both metaphors for the new age and testify to the fact that the intelligent curiosity and profound thought of the Renaissance found their most brilliant expression in visual imagery.

In the thirteenth and fourteenth centuries the works of the great classical thinkers had stimulated the scholarship of philosophers and theologians of the Middle Ages. In the next two centuries the works of classical artists and builders inspired the creative artistry of the masters of the Renaissance. The achievements of the medieval scholars were understood only in intellectual circles and the ivory towers of universities, but those of the artists of the Renaissance burst out into the public domain. The wealthy and powerful of the Italian city-states, especially Florence, awarded commissions and provided resources. Wanting to give legitimacy to their status and to display their knowledge of classicism, they were generous in their support of the arts and of architecture.

God, Man, and Proportion

The rational facades, colonnaded porticos and loggias, rectangular arcaded courtyards, and space-embracing domes of Renaissance architecture have Greek and Roman origins, but they reflect the refined energy of the new age. The structure commonly regarded as the first building of the Renaissance is the Ospedale degli Innocenti in Florence. Built as a hospital for abandoned infants and children, it is now a museum. The arcaded loggia of its facade was designed and built by Brunelleschi between 1419 and 1424. A detail is shown in Figure 5.2. The Corinthian columns affirm the classical origins of

Figure 5.2. Detail of the facade of Brunelleschi's Ospedale degli Innocenti, 1419–1424, Florence. Photo by Giacamo Augusto

Figure 5.3. Courtyard of the Palazzo della Cancellaria, the Vatican Chancellery, 1489–1513, Rome. Architect unknown. Possibly by Francesco di Giorgio. Incorrectly attributed to Donato Bramante. Photo by Emmanuel Brunner

the design of the structure, and the medallions over them give a hint of its purpose. Brunelleschi, by then fully engaged with the construction of the dome of the Santa Maria del Fiore, did not complete the hospital. This was done by architects who followed the stylistic tone that Brunelleschi had set. The arcaded courtyard in the interior of the building follows the form of the facade closely. Such arcaded courtyards, at times with two or three tiers of arcades, became a hallmark of Renaissance architecture. Figure 5.3 shows a later Roman example.

A critical link between Renaissance architecture and its Greek and Roman roots is the authoritative work *De Architectura* (*The Ten Books on Architecture*) by the Roman architect Marcus Vitruvius Pollio. Written in the first century B.C., this is the only surviving treatise about the architecture of antiquity. The topics it covers include basic building design ("a ground plan is made by the proper successive use of compasses and rule"), methods of construction (for instance, a discussion of the properties of Roman concrete), and urban planning. *De Architectura* was a bible for several great masters of the Renaissance. When they needed advice about the design of sacred and public buildings, as well as the relationships between their components, they turned to Vitruvius. One of Vitruvius's books explores the proportions of the human body. A drawing by Leonardo da Vinci, shown in Figure 5.4, illustrates a central idea. The extended hands and feet of a well-built man generate the circle and the square, two of the most basic and perfect geometrical shapes. To the master builders of the Renaissance, this simple picture revealed a deep and fundamental connection between God's creation, the human form, and

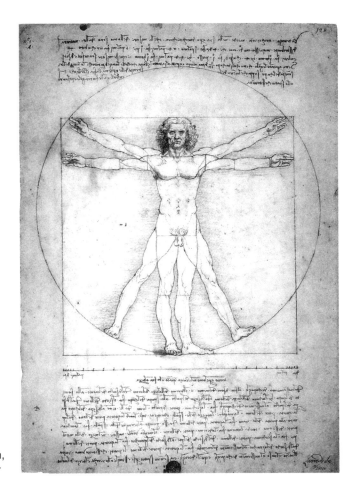

Figure 5.4. Leonardo da Vinci's *Homo Vitruvianus*, or *Vitruvian Man*, 1492. Pen, ink, watercolor, and metalpoint on paper

architecture. Humans are images of God and the proportions of their bodies are the result of divine will. Because architecture needs to be in tune with this cosmic order, buildings should reflect the proportions of the human body. This gives concrete form to the idea that "man is the measure of all things," one of the guiding principles of the Renaissance.

Leon Battista Alberti (1404–1472) was an architect, philosopher, mathematician, musician, horseman, and serious athlete. He was not only an imaginative, practicing architect, but his philosophical reflections about architecture established him as the leading architectural theorist of his time. Alberti saw architecture as much more than a discipline that deals with the practicalities of building. Because it shapes the physical environment, it was a framework for shaping society. Alberti had access to the circles of power in the Vatican and became one of the most influential thinkers of the Renaissance.

Alberti was commissioned to construct a new facade for a medieval church in Florence. The church Santa Maria Novella (shown in Figure 5.5) had been

Figure 5.5. Alberti's facade of the Santa Maria Novella, 1456–1470, Florence. Photo by Jebulon

built between 1279 and 1310 in the Gothic tradition (as one of only a few Gothic churches in Italy). The new facade needed to incorporate the six existing Gothic tombs as well as the two Gothic side doors of the old. True to the spirit of Vitruvius, Alberti's design made important use of both the circle and the square. Four circles, including the window, are prominent features of the upper facade. The lower part of the facade is a rectangle obtained from two identical squares (shown in gray in the figure). A square of the same dimension (also in gray) frames the central section of the upper facade, and the entire facade fits into a square (in black) of twice this size. The new facade of the Santa Maria Novella was completed between 1456 and 1470. Inspired by its design, a number of Renaissance architects developed variations on the theme of Alberti's composition.

Alberti, Music, and Architecture

Animated by Vitruvius, Alberti reflected about both the theoretical and practical aspects of building and collected his thoughts in a treatise on architecture. The initial manuscript was completed in 1452, but Alberti continued this work for the rest of his life. His *De Re Aedificatoria*, or *On the Art Building*, was not published until after his death. In the most influential parts of the book, Alberti formalized and codified the principal features of classical architecture and established a theory of proportion to guide the design of buildings and their components. Alberti promoted the idea that a building needs to be a harmonious whole and that each of its parts, interior and exterior, has to be integrated within its design. In order to achieve this, architects should be guided by a system of proportion with high aesthetic appeal. This could not be a system of the architect's own choosing, but a system that was rooted in a higher order, the harmonious order within God's universe. God's design of the universe, including that of the human form, conforms to certain mathematical principles. These principles needed to be comprehended and applied by architects in the designs of their structures.

Alberti drew on the understanding of the Pythagoreans that numbers explain everything in the universe and that the relationships that exist between two consonant musical tones and simple numerical ratios are an important expression of this. It became evident to Alberti that proportion in architecture needed to conform to these same simple numerical ratios. With reference to Pythagoras, he tells us that "the numbers by means of which the agreement of sounds affects our ears with delight, are the very same which please our eyes and our minds." Alberti continues, "we shall therefore borrow all our rules for harmonic relations from the musicians to whom this kind of numbers is extremely well known, and from those particular things wherein Nature shows herself most excellent and complete." This view became fundamental to the conception of proportion in the Renaissance. The architects of the Renaissance believed that musical consonances were the audible manifestations of a universal harmony that has binding implications for architecture. In expressing these in his designs, the architect is not simply translating musical ratios into architecture, but is applying universal principles that nature conforms to and that music reveals.

Music had a very special appeal because it was regarded to be a mathematical science. An unbroken tradition originating from classical times held that arithmetic (the study of numbers), geometry (the study of spatial relationships), astronomy (the study of the motion of the celestial bodies), and music (the study of the sounds perceived by the ear) formed the four important liberal arts. In contrast to these noble intellectual pursuits, painting, sculpture, and architecture were regarded to be crafts and were given much lower status. Providing architecture with a mathematical foundation would raise it from a manual art to an intellectual discipline.

Let's turn to the numerical relationships that arise in the analysis of musical tones. When a physical object vibrates, air molecules are set into motion. The vibration is transmitted into space in all directions as an alternating sequence of pressure fronts of denser and looser concentrations of air molecules. When such a sequence of fronts hits our ear, we hear a sound. Suppose that a string (think of the string of a string instrument) is under tension and that its ends are fixed. Figure 5.6 shows the string, its endpoints A and B, and the center P of the string. The string is elastic and can be displaced from its original position APB by pulling on P. When the string is released it will go back and forth in a moving wave. We'll call what was described pluck- ing the string. The sound that the vibrating string generates is a tone. The number of times per second that the wave moves back and forth—think of the point P going through one complete cycle of up, down, and back to its original position—is the frequency f of the tone. The frequency depends on the length of the string, the density of the material that it is made of, and the tension it is under. We hear the frequency as the pitch of the tone. The pitch can range from very high (think of the high tones of a violin) to very low (think of a bass fiddle). The frequency determines the pitch: the higher the frequency, the higher the pitch; the lower the frequency, the lower the pitch. If the string is plucked harder, the loudness of the tone increases but the frequency and pitch remain the same.

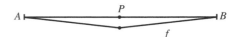

Figure 5.6

The Pythagoreans discovered the following interesting connection be- tween the length of a string and the frequency of the tone that it produces. Figure 5.7 considers two strings that are made of the same material and are under the same tension. Their lengths are L_1 and L_2 and the frequencies of the tones that they produce when plucked are respectively f_1 and f_2. The Py- thagoreans discovered that if the ratio of the lengths L_1 to L_2 of the strings are, respectively, 2 to 1, 3 to 1, 3 to 2, or 4 to 3, then the ratios of the frequen- cies f_1 to f_2 of the tones they generate are in the inverse ratios of 1 to 2, 1 to 3, 2 to 3, and 3 to 4, to each other. So the longer the string the lower the pitch, the shorter the string the higher the pitch. This relationship is numerically precise: *If the ratio of L_1 to L_2 is r, then the ratio of f_1 to f_2 is $\frac{1}{r}$.* In the four cases where the ratio of the frequencies is 1 to 2, 1 to 3, 2 to 3, or 3 to 4, the two tones sound pleasant together. They are consonant. It turns out that conso- nance is the exception, not the rule.

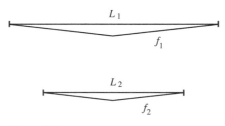

Figure 5.7

In addition to a string of length L, consider strings of lengths $\frac{1}{2}L$, $\frac{1}{3}L$, $\frac{1}{4}L$, $\frac{1}{5}L$ (and so on). All the strings are made of the same material and all are under the same tension. They are depicted in Figure 5.8. If f is the frequency of the tone generated by the string of length L, then by the inverse relation- ship between the ratios of length and frequency, the frequencies of the tones generated by the other strings are, respectively, $2f$, $3f$, $4f$, $5f$ (and so on). Each of these tones has a higher pitch than the tone with frequency f, but each is consonant with it. Now that we know this, we can refine an earlier discussion.

When a string is plucked, a blend of tones is generated. A tone of lowest frequency, say f, predominates. Its frequency or pitch is the fundamental fre- quency or pitch. It is called the first harmonic. But the tones of frequencies

Figure 5.8

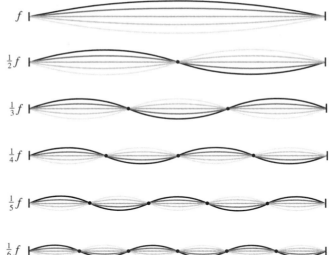

Figure 5.9

$$2f \quad 3f \quad 4f \quad 5f \quad \ldots$$

are generated as well. They are the second, third, fourth, and fifth (and so on) harmonics. The sound that is generated is a mix of all the harmonics. This mix is called the timbre of the sound. Geometrically, this means that after the string of Figure 5.6 is released, it changes its shape over time in a very complex way. The shape of the string is the composite of the waves of all its individual harmonics. The waves of the first six harmonics are shown in Figure 5.9. For each of the six frequencies, think of the string as moving smoothly and successively from the position depicted in black to those depicted in lighter grays, and back again, and forth. . . . (The fact that the sound that a vibrating string produces is a blend of harmonics tells us that tones of the pure frequencies shown in Figure 5.9 cannot in fact be produced by a vibrating string. Only electronic instruments can produce tones of just a single frequency or pitch. By capturing these, an oscilloscope can generate the waves of Figure 5.9. Nonetheless, the geometry of the string is a combination of such waves, and the sound it generates is a combination of the pure tones of its harmonics.)

In the discussion that follows, the frequency or pitch of the tone of a vibrating string is always understood to be the fundamental frequency or pitch. Again, all strings referred to are made of the same material and all are under the same tension.

Figure 5.10a shows a string of length L. We'll call the tone that it generates our base tone, and we'll let f be its frequency. Figures 5.10b, 5.10c, and 5.10d consider strings of lengths $\frac{3}{4}L$, $\frac{2}{3}L$, and $\frac{1}{2}L$. They generate tones of frequencies $\frac{4}{3}f$, $\frac{3}{2}f$, and $2f$, respectively. (The third of these tones is the second

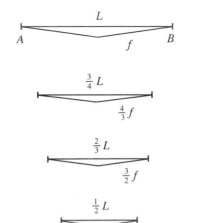

Figure 5.10

Figure 5.11

harmonic of the base tone.) If we designate the pitch of our base tone by 1, then the pitches of the other three tones are, respectively, $\frac{4}{3}$, $\frac{3}{2}$, and 2. The Pythagoreans knew that each of these tones is consonant with the base tone. The white rectangles of Figure 5.11 represents the four tones in increasing order of pitch.

Let's continue to add tones. Turn to the string of length $\frac{2}{3}L$ of Figure 5.10c. Figure 5.12 continues the strategy of Figure 5.10. It depicts strings of lengths $\frac{2}{3}(\frac{2}{3}L) = \frac{4}{9}L$, $2(\frac{4}{9}L) = \frac{8}{9}L$ and $\frac{2}{3}(\frac{8}{9}L) = \frac{16}{27}L$. Given that the pitch of the base tone is 1, the pitches of the tones that these three strings produce are, respectively, $\frac{9}{4}$, $\frac{9}{8}$, and $\frac{27}{16}$. In Figure 5.13, two of the new tones are added to the list. As in the case of Figure 5.10, the tones of the strings of Figures 5.12b and 5.12c are both consonant with the tone of the string in Figure 5.12a. However, it is not true that all the tones on the list of Figure 5.13 are consonant in pairs. Some of the pairs do not sound good together. This is to some extent a matter of taste, but generally two tones sound good together only if their frequencies are in simple numerical ratio. The tone of pitch $\frac{9}{8}$ is consonant with the tone of pitch $\frac{27}{16}$, because $\frac{9}{8}$ over $\frac{27}{16}$ is equal to $\frac{9}{8} \cdot \frac{16}{27} = \frac{2}{3}$. But the tone of pitch $\frac{27}{16}$ is not consonant with the base tone, because $\frac{16}{27}$ (1 divided by $\frac{27}{16}$) is a more complicated ratio.

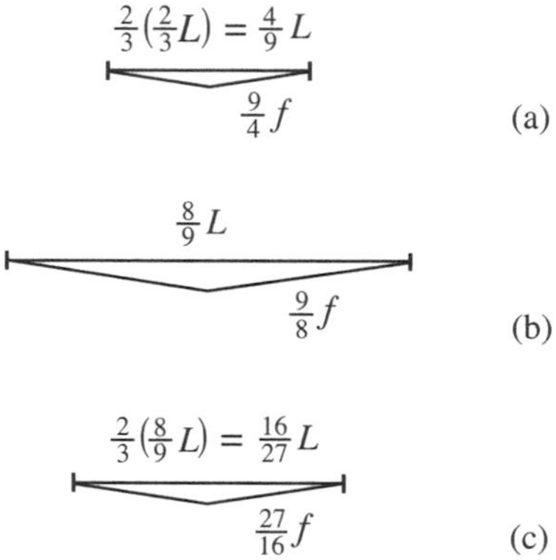

By filling in notes between the two successive harmonics, we have created a musical scale. If you learned the music lesson from the movie *The Sound of Music*, and if our base tone is tuned to give the note *do* (by providing the string with the correct tension), then you would be able to strum, or play as keys of a piano, the notes from Figure 5.13 (from left to right) as

do   re   (mi)   fa   so   la   (ti)   do.

The notes *mi* and *ti* are missing from our scale, but the process described above could have been continued to provide them. With the *mi* and *ti* included, there are eight notes from *do* to *do*, so that we speak of a musical octave.

Whether you play an instrument, sing, or neither, the fact is that the Pythagoreans' procedure has supplied the ratios 1 to 2, 2 to 3, 3 to 4, 8 to 9, and 16 to 27. These are the ratios that underlie Alberti's theory of proportion for architecture. From 2 to 3, he obtains the ratios 4 to 6, 8 to 12, 16 to 24, and

$$\frac{2}{3}\left(\frac{2}{3}L\right) = \frac{4}{9}L$$
$$\frac{9}{4}f \quad \text{(a)}$$

$$\frac{8}{9}L$$
$$\frac{9}{8}f \quad \text{(b)}$$

$$\frac{2}{3}\left(\frac{8}{9}L\right) = \frac{16}{27}L$$
$$\frac{27}{16}f \quad \text{(c)}$$

Figure 5.12

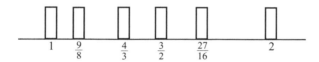

Figure 5.13

32 to 48. From 3 to 4, he gets the ratios 6 to 8, 12 to 16, 18 to 24, and 24 to 32. From 8 to 9, he obtains 16 to 18 and 32 to 36, and from 16 to 27, he gets 32 to 54. Linking these numbers gives him the sequence

1, 2, 3, 4, 6, 8, 9, 12, 16, 18, 24, 27, 32, 36, 48, and 54.

When an architect of the Renaissance uses the numbers of this list in the spatial layout of a building, he knows that he is imparting to his design the same harmonious order that God imparted to the universe.

Andrea Palladio, a master architect of the later Renaissance, used the basic geometric forms of Vitruvius as well as the numbers that Alberti gleaned from the musical ratios of the Pythagoreans to infuse the designs of his buildings with balance, proportion, and an overarching order.

Palladio's Villas and Churches

Andrea Palladio (1508–1580), sculptor and stone mason, came into the employ of a count from Vicenza, at that time a city in the Republic of Venice. The count was also a distinguished scholar and amateur architect. Palladio participated in the construction of a villa for the count and became a member of his intellectual circle. Mentored by the count in this environment, Palladio began to develop his talents as an architect. Informed by the study of ancient and Renaissance structures during visits to Rome, his designs matured swiftly. Palladio's buildings refine and rearticulate the practice of Renaissance architecture to create structures with components that are in harmonious relationship to each other and to the structure as a whole.

The purpose of Palladio's first public commission was to give the medieval market and town hall in Vicenza a new appearance. He constructed an elegant shell around the old structure that also served to buttress it. As Figure 5.14 shows, the dominant element of his design is a repeated bay consisting of a central arch supported by a pair of small columns with two circular openings on each side. The bays are separated by larger pilasters in the Ionic style. The bays at the corner do not have the circular openings, look more solid, and give the appearance of added strength. The structure was supplied with a new roof in the nineteenth century.

The Palazzo Chiericati, a later commission pictured in Figure 5.15, has a facade that consists of three sections, a dominant central section and two symmetric sections flanking the center. The ground level is defined by a Doric colonnade, and the second level by a parallel colonnade of Ionic columns. On the upper level, the exterior wall of the central section is moved forward. This adds to the size of the main salon of the palace and creates space on the sides for the two loggias (the open spaces set off by columns). Palladio's plan specifies the dimensions of the rooms on the ground floor

Figure 5.14. Palladio's shell for the basilica of Vicenza, a medieval market and council hall, 1549. Photo by GvF

Figure 5.15. Palazzo Chiericati, Vicenza, 1550–1552. Photo by Ivon Bishop

of the palace as 12 by 18, 18 by 18, and 18 by 30, in old Venetian feet (a unit slightly smaller than the foot we use today), and those of the central portico and the hall behind it as 16 by 54 Venetian feet. Notice that only the number 30 falls outside the sequence that Alberti drew from the musical ratios of the Pythagoreans.

In the middle of the sixteenth century, the Venetian Republic initiated a program of economic reforms as part of an effort to reverse the declining commercial fortunes that faced it. One aspect of this program called on able and well-to-do administrators to set up and manage agricultural estates in the Venetian countryside. The noblemen who took on this challenge needed a new kind of structure, one that combined a functional aspect with an air of elegance, one that was grand in style but not excessively expensive, and one that could house, all at once, their families, farm workers, farm equipment, and livestock. The villas that Palladio built were a response to this need. Nineteen of them survive. Palladio would build two or three of a similar design and then turn to a completely different style. The designs of all the villas are classical in aspect, balanced in composition, and elegant in appearance.

The Villa Emo illustrates the hierarchy of parts that Palladio's buildings often express. The central section is the dominant aspect of the structure and houses the principal spaces. Figure 5.16 shows that attention is drawn to it by the approach and ascent, the columns of the portico, and the triangular pediment. Figures 5.17 and 5.18 confirm that the central section rises elegantly above the simpler wings, but that it is tied to the wings by a proportional relationship. The Villa Emo exemplifies Alberti's principle that architectural proportions should follow musical ratios. Refer to Palladio's plan in Figure 5.19 and notice the sequence 12, 24, 48 that sets the lengths of the rooms of the arcaded wings and the 20 that provides their width (all in Venetian feet). Notice also the dimensions (again in Venetian feet) 16 by 16, 16 by 27, 27 by 27, and 12 by 16 (the 12 is mistakenly entered as a 2) of the rooms in the central block. All these numbers appear in the list that Alberti

Figure 5.16. The central block of the Villa Emo, north of Venice, 1555–1565. Photo by Marcok

Figure 5.17. Palladio's elevation for the Villa Emo from Palladio's *I Quattro Libri*. Marquand Library of Art and Archaeology, Princeton University Library

Figure 5.18. The Villa Emo, central block and arcaded wings. Photo by Hans A. Rosbach

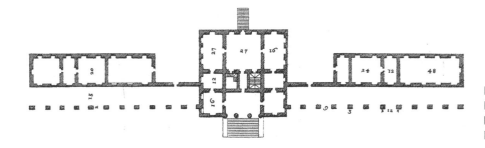

Figure 5.19. Palladio's plan for the Villa Emo from Palladio's *I Quattro Libri*. Marquand Library of Art and Archaeology, Princeton University Library

derived from Pythagorean musical ratios. The dimensions that Palladio provides in the plans and elevations of his buildings are ideal dimensions and differ from those of the executed buildings. Such differences are explained by the fact that Palladio needed to respond to the particular conditions that he encountered on the sites. However, surveys have shown that the various dimensions of the Villa Emo correspond closely to those specified in Palladio's plan. For example, the dimensions of the major rooms of the central section deviate from those of the plan by only a few inches.

The Villa Rotonda is Palladio's most famous villa. It crowns the summit of a hill just outside the city of Vicenza. Not a farm estate, it was built as a residence for a retired dignitary of the Church. At its core is a central, two-story, cylindrical space topped by a dome. The Italian word for round gives the villa the name La Rotonda. Figure 5.20 shows two of the porticos with

Figure 5.20. The Villa Rotonda, outside Vicenza, 1556–1567. Photo by Hans A. Rosbach

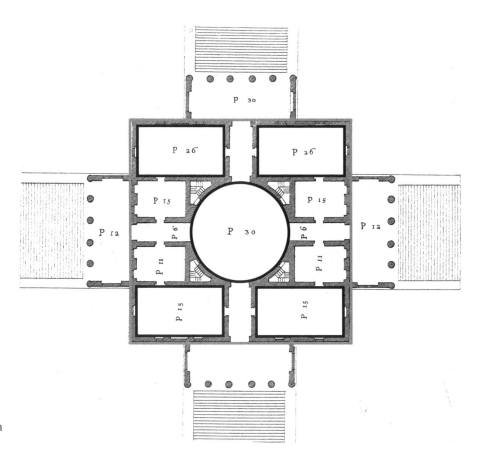

Figure 5.21. Plan of the Villa Rotonda from Palladio's *I Quattro Libri*. Marquand Library of Art and Archaeology, Princeton University Library

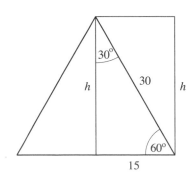

Figure 5.22. Plan of the corner room

their Ionic columns that flank the building on all four sides. It is a simple and powerful composition, inspired by the Roman Pantheon and copied many times over the centuries. For its proportional scheme, Palladio turned to the square and circle of Vitruvius. At the center of Palladio's plan of Figure 5.21 is the circle of the rotunda of 30 Venetian feet in diameter. The inner circle and the square of the exterior of the main structure are highlighted in black and gray, respectively. The four rectangular rooms at the corners are also highlighted in black. Their dimensions, listed as 15 by 26 in the plan, are also determined geometrically. Figure 5.22 considers an equilateral triangle of side length 30 along with its height h. By the Pythagorean Theorem, $h^2 = 30^2 - 15^2 = (2 \cdot 15)^2 - 15^2 = 3 \cdot 15^2$. So $h = 15\sqrt{3} \approx 25.98 \approx 26$ and it follows that these corner rooms correspond in their design to the rectangle constructed in the figure.

The architects of the Renaissance believed that each category of building—private buildings, civic buildings, palaces, and churches—has its own inner logic and rhythm, and that there is an ascending order of values.

Figure 5.23. The church of Christ the Redeemer (Il Redentore), Venice, 1576–1592. Photo by Hans A. Rosbach

Churches were at the top of the pyramid. The planning and building of churches was the architect's most noble task. Their design had to express symbols of eternal validity in concept as well as in every detail. When Palladio says that he designs his churches "in such a manner and with such proportions, that all the parts together may convey a harmony to the eyes of the beholders," he refers to precise spatial relationships provided by universally valid ratios. In Venice, Palladio was commissioned to build the two churches San Giorgio Maggiore (named in honor of Saint George, the Martyr) and Il Redentore (Christ the Redeemer). The design of each features the nave of a basilica with a dome over the crossing and a short transept formed by two rounded bays on each side of the nave. In the church Il Redentore, the crossing and the two bays form a round, theaterlike space in front of the altar (a sacred space appropriate for the celebration of the annual commemorative mass for the doge and his entourage). Both churches stand prominently on banks of canals. This meant that each of their facades had to be designed to respond to both an external and an internal role. On the one hand, it had to represent the church in the "public square," and on the other, it had be in tune with the tall nave and lower sides of its interior. In his response, Palladio again turned for inspiration to the Pantheon, both to elements of its exterior (Figure 2.42) and interior (Plate 5). The facades of the two churches are very similar. Figure 5.23 shows that of Il Redentore. The two minaretlike spires that flank the dome of Il Redentore point credibly to Islamic influences (Ottoman influences in particular). The construction of the two churches commenced under Palladio's supervision, San Giorgio Maggiore in 1565 and Il Redentore in 1576. Both were finished long after Palladio's death.

Palladio published his treatise *I Quattro Libri dell' Architettura* (*The Four Books of Architecture*) in Venice in 1570. It discusses the ancient traditions of architecture and makes use of ideas of Alberti and Palladio's contemporaries, but it focuses its presentation and illustrations on his own designs. Figures 5.17, 5.19, and 5.21 show woodcuts from *I Quattro Libri*. The influential treatise also sets out the orders of architecture (see Discussion 5.1) and deals with aspects of design and engineering, as well as city planning. It is addressed to the practicing architect and emphasizes practical know-how. So when *I Quattro Libri* tells us that "architecture, like all other arts, imitates nature," "nature" refers both to a rationality of structure and to a practicality of design.

Palladio, both with his treatise and with his beautifully conceived and executed buildings, had an enormous influence on the development of Western architecture. Many houses, public buildings, and churches all over the world are dominated by neoclassical elements of symmetrical, columned facades that derive their design from his work.

Da Vinci and Bramante: Churches with Central Plan

Leonardo da Vinci (1452–1519) was a master painter, a military engineer, an inventor of mechanical devices, and a penetrating observer and recorder of nature in action. The splendid illustrations of mechanical devices, turbulent waters, geologic formations, birds in flight, horses in motion, and studies of the human anatomy (arrived at by careful dissections of cadavers) that fill his *Notebooks* demonstrate that he understood how these complex structures and organisms functioned. The *Notebooks* include sketches of the ingenious hoists and cranes that Brunelleschi invented to facilitate the construction of the dome of the Santa Maria del Fiore (with the consequence that da Vinci is often given credit for their invention). In the 1480s and 1490s, da Vinci was in the service of the duke of Milan as a sort of expert in residence. His work for the duke ranged from the design of costumes and stage machinery for the elaborate feasts of the court, to studies for a huge equestrian statue, to detailed plans for casting it in bronze. (The statue was never made. With an attack by France on the horizon, the duke used the tens of tons of bronze for a cannon instead.) During his stay in Milan, da Vinci also found time to paint the famous *Last Supper* and to extend the remarkable reach of his talents to mathematics and architecture.

Da Vinci's interest in architecture was stimulated by two other famous men whom the duke had called to Milan. One of them was Francesco di Giorgio (1439–1502), a painter, sculptor, architect, and engineer from Siena.

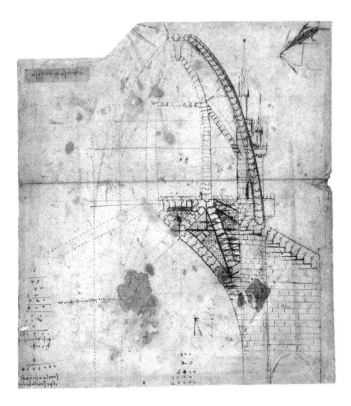

Figure 5.24. Detail from Leonardo da Vinci's study for the Dome of Milan, 1487. Codex Atlanticus, fol. 850r. © Veneranda Biblioteca Ambrosiana, Milan/De Agostini Picture Library

He is thought to have contributed to the construction of the Chancellery of the Vatican (see Figure 5.3). The other was the architect Donato Bramante (1444–1514), who had been commissioned to complete an addition to one church in Milan and the reconstruction of another. The three men became acquainted and shared their ideas about architectural design. All three were consulted about the vault over the crossing of the Duomo of Milan, the great Gothic cathedral still under construction at that time. (See Chapter 3, "From the Annals of a Building Council.") Da Vinci's design for the vault, shown in Figure 5.24, gives the sense that he understood thrusts and how they needed to be channelled downward. Da Vinci's design was not accepted. Nor were the ideas of his friends.

Like Alberti before them, da Vinci, di Giorgio, and Bramante believed that the ideal design for a church is not the traditional rectangular basilica plan, but a circular plan where the principal structural elements radiate from the center. As a symbol of perfection, they regarded the circle to be a geometric expression of the perfection of God.

Di Giorgio drew up a template for the design of a variety of such churches. His construction is analyzed in Figure 5.25. It starts with a square and its subdivision—by three vertical and three horizontal lines—into 16 equal

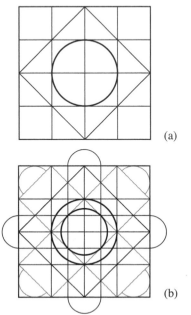

(a)

(b)

Figure 5.25. Francesco di Giorgio, a template for designing centrally planned churches, 1489–1492

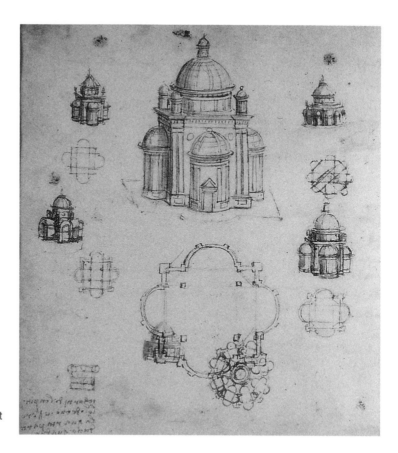

Figure 5.26. DaVinci's designs for churches with central plan, around 1490. Bibliothèque de l'Institut de France, Institut de France, Paris.

squares. The square and the circle that this subdivision determines in the interior are shown in Figure 5.25a. The four squares in the corners determine two intersecting rectangles along the diagonals of the original square. Di Giorgio caps these two diagonal rectangles with semicircles. These capped rectangles are shown in gray in Figure 5.25b. Di Giorgio then draws in a horizontal and a vertical rectangle and caps each with two semicircles (that fall outside the original square) to form the central cross of Figure 5.25b (again shown in gray). The inner circle that the cross determines completes DiGiorgio's template. The two circles locate the drum of the dome of the church. Along with the innermost square, they also position the four supporting piers and set their width. Geometry again serves as a guide for important structural aspects just as it did in Gothic construction. Several of da Vinci's designs of centrally planned churches are shown in Figure 5.26. Notice the similarity between some of da Vinci's plans and di Giorgio's template.

When the monk and mathematician Luca Pacioli (1445–1517) came to teach at the court of Milan at the invitation of the duke, da Vinci had an opportunity to deepen his understanding of mathematics. Pacioli had already published his *Summa*, a comprehensive summary of the mathematics available at the time. The work contained little that was not already in Euclid and Fibonacci (see Chapter 4, "A Line of Numbers"), but it was widely circulated. Guided by Pacioli, da Vinci pushed his way through all 13 books of Euclid's *Elements* and filled two of his *Notebooks* with commentaries on the text. At this time, Pacioli was at work on a volume that developed the divine proportion (see Chapter 2, "Gods of Geometry") and its connection with the human body. He followed Alberti when he insisted that architects needed to use the proportions exhibited by the human body in their design of sacred buildings. Published later with the title *Divina Proportione*, it also discussed regular polygons and the five Platonic solids (see Discussion 4.1). It contained more than 60 illustrations attributed to da Vinci (see Problem 14), thought to be the only of his drawings to be published in his lifetime.

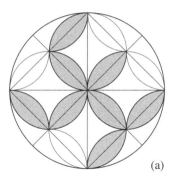

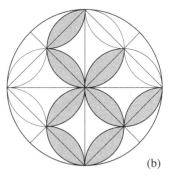

Figure 5.27. Da Vinci's changing patterns. Codex Atlanticus, Biblioteca Ambrosiana, Milan, adapted from fol. 455

Da Vinci also pursued his own mathematics. The circle and square that he captured in the Vitruvian Man (see Figure 5.4), and that formed the basis of his designs of centrally planned churches, also became objects of his mathematical investigations. His *Notebooks* contain many examples of diagrams built with intersecting circles and squares. There are rows and rows of varying forms of crescents, rosettes, and floral designs, perhaps meant to suggest the phenomenon of nature's ever-changing patterns and rhythms. An illustration of his studies follows.

Da Vinci considers a circle of radius r, places four circles of radius $\frac{1}{2}r$ into it, and considers patterns of leaves that the circles determine. Figure 5.27 shows two variations. Notice that the total area of the leaves in Figures 5.27a and 5.27b are the same. Da Vinci computes this area by thinking along the following lines. Turn to Figure 5.28. Let G and W be the respective areas of the gray flower and the white region in Figure 5.28a. Notice that

$$\frac{1}{2}G + \frac{1}{4}W = \pi\left(\frac{1}{2}r\right)^2.$$

It follows that $G + G + W$ is equal to the area of four circles of radius $\frac{1}{2}r$. Consider Figure 5.28b and let D be the area of the dark gray region. A look at the diagram tells us that $G + W + D$ is equal to the area of the circle of radius r. By combining this information, da Vinci gets

$$G + G + W = 4\left(\pi\left(\frac{1}{2}r\right)^2\right) = \pi r^2 = G + W + D.$$

It follows that $G = D$. Now take the eight half-petals of the gray flower and move them as shown in Figure 5.29a. The light and dark gray regions together have area $G + D = 2G$. Figure 5.29b and the Pythagorean Theorem inform da Vinci that

$$2G = G + D = \pi r^2 - s^2 = \pi r^2 - 2r^2 = (\pi - 2)r^2.$$

This is the area of the patterns of leaves of Figures 5.27a and 5.27b.

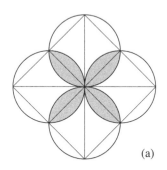

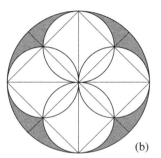

Figure 5.28. Da Vinci's leaf geometry

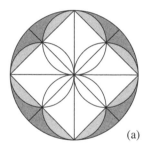

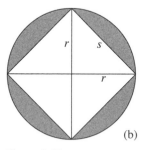

(a)

(b)

Figure 5.29

The brilliant men of the duke's circle in Milan would soon disperse. Francesco di Giorgio returned to Siena where he created two magnificent bronze angels for the high altar of the Siena Cathedral. An invasion by the French a little later in 1499 put an end to the reign of the duke and drove Bramante, da Vinci, and Pacioli from the city. Da Vinci and Pacioli fled together, first to Venice, then to Florence. But our story follows Bramante to Rome. He had been commissioned by Queen Isabella and King Ferdinand of Spain to build a monument on the site where the martyrdom of Saint Peter was thought to have taken place. Bramante's response was the Tempietto. Figure 5.30 tells us that the Tempietto is a small monument built on a circular plan, surrounded by a circular colonnade of Doric columns (with a diameter of only 25 feet), and topped by a dome. It is remarkable for its elegantly simple interpretation of classical forms. The different sculptural elements shape light and shade and give depth to the structure. The dynamic emphasis on space and mass exemplifies Bramante's designs. We will see in Chapter 6 how influential the Tempietto has been to the development of architecture.

By the middle of the fifteenth century, the Basilica of St. Peter (see Figure 3.13) had stood virtually unaltered for over 1000 years and was in bad shape. Alberti, then architect of the pope, reported that one of the walls was in danger of collapse. In response to this assessment, plans for an extensive restoration and expansion of the old church were drawn up. Soon the existing

Figure 5.30. Bramante's Tempietto in Rome, 1502. Photo by marpet

apse had been replaced by an extended vaulted apse and chancel. Unfortunately—as was to happen again and again in the history of St. Peter's—when the pope died, the energy to continue the construction died with him. Another half-century went by before major work on St. Peter's resumed. When Pope Julius II took charge of Rome's religious and civic affairs in the early part of the sixteenth century, the city was on its way to become the artistic center of the Italian peninsula and Europe. The pope quickly turned his attention to St. Peter's. The assumption had been that the core of the old church would be left intact. It was the sacred home of the tomb of Saint Peter and a holy relic of the founding of Christianity. So it was astonishing when Julius II made the decision to replace Old St. Peter's with a completely new church. It was a decision that was severely criticized. But the pope was adamant that there should be a new St. Peter's to rival or surpass the greatest monuments of ancient Rome. The pope had been impressed by the work of Bramante and put him in charge of the St. Peter's project.

Bramante conceived a structure that combined the symbolism of the cross with a centralized geometry. His design united the four equal perpendicular arms of the Greek cross with the square and the circle. Four apses would radiate outward from the center, reaching out symmetrically and symbolically to the four corners of the Earth. A great dome in the spirit of that of the Pantheon was to soar above the crossing. Bramante's plan for St. Peter's is depicted in Figure 5.31 in black. Superimposed in gray is di Giorgio's template of Figure 5.25b. The close correspondence between the two is consistent with the suggestion that Bramante's design was based on di Giorgio's template.

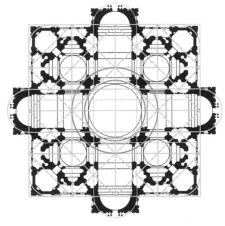

Figure 5.31. Bramante's plan for St. Peter's in Rome, 1505 (with Francesco di Giorgio's template superimposed)

Bramante soon realized that the main piers of his design for the new St. Peter's would not be strong enough to support the massive dome that he envisioned. The revised plan of Figure 5.32 shows, outlined in white, three of Bramante's much more powerful piers (the circles inside the cross sections represent spiral staircases) and the fourth pier still in its original size. It also shows, framed in white, some components of Bramante's original plan of Figure 5.31 and the large semicircles of a grander concept. Finally, it depicts, in dashed white outline, a plan of Old St. Peter's as well as the fifteenth-century apse and chancel.

Construction began in 1506. The tomb of Saint Peter was sacred and church services needed to continue, so Bramante covered the main altar above the tomb with a protective structure. But the destruction of old tombs and venerated murals and mosaics was inevitable, and Bramante came to be known as "Il ruinante" (Italian for "the destroyer"). Progress on the new St. Peter's was rapid. In just a few years, Bramante raised the four giant main piers and linked them with soaring coffered arches. This was the central core of the new church. The stability of the massive dome would depend on this core and the buttressing provided by the vaults surrounding it. The work

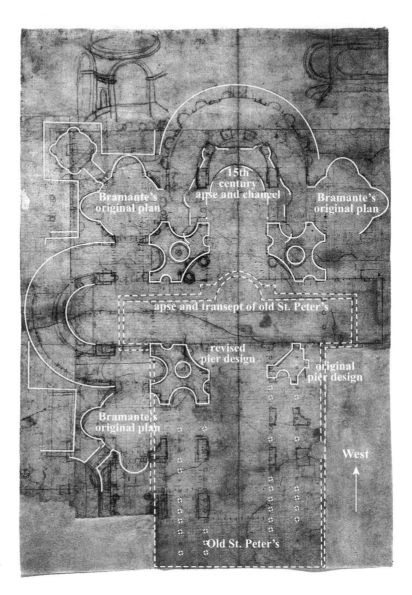

Figure 5.32. Bramante's revised plan, around 1505

of each of the subsequent architects had to be consistent with and flow out from Bramante's central structure. Unlike the construction of the nave of a Gothic cathedral, the building of the new St. Peter's could not proceed one bay at a time.

Projects of this size and complexity required organizations that we would call construction companies today. In the case of St. Peter's this was the Fabbrica di San Pietro. Cardinals who reported directly to the pope had the oversight. The capomaestro was responsible for executing the project and managing the work of the master masons, craftsmen, technical staff, and laborers. The Fabbrica facilitated and documented the continuous and informed evolution of the construction.

Michelangelo's St. Peter's

After Bramante's death in 1514, Raphael became the new capomaestro. It is likely that it was Bramante's vision for the new St. Peter's that Raphael chose as the setting for the meeting of the great minds of his *School of Athens.* See Plate 20. In his *Lives of The Artists,* the sixteenth-century painter, architect, historian, and biographer Giorgio Vasari (1511–1574) maintains that Bramante actually designed the architectural setting of Raphael's fresco. Raphael had little impact on the St. Peter's project. Work was brought to a standstill by the devastating occupation of Rome by the troops of the Spanish Habsburgs (engaged in a power struggle with the French over control of the Italian city states) as well as the ongoing difficulties with the financing of the enormous building costs. Artists' sketches of the construction site made in the 1530s show very little progress beyond what Bramante had already built. Raphael did influence the practice of design in architecture. In his drawings of buildings, he provided ground plans, elevations, and sections. This became standard practice in architecture only later.

Antonio Sangallo (1484–1546), known more for his technical expertise than his creative ability, took over as capomaestro in 1534. With Sangallo in charge, major construction resumed. Sangallo realized that even Bramante's more powerful main piers would be too weak and strengthened them by walling up their large niches. By adding four pendentives to Bramante's piers and connecting arches, he completed the central structure from which the drum and the dome of St. Peter's would later rise. Sangallo also built the secondary piers to the east and south and the barrel vaults connecting them to the central structure. The semicircular wall of the southern apse—such a wall is called a hemicycle (in Greek, *hemi* = half, *kyklos* = circle or cycle)— was finished to the top of the first story. Sangallo's vision for St. Peter's was even more grandiose and elaborate than Bramante's. It called for chapels and corridors in new semicircular sections beyond Bramante's plan of Figure 5.32. The walls that Sangallo proceeded to build for these new spaces began to encroach on other buildings of the Vatican. By the time of Sangallo's death in 1546, the projected size and complexity as well as the actual cost of the structure had spun completely out of control.

The pope asked an unwilling Michelangelo to take charge of the construction. Michelangelo could not refuse the request of a pope, but on his insistence was granted complete authority over the project. The new capomaestro brought rationality to both the design of the new St. Peter's and its execution. He strengthened the wall structures of Bramante's plan and also

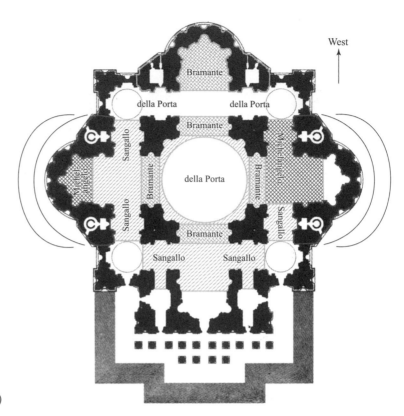

Figure 5.33. Michelangelo's plan for
St. Peter's in Rome, from around 1546
(with shading and crosshatching added)

broke it open by simplifying the interior into a clear, continuous space. By removing Sangallo's outer walls, he brought direct light into all parts of the interior, reduced the size of the church, and saved large amounts of money and construction time. With a few masterful strokes, Michelangelo had replaced Bramante's complicated configuration of Figure 5.31 and reigned in Sangallo's grandiose vision. Michelangelo's cross and square plan was cohesive and rational. It could be realized, and it was to be built.

Figure 5.33 shows Michelangelo's plan. The shading, crosshatching, and outlines of the three hemicycles are superimposed on the plan to identify the architects who undertook the construction. Areas crosshatched in light gray highlight Bramante's coffered arches that connect his main piers. The western apse and chancel are also outlined and crosshatched in light gray. They were begun in the fifteenth century and completed by Bramante. The light gray, singly hatched sections identify Sangallo's vaults and his pendentives around the dome. The two arcs to the left and right of the plan refer to the outer walls that Sangallo started and Michelangelo removed. The outlines and crosshatching in dark gray highlight the work of Michelangelo. He completed the southern hemicycle, built the northern hemicycle up to the second story, and added vaults. The construction of the drum and the design

Figure 5.34. Michelangelo's design of the exterior wall of the southern hemicycle of St. Peter's with papal seal of approval, engraving published by Vincenzo Luchino, 1564. From Henry A. Millon and Craig Hugh Smyth, "Michelangelo and St. Peter's I: Notes on a Plan of the Attic as Originally Built on the South Hemicycle," *Burlington Magazine*, vol. 111, no. 797 (Aug. 1769)

of the dome above it are also the work of Michelangelo. (The dome and the remaining vaults, except those on the eastern end, were executed by Michelangelo's assistant, Giacomo della Porta, in the 1580s and 1590s. Della Porta also rebuilt the hemicycle of the western apse to conform to Michelangelo's southern hemicycle.)

It was pointed out in the concluding paragraph of "The Roman Arch" section of Chapter 2 that the Romans built many of their large structures with concrete. Concrete walls were often provided with brick facing in order to solidify and protect the exterior surfaces. Construction in the Middle Ages relied on stone block (and brick as inexpensive substitute), but did not use concrete in the advantageous way the Romans did. The fact that the wall structures in Bramante's plan of St. Peter's (see Figure 5.31) have many indentations and irregularities is evidence that he had rediscovered the Roman method of building with concrete. The strength of concrete, and the ease with which it can be given shape and form, made the execution of these complex wall structures practical. Michelangelo's facade of the southern apse is depicted in Figure 5.34. Not only does this facade curve (in following the curving hemicycle) but its pilasters, windows, and decorative elements lie in different planes and project at various angles. It would not have been feasible for Michelangelo to achieve such a rich geometry without the flexibility and plasticity (not to mention economy) offered by concrete-brick construction.

The great dome with its supporting drum was to be the crowning feature of the new St. Peter's from the beginning. It was also the most challenging

Figure 5.35. Michelangelo, section and elevations for the drum and dome of St. Peter's, Musée des Beaux-Arts, Lille, France. Sketched sometime during the years 1546–1557

Figure 5.36. Giovanni Antonio Dosio, Michelangelo's drum under construction around 1562. (The structure Bramante built to protect the altar can be seen seen under the arching vault). From Charles B. McClendon, "The History of the Site of St. Peter's Basilica, Rome," *Perspecta*, vol. 25, 1989

aspect of the construction. Michelangelo's concept was strongly influenced by Brunelleschi's dome of the Santa Maria del Fiore. It was the only prototype on a comparable scale and Michelangelo took the basic structural aspects from it: the drum with its large windows, the double shell construction, the ribs, and the crowning lantern. But the specifics of the design of the dome and its drum needed to evolve and gain definition from the structure as a whole. Figure 5.35 is a study undertaken by Michelangelo (or an assistant under his supervision). Notice how the double columns of the design of the exterior of the drum parallel the large paired pilasters of the facade in Figure 5.34. It was Michelangelo's sense that the strong vertical accents of the facade be continued upward through the paired columns of the drum, along curving external ribs of the dome, to converge at the lantern, the focal point of the composition. When Michelangelo doubled the eightfold symmetry of the drum and ribs of the dome of the Santa Maria del Fiore to a sixteenfold symmetry in his design for St. Peter's, he enhanced these vertical accents. The drum of the dome was started in 1557 and Figure 5.36 shows it under construction in 1562. The configuration of cords seen in the sketch was used

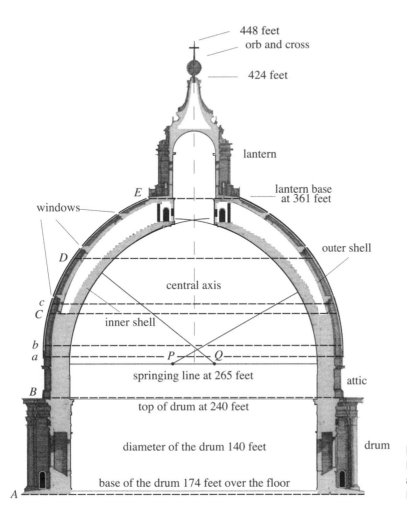

448 feet
orb and cross

424 feet

lantern

lantern base
at 361 feet

E

windows

outer shell

D

central axis

c
C

inner shell

b
a

P *Q*

springing line at 265 feet

attic

B

top of drum at 240 feet

diameter of the drum 140 feet

drum

base of the drum 174 feet over the floor

A

Figure 5.37. Cross section of della Porta's dome of St. Peter's, based on an engraving published by Hieronymus Frezza, 1696

to guide the cylindrical geometry. In 1564, with the drum nearly complete, Michelangelo died and work on St. Peter's came to a halt once again.

When Giacomo della Porta (1533–1602), architect and sculptor, took over in 1573, things began to move forward again. Della Porta would work on many important buildings in Rome and become Rome's leading architect of the latter part of the sixteenth century. Della Porta rebuilt the western apse of St. Peter's and constructed the vaults on the west side (refer to Figure 5.33). When Sixtus V became pope, the project went into overdrive. Known as the greatest builder among popes, he approved della Porta's plans for the dome and pushed for its construction. Della Porta retained the essence of Michelangelo's design, but raised the dome's profile. The greater height made the dome more impressively visible over the main structure of the church. A comparison of Figures 5.35 and 5.37 shows that della Porta achieved the extra height primarily by extending the drum and adding an attic above it, and only to a small extent by raising the geometry of the dome itself. The vertical cross section of the interior of the inner shell is a pointed arch defined by two circular arcs. Because the centers of the two circles—the points P and Q in Figure 5.37—are relatively close to each other, this cross section is close to being a semicircle. In fact, the ratio of the interior height of the dome to the diameter of the dome at the springing line is about 0.58. This is not appreciably more than the height to base ratio of 0.5 for a dome with a semicircular cross section. (Refer to Problem 9.) The exterior cross section of the outer shell of the dome also follows a pointed arch. It rises slightly more steeply than that of the inner shell.

The pope appointed the architect and engineer Domenico Fontana (1543–1607) to assist della Porta, and the construction of the dome began in 1588. Work progressed nonstop with 600 to 800 men on the site at a given time. The attic continued the cylindrical shape of the drum. Sixteen massive masonry ribs grew from the top of the attic. They were equally spaced and curved upward and inward. The ribs were built to taper as they rise, but to increase in thickness horizontally—from about 6 feet at the base to about 16 feet at the top—in the direction of the dome's vertical axis. Figure 5.38 shows how they shape the dome. The inner and outer shells of the dome rose simultaneously in brick, travertine block, and mortar each in 16 curving panels between the bracing ribs. The horizontal sections were not given the geometry of a regular 16-gon (as the octagonal cross sections of Brunelleschi's dome would have suggested), but were rounded out to follow the circular cross sections of the drum and attic. The inner shell was made about $6\frac{1}{2}$ feet thick and the outer shell about 3 feet thick. Three rings of iron chains were walled into and around the circumference of the dome, two slightly above the attic and another slightly above the point of separation of the two shells. These chains are about $2\frac{1}{2}$ by $1\frac{1}{2}$ inches in cross section and served to contain the enormous outward thrusts. They are represented in Figure 5.37 as the dashed lines labeled a, b, and c. (The other chains that are shown were added much later.)

Incredibly, the dome was finished in less than two years. Wooden centering carried by beams slanting upward from supports on the inner wall above the attic supported the construction of the shells. Recall that Brunelleschi had not relied on centering when he erected the dome in Florence. Why the departure from this precedent? Della Porta knew that his dome was flatter than the dome in Florence and that it would be built at a much greater speed. The greater steepness of Brunelleschi's dome meant that the successive octagonal rings of bricks of its shells received more support from the finished structure below it. The 16 years needed to complete his dome gave the mortar binding a ring of bricks ample time to cure before it was subjected to great loads. Della Porta decided to lessen the risk posed by the breakneck speed of the construction of his flatter dome by supporting its rising shells with a centering structure.

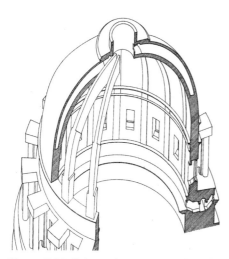

Figure 5.38. Schematic representation of the structure of the dome by F. Nespoli

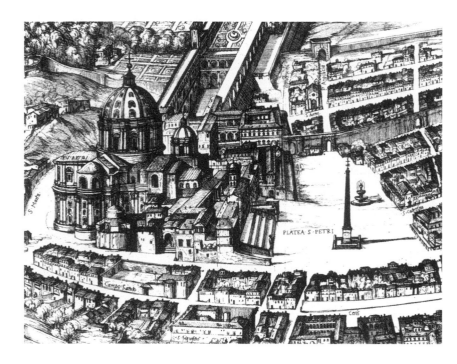

Figure 5.39. St. Peter's from the south. Detail of *View of Rome*, an etching published by Antonio Tempesta, 1593

The lantern took two more years to complete and before the year 1593 was over, the orb and cross were in place at the very top. Figure 5.39 pictures an etching showing della Porta's dome and lantern high over Michelangelo's southern hemicycle. The cluster of structures to the east of the dome includes remnants of Old St. Peter's. Visible in the middle of St. Peter's square is an ancient Egyptian obelisk. An extraordinary effort had moved it to its central position in the square only a few years before the completion of the dome. The story of this obelisk is told in Discussion 5.2.

St. Peter's was completed during the next two decades. The question as to whether St. Peter's was to be a church with a central or a basilica plan was finally resolved. Instead of the central plan envisioned by both Bramante and Michelangelo, the pope gave final approval to a basilica design with an elongated nave and a wide entrance area. The nave was to consist of a coffered barrel vault in the same form as the supporting vaults of the dome. It was to be segmented into three bays, supported by arches and piers, that open to an aisle on each side. The older structures to the east were demolished and Carlo Maderno (1556–1629), the new capomaestro, finished both the nave and the entrance area with its facade by 1612.

Codazzi's painting of Figure 5.40 shows St. Peter's a few years after its completion. It depicts the dome and its drum in strangely dark tones and includes two bell towers that would become embroiled in controversy (and no longer exist today). The painting shows that Maderno's facade reflects the accents

Figure 5.40. Viviano Codazzi, *St. Peter's, Rome*, c. 1630. Oil on canvas, Museo del Prado, Madrid

that Michelangelo had set with his southern hemicycle of Figure 5.34. These include the large Corinthian pilasters, the sequence of wider and narrower bays, the balconies framed by Ionic columns, the alternating triangular and curved pediments, as well as the vertical expression of the composition as a whole. But Codazzi's painting also shows that the almost solid attic of Maderno's facade, broken only by small rectangular windows, slows the upward flow of the composition. By contrast, the large arched openings in the attic of Michelangelo's southern hemicycle facilitate this flow. While the texture of the exterior as well as the dome of St. Peter's express Michelangelo's vision, the elongated nave and broad facade of the basilica do not.

The construction of St. Peter's had taken 120 years. The greatest architects of their time contributed their talents to make the new church the spectacular monument to the Christian faith that it is today. Bramante gave the new church its central core, Michelangelo shaped the essential aspects of its structure, della Porta raised its huge dome, and Maderno lengthened its nave to give it its basilica form. It remained for Gian Lorenzo Bernini to give the interior of the basilica its Baroque expression and to construct the colonnade that embraces St. Peter's square.

Bernini's Baroque Basilica

The Roman Catholic Church had gathered at the Council of Trent in 1545 to embark on reforms to counter the Protestant reformation. One of the initiatives called on art and architecture to promote the influence of the Church and to connect the faithful directly to the narrative and symbols of their religion. The new Baroque style that emerged retained the forms and structures of classical and Renaissance architecture but embellished them with opulent, decorative details. (The word baroque may have been derived from the Portuguese *barocco* or the Spanish *barueco* that both refer to an irregularly shaped pearl.) Defined by frescos, stuccoed surfaces, sculpture, sophisticated use of light and color, and the interplay of different surfaces, the new architecture created forms and spaces of theatrical grandeur.

In the same way that Florence came to exemplify the artistry of the Renaissance, Rome became the center of Baroque art and architecture. Gian Lorenzo Bernini (1598–1680), successor of Maderno and heir to Bramante and Michelangelo, dominated the development of the new style in the seventeenth century. Much of the character of the interior of St. Peter's is the result of the inspiration of Bernini and his workshop. This included not only the floor of the nave and the decoration of the piers, as illustrated in Panini's painting seen in Plate 21, but also the addition of groups of sculptures around the altars and tombs and decorative elements for the crossing and the main apse. To reduce the scale of the vast space under the dome, Bernini designed a monumental baldacchino that fuses architecture and sculpture in gilded bronze. Erected by an army of craftsmen from 1624 to 1634, this symbolic protective canopy towers some 90 feet over the main altar. Its four bronze columns spiral sinuously upward in the new ornate language of the Baroque. The bronze had been stripped from the Pantheon, melted, and recast. Plate 21 provides a glimpse of the baldacchino at the far end of the nave.

By 1656 Bernini was at work on St. Peter's square. The scale of the great new church of Latin Christendom called for a grand exterior setting. Bernini created a colonnade that moves out from Maderno's facade to expand—four columns abreast—into a large oval centered at the obelisk. An engraving by Giovanni Piranesi pictured in Figure 5.41 shows how its pair of arms reach out in a symbolic majestic embrace of the faithful. More than 250,000 people

Figure 5.41. Giovanni Battista Piranesi, *St. Peters with Forecourt and Colonnades*, 1775. Etching from the series *Views of Rome*. Courtesy René Seindal

can gather in the square for the papal blessing. Statues of saints—140 of them—overlook the square from their perch on top of the colonnade. Piranesi's prints in the series *Vedute di Roma (Views of Rome)* are executed with great accuracy as well as technical mastery. They were very influential in shaping the understanding and spreading the fame of the eternal city.

To lay out the oval geometry of his colonnade, Bernini turned to a strategy of the Romans. The two isosceles triangles of Figure 5.42 with vertices the points 1, 2, and 3, and 1, 3, and 4 determine the concentric ovals of the colonnade exactly as illustrated in Figure 2.39 and described in "The Colosseum" section in Chapter 2. In view of the gaps in the colonnade, the upper and lower parts of the ovals only play a minor role. Figure 5.42 shows how the two pairs of concentric circular arcs centered at the points 1 and 3 and the radii emanating from these two points determine the positions of most of the columns and the spacing between them. The circle centered at point 2 (and its radii) determines the position of the two sets of columns farthest from the facade.

Not everything that Bernini touched turned to gold. The two outermost bays of Maderno's facade were wider than the others and were intended to

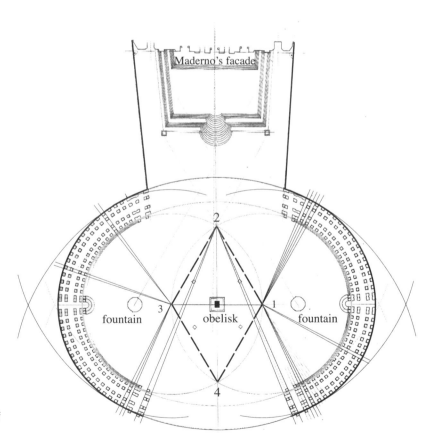

Figure 5.42. The oval geometry of Bernini's colonnade

Figure 5.43. The interior of the dome of St. Peter's as it looks today. Photo by Jay Berdia

support high bell towers. This intention became problematic after the facade experienced considerable uneven settlement during construction. The subsoil under the facade was of uneven quality and the drainage was poor. The piles that Maderno had sunk into the ground and the wide footed foundations above them were not strong enough. As a consequence, the facade of St. Peter's—a block 380 feet long, 75 feet wide, and 160 feet high—rotated slightly toward the southern end causing cracks in some walls and vaults in the process. When Bernini continued the construction of the towers that Maderno had started, their growing weight magnified the problem. After much criticism of Bernini, the tower project was abandoned. (Figure 5.40 tells us that short versions of such towers were in place at one time.)

Given the massive size and weight of the dome of St. Peter's (a later study provided the estimates of 41 million pounds for the weight of the ribs and shells, 3 million pounds for the lantern, and over 100 million pounds for the entire dome), it is not surprising that there would be structural problems. The great weight of the masonry generated hoop stresses that the three iron chains della Porta had placed around the dome could not contain. Serious cracks extending from the springing line up to the lantern developed along meridians of the inner shell. They were more severe than those of Brunelleschi's dome in Florence. The dome was thoroughly restored in the middle of the eighteenth century. This included the placement of five more iron chains around the dome at various elevations and the repair of one of the three original chains. (This story is taken up in Chapter 6.) The additional chains are labeled *A, B, C, D,* and *E* in Figure 5.37 (and *c* labels the repaired

Figure 5.44. The dome and facade of St. Peter's. Photo by Wolfgang Stuck

chain). Chains *A* and *B* strengthened the weak interconnection between the outer ring of paired columns of Michelangelo's drum and the inner ring of piers between its windows. These corrective measures have stabilized the dome and no major structural problems have been observed since. Figure 5.43 shows the interior of the drum and inner shell of St. Peter's and Figure 5.44 shows the facade and the exterior of the dome as they look today.

Brunelleschi and Perspective*

Let's turn to Raphael's *School of Athens* in Plate 20. The scene moves from the stone floor in the front up the stairs toward the facade of the building and the passageway that leads to the arch. The human figures are depicted larger in the foreground and become smaller in the background. Raphael has created a sense of depth as the picture unfolds in several planes from front to back. Viewers experience the scene as they would see it "live." How does a painter or graphic artist achieve such realistic depictions? How did Panini go about the representation of the receding nave, aisles, vaults, and arches of St. Peter's on the flat two-dimensional surface of his painting in Plate 21? The short answer to these questions is that painters and graphic

---

*This section and the next section undertake a mathematical study of perspective. They are technical, do not impact any other parts of the book, and may be skipped.

artists simply draw what they see. But is there a way to formulate in precise geometric terms how artists can transform the three-dimensional space that they look out on to the plane of their canvas or paper?

It is none other than Filippo Brunelleschi who is given credit for providing the answer. Brunelleschi's ideas are developed in two of Alberti's treatises on painting. The first was the theoretical work *De Pictura* published in Latin in 1435. The *Della Pittura* appeared a year later. It was dedicated to Brunelleschi and was addressed to any citizen interested in art. It is important to point out that perspective is not only important to painters and graphic artists, but to architects as well. The architect is not only interested in the building *as it is*, but also *how it looks* from important vantage points. How a building appears from different angles is a matter that belongs to the domain of perspective.

We see an object because light from every point on the part of its surface within our field of vision travels to strike our eye where the light is processed by the mechanism that allows us to see. To understand perspective we focus on a ray of light from the object to the eye and study it abstractly. Before we do, we'll need to learn more about abstract lines in the plane and in space. We will use information developed in Chapter 4, "The Coordinate Plane" and "Coordinate System in Three Dimensions."

Let's consider an xy-coordinate plane and let L be a nonvertical line in the plane. Take two distinct points $P_1 = (x_1, y_1)$ and $P_2 = (x_2, y_2)$ on L and recall that $y - y_1 = \frac{y_2 - y_1}{x_2 - x_1}(x - x_1)$ is an equation of L in two-point form. Rewrite this equation as $\frac{y - y_1}{y_2 - y_1} = \frac{x - x_1}{x_2 - x_1}$ and set $\frac{y - y_1}{y_2 - y_1} = \frac{x - x_1}{x_2 - x_1} = t$. So $x - x_1 = t(x_2 - x_1)$ and $y - y_1 = t(y_2 - y_1)$, and therefore

$$x = x_1 + t(x_2 - x_1) \quad \text{and} \quad y = y_1 + t(y_2 - y_1).$$

These equations are parametric equations for L in the parameter t. (The word *parameter* comes from the Greek. Here it means determining factor, as a given t determines both x and y.) The coefficient $y_2 - y_1$ of t in the equation for y over the coefficient $x_2 - x_1$ of t in the equation for x is the slope of L. Any pair of numbers x and y determined by a value of t and the two equations satisfies $\frac{y - y_1}{y_2 - y_1} = \frac{x - x_1}{x_2 - x_1}$. So the point $P = (x, y)$ that t determines lies on L. For instance, for $t = 0$, the point obtained is $P_1 = (x_1, y_1)$ and for $t = 1$, the point is $P_2 = (x_2, y_2)$. As t runs through the real numbers, the corresponding points $P = (x, y)$ trace out the line L. A look at the form of the two equations tells us that different choices of points P_1 and P_2 give rise to different pairs of parametric equations for L. Even though they were derived only for a nonvertical line, the parametric equations above apply to a vertical line as well. In this case, $x_1 = x_2$, so that $x = x_1$. Since $y = y_1 + t(y_2 - y_1)$ (and $y_2 \neq y_1$), all the points (x_1, y) on the line arise as t varies.

Let's look at an example. Consider the line L that has the points $P_1 = (-2, 3)$ and $P_2 = (4, 1)$ on it. See Figure 5.45. Check that $y = -\frac{1}{3}x + \frac{7}{3}$ is the slope-intercept form of the equation for L. By substituting the coordinates

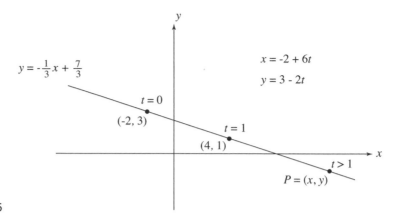

Figure 5.45

of P_1 and P_2 into the general form of the parametric equations, we get the equations $x = -2 + t(4 - (-2))$ and $y = 3 + t(1 - 3)$, or

$$x = -2 + 6t \quad \text{and} \quad y = 3 - 2t$$

for L. Think of the parameter t as representing time. Because the location of the point $P = (x, y)$ that t determines changes as t changes, the point will move. At time $t = 0$, the point P is at $(-2, 3)$. As t increases, x increases so that P moves to the right. When $t = 1$, the point P has arrived at $(-2 + 6, 3 - 2) = (4, 1)$. For increasing $t > 1$, the point P continues to move to the right beyond the point $(4, 1)$. For what range of t does the point P trace out the part of L that lies to the left of $(-2, 3)$? If, in the determination of the two coefficients of t, the points $P_1 = (-2, 3)$ and $P_2 = (4, 1)$ are taken in the opposite order, different parametric equations, namely $x = -2 - 6t$ and $y = 3 + 2t$, are obtained for the line L. How does the point $P = (x, y)$ move along the line L in this case?

We turn to consider an xyz-coordinate system and a line L in space. Let $P_1 = (x_1, y_1, z_1)$ and $P_2 = (x_2, y_2, z_2)$ be two distinct points on L. We now let $P = (x, y, z)$ be any point and ask for the conditions on x, y, and z that place the point P on the line L. The discussion that follows makes use of Figure 5.46. Push the points P_1 and P_2 into the xy-plane—by going parallel to the z-axis— to get the points (x_1, y_1) and (x_2, y_2) in that plane. In the process, L is pushed to the line L_{xy} that (x_1, y_1) and (x_2, y_2) determine. In the same way, push P to (x, y). If P is on L, then (x, y) is on the line L_{xy}. It follows from the earlier discussion that $\frac{y - y_1}{y_2 - y_1} = \frac{x - x_1}{x_2 - x_1}$. Now start over and push P_1, P_2, and L into the xz-plane by going parallel to the y-axis. If P is on L, then (x, z) is on L_{xz} and $\frac{z - z_1}{z_2 - z_1} = \frac{x - x_1}{x_2 - x_1}$. It follows that if $P = (x, y, z)$ is on the line determined by $P_1 = (x_1, y_1, z_1)$ and $P_2 = (x_2, y_2, z_2)$, then $\frac{x - x_1}{x_2 - x_1} = \frac{y - y_1}{y_2 - y_1} = \frac{z - z_1}{z_2 - z_1}$. Setting $\frac{x - x_1}{x_2 - x_1} = \frac{y - y_1}{y_2 - y_1} = \frac{z - z_1}{z_2 - z_1} = t$, we get

$$x = x_1 + t(x_2 - x_1), \, y = y_1 + t(y_2 - y_1), \quad \text{and} \quad z = z_1 + t(z_2 - z_1).$$

We have shown that if a point $P = (x, y, z)$ is on L, then its coordinates satisfy these equations—they are again called parametric equations—for some real

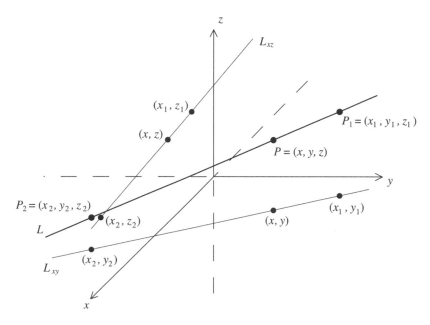

Figure 5.46

number t. This is also true in the other direction. Namely, if the x-, y-, and z-coordinates of a point P satisfy these equations for some real number t, then the point P is on L. (We'll skip the verification of this.)

Take $P_1 = (3, -4, 2)$ and $P_2 = (-5, 6, 1)$, for example. It follows from the discussion above that a point $P = (x, y, z)$ is on the line determined by P_1 and P_2 precisely if

$$x = 3 - 8t, \; y = -4 + 10t, \quad \text{and} \quad z = 2 - t$$

for some real number t. Letting t be equal to -2 and then 3 tells us that the points $(3 - 8(-2), -4 + 10(-2), 2 - (-2)) = (19, -24, 4)$ and $(3 - 8 \cdot 3, -4 + 10 \cdot 3, 2 - 3) = (-21, 26, -1)$ are on this line.

Coordinate geometry—developed about 200 years after Alberti's publications—is tailor-made to explain what Alberti learned from Brunelleschi. Alberti considered a horizontal tile floor paved with square tiles, focused on a 6 by 6 square of tiles, and described a strategy for drawing this square from the vantage point of a painter.

Figure 5.47 considers three-dimensional space and an xyz-coordinate system within it. The figure pictures the 6 by 6 arrangement of tiles on the xy-plane of the floor. Each tile is a 1 by 1 square, the y-axis runs through the center of the arrangement, and its front edge is a distance h from the x-axis. The artist's rectangular canvas is placed in the xz-plane. The artist's eye is fixed at a height e directly above the y-axis at a distance d behind the canvas. Notice that the artist's eye is at the point $E = (0, -d, e)$. Suppose, for the sake of the upcoming explanation, that the artist is working with a see-through canvas.

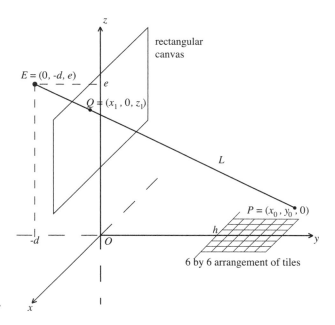

rectangular canvas

$E = (0, -d, e)$

$Q = (x_1, 0, z_1)$

L

$P = (x_0, y_0, 0)$

h

6 by 6 arrangement of tiles

Figure 5.47

The answer to the following question is the key. Given a point P on the floor (on the arrangement of tiles or not), at precisely what point Q will the artist see the point on the plane of his canvas? Because P lies in the xy-plane, $P = (x_0, y_0, 0)$ for some x_0 and y_0. To pinpoint Q in terms of its coordinates, we'll find a set of parametric equations for the line L that joins E and P, and then determine the intersection of L with the xz-plane of the canvas. Form the respective differences $0 - x_0$, $-d - y_0$, and $e - 0$ between the coordinates of E and P, and recall from the discussion about lines in three space that the points on L are of the form (x, y, z) with $x = x_0 - tx_0$, $y = y_0 - t(d + y_0)$, and $z = 0 + te$, where t ranges over the real numbers. For $t = 0$, $(x, y, z) = (x_0, y_0, 0) = P$ and for $t = 1$, $(x, y, z) = (0, -d, e) = E$. At the intersection of L with the xz-plane, the y-coordinate $y_0 - t(d + y_0) = 0$. It follows that $t = \frac{y_0}{d + y_0}$. By substituting this t into the parametric equations for L, we find that the required point on the canvas is $Q = ((x_0 - \frac{y_0 x_0}{d + y_0}), 0, \frac{e y_0}{d + y_0})$. Because $x_0 - \frac{y_0 x_0}{d + y_0} = \frac{x_0(d + y_0) - y_0 x_0}{d + y_0} = \frac{dx_0}{d + y_0}$, this is the point $Q = (x_1, 0, z_1)$ where $x_1 = \frac{dx_0}{d + y_0}$ and $z_1 = \frac{e y_0}{d + y_0}$.

So Alberti's instruction to the painter, expressed within the framework of the given coordinate system, is this rule: A point P with coordinates $P = (x_0, y_0)$ at any location in the xy-plane with positive y-coordinate (including points beyond the square of tiles) should be drawn at the point

$$Q = (x_1, z_1), \text{ where } x_1 = \frac{dx_0}{d + y_0} \text{ and } z_1 = \frac{e y_0}{d + y_0},$$

in the xz-plane of the canvas.

An exploration of the consequences of Alberti's rule for the placement of the tile floor on the canvas follows. Figure 5.48a shows the outline of the 6 by

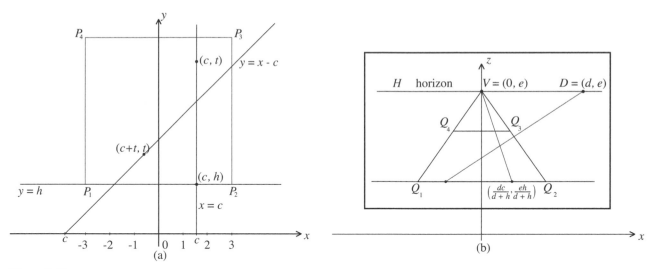

Figure 5.48

6 square arrangement of tiles along with the four points P_1, P_2, P_3, and P_4 at the corners. The points Q_1, Q_2, Q_3, and Q_4 in Figure 5.48b are the images of these points on the canvas as Alberti's rule provides them. The placement on the canvas of vertical (in this discussion the term "vertical" derives its meaning from Figure 5.48a) and diagonal lines on the floor is decisive and is studied next. Figure 5.48a considers a point c on the x-axis and the line $x = c$ on the floor. Notice that the points on this line all have the form $P = (c, t)$, for a real number t. By Alberti's rule, the point $P = (c, t)$ needs to be drawn at $Q = (\frac{dc}{d+t}, \frac{et}{d+t})$ on the canvas. Taking $t = h$ tells the artist that the point (c, h), at the front edge of the tile floor, falls on the point $Q = (\frac{dc}{d+h}, \frac{eh}{d+h})$ of the canvas. Think of $t \geq 0$ as designating time and think of the point $P = (c, t)$ as moving on the line $x = c$ in the upward direction. As t becomes larger and larger, the point P on the floor recedes farther and farther from the artist. Rewriting $Q = (\frac{dc}{t+d}, \frac{et}{t(1+\frac{d}{t})}) = (\frac{dc}{t+d}, \frac{e}{1+\frac{d}{t}})$ tells the painter that the point Q on the canvas moves to the point $V = (0, e)$. What has been demonstrated is that the line $x = c$ on the xy-plane of the floor has to be sketched on the xz-plane of the canvas of Figure 5.48b as the line from $(\frac{dc}{d+h}, \frac{eh}{d+h})$ to the point $V = (0, e)$. It follows that *all* vertical lines on the floor—including the seven vertical lines through the boundary points of the tiles at the front edge—meet at the point $V = (0, e)$ when drawn on the canvas. The point $V = (0, e)$ on the canvas is called the vanishing point. The conclusion just reached is confirmed by our experience. Think of two perfectly straight and parallel horizontal railroad tracks. Think of yourself as standing between them looking in the direction of the tracks. The two tracks will appear to meet at the horizon.

The last piece of the puzzle concerns the position on the canvas of lines on the xy-plane of the floor with slope 1, in particular the diagonal line through P_1 and P_3. Let such a line start at the point c on the x-axis. Because $(c, 0)$ is on the line, its equation is $y = x - c$ (using the point-slope form). It is sketched in Figure 5.48a. Notice that any point on this line has the form $P = (t + c, t)$ for a real number t. (This is equivalent to saying that $x = t + c$ and $y = t$ are parametric equations for the line.) By Alberti's rule, the corresponding point on the canvas is $Q = (\frac{d(t+c)}{d+t}, \frac{et}{d+t})$. Again think of $t \geq 0$ as time and of $P = (t + c, t)$ in motion up along the line. As t becomes larger and larger, the point P on the floor recedes into the distance. Because $Q = (\frac{dt(1+\frac{c}{t})}{t(1+\frac{d}{t})}, \frac{et}{t(1+\frac{d}{t})})$ $= (\frac{d(1+\frac{c}{t})}{1+\frac{d}{t}}, \frac{e}{1+\frac{d}{t}})$, the corresponding point Q on the canvas moves to $D = (d, e)$ (because $\frac{c}{t}$ and $\frac{d}{t}$ both go to 0). It follows that *any* line on the xy-plane of the floor with slope 1—in particular the line through P_1 and P_3—converges to the point D when drawn on the xz-plane of the canvas.

The artist can now draw Alberti's 6 by 6 tile floor as follows. He starts by drawing in the points $V = (0, e)$ and $D = (d, e)$ and the horizontal line H that they determine. This horizontal line corresponds to the horizon as the artist would see it (if there were no obstructions). The artist considers the vertical line $x = c$ of Figure 5.48a for each of the two values $c = -3$ and $c = 3$. Representing these two lines as Figure 5.48b prescribes, provides the lines from Q_1 to V and from Q_2 to V shown on the canvas of Figure 5.49b. The segment between the starting points Q_1 and Q_2 of these two lines is the lower boundary of the image of the floor. The artist puts in five points between Q_1 and Q_2, spacing them equally, and draws a segment to V from each of them. These lines are the images on the canvas of the vertical boundaries between the tiles of the 6 by 6 arrangement. Next, the artist considers the diagonal that the points P_1 and P_3 determine on the floor. When drawn on the canvas,

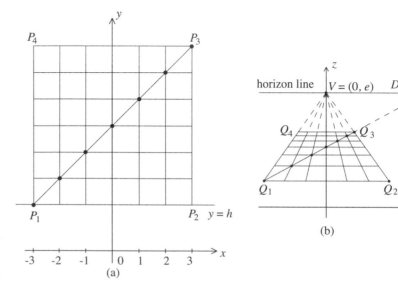

Figure 5.49

this is the line from Q_1 to Q_3 to the point $D = (d, e)$. The seven points highlighted in Figure 5.49a are the points of intersection of the seven vertical lines with the diagonal. On the canvas they are the points of intersection between the seven lines to V and the diagonal from Q_1 to D. Notice that these seven points tell the artist where to draw the horizontal lines of separation between the tiles. The sketch of Alberti's tile floor as shown in Figure 5.49b is now complete.

Floors are a good start, but how are objects that have height to be drawn? How are curves to be dealt with? The section "From Circle to Ellipse" and Discussion 5.3 below will inform us that the explanation of perspective drawing just provided can be extended to answer such questions.

The sculptors Lorenzo Ghiberti (1378–1455) and Donatello (1386–1466) and the painter Masaccio (1401–1428) were among the first artists to put to deliberate use what Alberti later explained mathematically in his *De Pictura*. Ghiberti and Donatello applied linear perspective in their delicately artistic bronze relief panels such as the one shown in Figure 5.50. One of the first paintings that made use of Brunelleschi's principles of perspective is the fresco *Trinity* by Masaccio depicted in Figure 5.51. It was painted in the same Santa Maria Novella for which Alberti would later build the new facade pictured in Figure 5.5. Not surprisingly, Masaccio's work seems much less confident in its execution than Raphael's *School of Athens* painted about eight decades later.

The discussion of perspective presented in this section relies on Figure 5.47 and therefore on the assumption that the object (in this case the tile floor) is being looked at with one eye. The fact that we look out on things with two eyes means that what we actually see is a combination of two slightly different images from slightly different angles that are merged into a single image. This imparts a perception of depth to what we see.

Figure 5.50. Donatello, *Feast of Herod*, c. 1425. Bronze relief panel of the baptismal font of the Siena Cathedral

Figure 5.51. Masaccio, *The Trinity*, 1427–1428. Fresco in the Santa Maria Novella

From Circle to Ellipse[*]

Have a look at the depiction of the rosette window of the Chartres Cathedral provided in Figure 3.24. We know it to be circular, but it is represented as an oval. Panini's painting of the interior of St. Peter's in Plate 21 represents the circular arches along the nave as arcs that are no longer circular. These depictions raise a question. What is the precise shape of a circle when it is viewed at an angle? Is it an ellipse? Or some other oval?

We will begin the answer to these questions by considering Apollonius's insights about the conic sections as they were collected in Chapter 4, "Remarkable

[*]This section depends on the previous section, "Brunelleschi and Perspective," but does not impact any other material in the book.

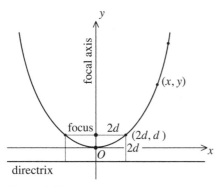

Figure 5.52

Curves and Remarkable Maps," and recasting them within the framework of a coordinate plane. Let a plane with an xy-coordinate system be given.

Consider any parabola in the plane and let c be the distance between its focus and directrix. Let $d = \frac{1}{2}c$. So $c = 2d$. Now move the parabola in such a way that its focal axis lies on the y-axis and the intersection of the parabola with its focal axis is the origin O. Figure 5.52 shows the parabola in this position. Because O is on the parabola, the distances from O to the focus and from O to the directrix are the same. So both are equal to d. Consider the point of intersection of the parabola and the line through the focus parallel to the x-axis. Because this point is on the parabola, the distance between it and the focus is equal to $2d$. So this point of intersection is $(2d, d)$. Now take any point (x, y) on the parabola. By a direct application of Apollonius's Proposition P2, we get that $\frac{y}{d} = \frac{|x|^2}{4d^2}$. Rearranging things provides the standard equation

$$y = \frac{1}{4d}x^2$$

of the parabola.

Next consider any ellipse in the plane. The ellipse is determined by two focal points and a positive constant k. Let $a = \frac{1}{2}k$. So $k = 2a$. Let c be the distance between the center of the ellipse and a focal point. The distance between the two focal points is $2c$. It follows from the requirement $k > 2c$ that $a > c$. Move the ellipse in such a way that its focal axis is the x-axis and its center is the origin O. Figure 5.53 shows the ellipse in this position. The points E and D are the points of intersection of the ellipse with the positive x- and y-axes. Because D is on the ellipse, twice the distance from D to a focus is equal to k, so that the distance between D and a focus is a. Let d be the distance between E and the focus on the right. Because E is on the ellipse, the sum of the distances d and $d + 2c$ is equal to k. So $2c + 2d = k = 2a$, and hence $c + d = a$. So E is the point $(a, 0)$ and a is the semimajor axis of the ellipse. Letting b be the semiminor axis completes the information displayed in the figure. Let (x, y) be any point on the ellipse. By an application of Apollonius's Proposition E2 to the two cases $A = O$ and $A = (x, 0)$, we get that $\frac{a^2}{b^2}$ and $\frac{|-a-x||a-x|}{|y|^2}$ are equal.

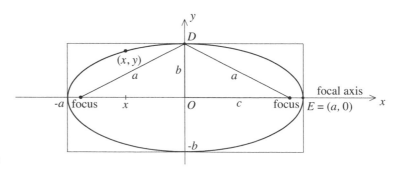

Figure 5.53

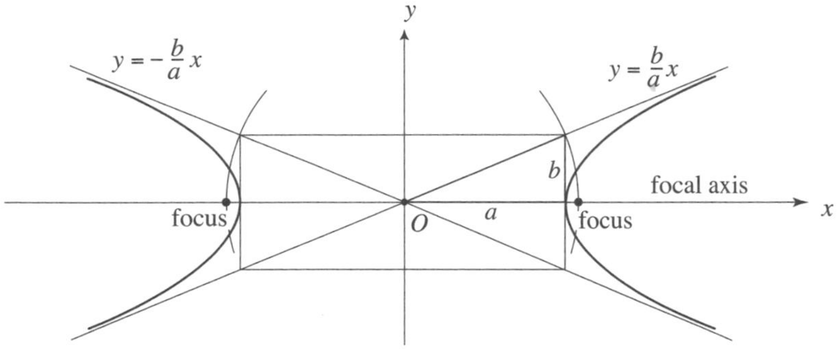

Figure 5.54

Because $|-a-x||a-x|=|a+x||a-x|=a^2-x^2$, this tells us that $\frac{a^2}{b^2}=\frac{a^2-x^2}{y^2}$. So $\frac{y^2}{b^2}=\frac{a^2-x^2}{a^2}$, and we have arrived at the standard equation

$$\frac{x^2}{a^2}+\frac{y^2}{b^2}=1$$

of the ellipse. Because the box surrounding the ellipse determines the constants a and b, it determines the ellipse. Therefore there is only one ellipse for a given box.

Apollonius was also familiar with hyperbolas. We will summarize the basic facts. The shape of a hyperbola is determined by a box and the extension of its diagonals. Figure 5.54 shows a typical hyperbola, the box that shapes it, and the constants a and b that the box determines. The hyperbola has two focal points. They are given, as shown, by the box and a circle with center at the center of the box. The line through the two focal points is the focal axis of the hyperbola. In Figure 5.54, the hyperbola has been moved so that the focal axis is the x-axis and the origin O is at the center of the box. The slopes of the diagonal lines are $\frac{b}{a}$ and $-\frac{b}{a}$ respectively. Their equations are displayed in the figure. The standard equation of the hyperbola is

$$\frac{x^2}{a^2}-\frac{y^2}{b^2}=1.$$

Let's consider the three equations

$$y=2x^2, \quad x^2+3y^2=12, \quad \text{and} \quad 2x^2-5y^2=10.$$

The following can be concluded from the discussion above. Set $\frac{1}{4d}=2$ and notice that $d=\frac{1}{8}$. So the graph of $y=2x^2$ is the parabola of Figure 5.52 with $d=\frac{1}{8}$. The graph of $x^2+3y^2=12$ is the same as the graph of $\frac{x^2}{12}+\frac{y^2}{4}=1$. This equation can be rewritten as $\frac{x^2}{(\sqrt{12})^2}+\frac{y^2}{2^2}=1$, and it follows that the graph is the ellipse of Figure 5.53 with $a=\sqrt{12}$ and $b=2$. The graph of $2x^2-5y^2=10$ is the same the graph of $\frac{x^2}{5}-\frac{y^2}{2}=1$. This can be rewritten as $\frac{x^2}{(\sqrt{5})^2}-\frac{y^2}{(\sqrt{2})^2}=1$, so that the graph is the hyperbola of Figure 5.54 with $a=\sqrt{5}$ and $b=\sqrt{2}$.

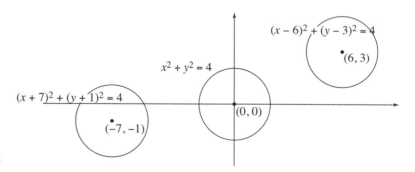

Figure 5.55

Consider the circle $x^2 + y^2 = 4$ of radius 2 and center the origin $(0,0)$. Replacing x by $x - 6$ and y by $y - 3$ gives the equation $(x - 6)^2 + (y - 3)^2 = 4$. This circle also has radius 2, but (by facts that conclude "The Coordinate Plane" section in Chapter 4) its center has been shifted to the point $(6,3)$. In the same way, if we replace x and y by $x + 7$ and $y + 1$, respectively, then the center of the circle is shifted from $(0,0)$ to $(-7,-1)$. Figure 5.55 shows all three circles.

The same considerations apply to any graph. For example, the graph of $y - 2 = 2(x + 3)^2$ has the same shape and orientation in the coordinate plane as the graph of $y = 2x^2$. The graph of $y - 2 = 2(x + 3)^2$ is obtained by a translation—this is a shift or slide of a figure in the plane to a new position *without* rotating it—of the parabola $y = 2x^2$ in such a way that the bottom of the parabola (originally at the origin $(0,0)$) ends up at the point $(-3,2)$. In the same way, the graph of $(x - 5)^2 + 3(y + 7)^2 = 12$ is obtained by translating the ellipse $x^2 + 3y^2 = 12$ so that its center (at the origin $(0,0)$) ends up at the point $(5,-7)$. Finally, the graph of $2(x + 9)^2 - 5(y + 11)^2 = 10$ is obtained by translating the hyperbola $2x^2 - 5y^2 = 10$ in such as way that the point of intersection (the origin $(0,0)$) of the two diagonal lines ends up at the point $(-9,-11)$.

Consider an equation of the form

$$Ax^2 + Bxy + Cy^2 + Dx + Ey + F = 0, \qquad (*)$$

where A, B, C, D, E, and F are constants. Notice that the equations of the circles, parabolas, ellipses, and hyperbolas considered in this section can all be written in this form. We will call the equation $(*)$ degenerate if its graph has either no points on it, consists of a single point, or is a line or a combination of two lines. The equation $x^2 + y^2 + 1 = 0$ is degenerate because it has no solutions. The equation $x^2 + y^2 = 0$ is degenerate because the only solution is $x = 0$, $y = 0$. The equation $x^2 - y^2 = 0$ can be factored as $(x - y)(x + y) = 0$. So the graph of $x^2 - y^2 = 0$ is a combination of the two lines given by $y = x$ or $y = -x$. It follows that $x^2 - y^2 = 0$ is degenerate. The equation

$$2x^2 - 7xy + 6y^2 + 7x - 11y + 3 = 0$$

is also degenerate. Check that $2x^2 - 7xy + 6y^2 + 7x - 11y + 3 = (x - 2y + 3)(2x - 3y + 1)$. Since $(x - 2y + 3)(2x - 3y + 1) = 0$ means that either $x - 2y + 3$

$= 0$ or $2x - 3y + 1 = 0$, it follows that the graph of the equation is a combination of the graphs of the lines $y = \frac{1}{2}x + \frac{3}{2}$ and $y = \frac{2}{3}x + \frac{1}{3}$.

We will summarize some basic facts about conic sections and their equations. Descartes, one of the two French creators of coordinate geometry, understood them in substance, but they were established in definitive terms later in the seventeenth century. Today, they are standard fare in any comprehensive calculus book.

Basic Fact 1. Any conic section in the xy-plane is the graph of an equation of the form $Ax^2 + Bxy + Cy^2 + Dx + Ey + F = 0$ that is (by definition) not degenerate.

Consider a quadratic equation $ax^2 + bx + c = 0$, where a, b, and c are constants with $a \neq 0$, and recall that its solutions are given by the quadratic formula

$$x = \frac{-b \pm \sqrt{b^2 - 4ac}}{2a}.$$

Observe that the term $b^2 - 4ac$ controls the solutions. If $b^2 - 4ac = 0$, then there is only one solution, namely $x = -\frac{b}{2a}$. If $b^2 - 4ac > 0$, then there are two solutions. Finally, if $b^2 - 4ac < 0$, then there are no solutions. Notice that if $b = 0$, then the solutions have the simple form $x = \pm\sqrt{\frac{-c}{a}}$, where $\frac{-c}{a} \geq 0$.

A criterion that is similar in flavor provides information about equations of form (∗) that are not degenerate.

Basic Fact 2. If $Ax^2 + Bxy + Cy^2 + Dx + Ey + F = 0$ is not degenerate, then its graph is a conic section. The graph is a parabola if $B^2 - 4AC = 0$, an ellipse if $B^2 - 4AC < 0$, and a hyperbola if $B^2 - 4AC > 0$.

Because the proofs of Basic Fact 1 and Basic Fact 2 are beyond the intentions of this text, we'll only make brief comments about them. Observe that the conic sections depicted in Figures 5.52, 5.53, and 5.54 have equations that can be rearranged to satisfy Basic Fact 1. It is easy to check that these rearranged equations conform to Basic Fact 2. The fact that any conic section in the xy-plane can be moved (by a combination of a translation and a rotation) to coincide with one of the conic sections in Figures 5.52, 5.53, and 5.54 can be exploited to supply the proofs in general.

Basic Fact 3. If $Ax^2 + Bxy + Cy^2 + Dx + Ey + F = 0$ is not degenerate, then $B = 0$ precisely if the graph of the equation is a conic section with focal axis parallel to either the x- or y-axis.

Important in the verification of Basic Fact 3 is the completing of squares procedure explained in Discussion 1.1. Let's have a look at the equation

$3x^2 + 5y^2 + 42x + 10y + 137 = 0$, for example. Notice that $B^2 - 4AC = 0 - 15 = -15$, so if the equation is not degenerate, then its graph is an ellipse by Basic Fact 2. Because $B = 0$, it is possible to say more. By completing the square for the equation twice (once for each of the variables), we get

$$3x^2 + 42x + 5y^2 + 10y + 137 = 3(x^2 + 14x) + 5(y^2 + 2y) + 137$$
$$= 3(x^2 + 14x + 7^2 - 7^2) + 5(y^2 + 2y + 1 - 1) + 137$$
$$= 3(x^2 + 14x + 7^2) + 5(y^2 + 2y + 1) - 3 \cdot 7^2 - 5 \cdot 1 + 137$$
$$= 3(x + 7)^2 + 5(y + 1)^2 - 15.$$

So the graph of $3x^2 + 42x + 5y^2 + 10y + 137 = 0$ is the same as that of $3(x + 7)^2 + 5(y + 1)^2 - 15 = 0$. After dividing both sides of $3(x + 7)^2 + 5(y + 1)^2 = 15$ by 15, we get $\frac{(x+7)^2}{5} + \frac{(y+1)^2}{3} = 1$, and hence $\frac{(x+7)^2}{(\sqrt{5})^2} + \frac{(y+1)^2}{(\sqrt{3})^2} = 1$. It follows that the graph of $3x^2 + 42x + 5y^2 + 10y + 137 = 0$ is the ellipse $\frac{x^2}{a^2} + \frac{x^2}{b^2} = 1$ of Figure 5.53 (with $a = \sqrt{5}$ and $b = \sqrt{3}$), translated so that its center at the origin $(0,0)$ is shifted to the point $(-7, -1)$. Its focal axis is parallel to the x-axis.

Let's now return to Alberti's tile floor and place a circle in the middle of it. Refer to Figure 5.56. How exactly should this circle be drawn in perspective on the canvas of Figure 5.49b? Since the center of the circle is $(0, h + 3)$ and its radius is 3, any point $P = (x_0, y_0)$ on the circle satisfies $x_0^2 + (y_0 - (h + 3))^2 = 3^2$. What does this tell us about the corresponding point $Q = (x_1, z_1)$ on the xz-plane of the canvas that Alberti's rule provides? Here is how we might try to proceed. If we could express x_0 and y_0 in terms of x_1 and z_1, then a substitution of these expressions into the equation of the circle would provide a connection between x_1 and z_1.

The expression we are looking for is the answer to a question about Figure 5.47. Given a point $Q = (x_1, 0, z_1)$ on the canvas, precisely what point $P = (x_0, y_0, 0)$ on the floor does it depict? To respond, let's find the line that $E = (0, -d, e)$ and $Q = (x_1, 0, z_1)$ determine, and then see what its point of intersection with

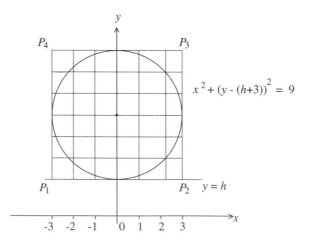

Figure 5.56

the xy-plane of the floor is. The differences between the coordinates of E and Q are $0 - x_1 = -x_1, -d - 0 = -d$, and $e - z_1$, so that by facts from the preceding section, a point (x, y, z) is on this line if $x = x_1 - tx_1$, $y = 0 - td$, and $z = z_1 + t(e - z_1)$ for some t. To get the intersection with the xy-plane, set $z = z_1 + t(e - z_1) = 0$. So $t = \frac{-z_1}{e - z_1}$, and it follows that $x_0 = x_1 + x_1 \frac{z_1}{e - z_1}$ and $y_0 = \frac{dz_1}{e - z_1}$. After simplifying, $x_0 = \frac{x_1(e - z_1) + x_1 z_1}{e - z_1} = \frac{x_1 e}{e - z_1} = \frac{ex_1}{e - z_1}$.

The substitution of $x_0 = \frac{ex_1}{e - z_1}$ and $y_0 = \frac{dz_1}{e - z_1}$ into the equation $x_0^2 + (y_0 - (h + 3))^2 = 3^2$ and some cumbersome algebra provides the steps below:

$$\left(\frac{ex_1}{e - z_1}\right)^2 + \left(\frac{dz_1}{e - z_1} - (h + 3)\right)^2 = 3^2$$

$$\frac{(ex_1)^2}{(e - z_1)^2} + \frac{(dz_1 - (h + 3)(e - z_1))^2}{(e - z_1)^2} = 3^2$$

$$e^2 x_1^2 + d^2 z_1^2 - 2dz_1(h + 3)(e - z_1) + (h + 3)^2(e - z_1)^2 = 3^2(e - z_1)^2$$

$$e^2 x_1^2 + d^2 z_1^2 - 2d(h + 3)(ez_1 - z_1^2) + (h + 3)^2(e - z_1)^2 - 3^2(e - z_1)^2 = 0$$

$$e^2 x_1^2 + d^2 z_1^2 - 2ed(h + 3)z_1 + 2d(h + 3)z_1^2 + (h^2 + 6h)(e - z_1)^2 = 0$$

$$e^2 x_1^2 + d^2 z_1^2 - 2ed(h + 3)z_1 + 2d(h + 3)z_1^2 + (h^2 + 6h)e^2 - 2(h^2 + 6h)ez_1 + (h^2 + 6h)z_1^2 = 0$$

$$e^2 x_1^2 + [d^2 + 2d(h + 3) + h^2 + 6h]z_1^2 - 2[ed(h + 3) + e(h^2 + 6h)]z_1 + (h^2 + 6h)e^2 = 0$$

$$e^2 x_1^2 + [h^2 + 2hd + d^2 + 6(h + d)]z_1^2 - 2e[d(h + 3) + (h^2 + 6h)]z_1 + (h^2 + 6h)e^2 = 0$$

$$e^2 x_1^2 + [(h + d)^2 + 6(h + d)]z_1^2 - 2e[d(h + 3) + (h^2 + 6h)]z_1 + (h^2 + 6h)e^2 = 0.$$

It's now been verified that if $P = (x_0, y_0)$ is a point on the circle $x^2 + (y - (h + 3))^2 = 3^2$, then the coordinates of the corresponding point $Q = (x_1, z_1)$ on the canvas satisfies the equation

$$e^2 x^2 + \big((h + d)^2 + 6(h + d)\big)z^2 - 2e\big(d(h + 3) + (h^2 + 6h)\big)z + (h^2 + 6h)e^2 = 0.$$

Because the circle on the floor is not seen as a point, a line, or a combination of two lines, this equation is not degenerate. Since $-4e^2((h + d)^2 + 6(h + d)) < 0$, it follows from Basic Fact 2 that it is the equation of an ellipse. So the perspective image of the circle on Alberti's floor is an ellipse. It is sketched in Figure 5.57 inside the trapezoid $Q_1 Q_2 Q_3 Q_4$ of Figure 5.49b.

We will now study the ellipse of Figure 5.57 and the trapezoid that surrounds it. An application of Alberti's rule to the points $P_1 = (-3, h)$ and $P_2 = (3, h)$ tells us that $Q_1 = (\frac{-3d}{d + h}, \frac{he}{d + h})$ and $Q_2 = (\frac{3d}{d + h}, \frac{he}{d + h})$. So the base of the trapezoid is $w = \frac{6d}{d + h}$. Applying Alberti's rule twice more, we get $Q_3 = (\frac{3d}{d + h + 6}, \frac{(h + 6)e}{d + h + 6})$ and $Q_4 = (\frac{-3d}{d + h + 6}, \frac{(h + 6)e}{d + h + 6})$. It follows that the length of the top side of the trapezoid is $u = \frac{6d}{d + h + 6}$ and that its height is

$$v = \frac{e(h + 6)}{d + h + 6} - \frac{eh}{d + h} = \frac{e(h + 6)(d + h) - eh(d + h + 6)}{(d + h + 6)(d + h)}$$

$$= \frac{e[h(d + h) + 6(d + h) - h(d + h) - 6h]}{(d + h + 6)(d + h)} = \frac{6ed}{(d + h)(d + h + 6)}.$$

Let's turn to the ellipse. Notice from Figure 5.57 that its semiminor axis is $b = \frac{1}{2}v = \frac{3ed}{(d + h)(d + h + 6)}$. Adding $\frac{1}{2}v$ to the z-coordinate $\frac{he}{d + h}$ of Q_1 provides the

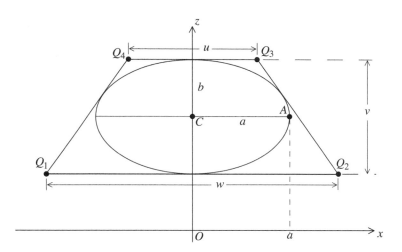

Figure 5.57

z-coordinate of the center C of the ellipse. After taking common denominators, we find that $C = (0, \frac{e[h(d+h+6)+3d]}{(d+h)(d+h+6)})$. A comparison of Figures 5.49a and 5.49b reveals that C is *not* the perspective image of the center of the circle. One last look at Figure 5.57 tells us that the semimajor axis a of the ellipse is the x-coordinate of the point A. So a can be found by plugging the z-coordinate of A (it is equal to the z-coordinate of C) into the equation of the ellipse and solving for x. This is an unpleasant computation that we will omit.

Example. Refer to Figure 5.47 and take $e = 6$, $d = 2$, and $h = 12$, all in feet. How large on the canvas are the trapezoid and the ellipse that respectively depict the 6 by 6 foot tile floor and the circle of radius 3 feet? The bottom edge of the trapezoid is $w = \frac{6 \cdot 2}{2+12} = \frac{6}{7}$ feet, or about 10.3 inches long. The top edge of the trapezoid is $u = \frac{6 \cdot 2}{2+12+6} = \frac{3}{5}$ feet, or 7.2 inches long. The height of the trapezoid is $v = \frac{6 \cdot 6 \cdot 2}{(2+12)(2+12+6)} = \frac{9}{35}$ feet, or about 3.1 inches. The semiminor axis of the ellipse is $b = \frac{1}{2} v = \frac{9}{70}$ feet, or about 1.5 inches. The equation of the ellipse derived earlier simplifies to

$$9x^2 + (7 \cdot 10) z^2 - 2 \cdot 3 (15 + 6 \cdot 18) z + 9 \cdot 12 \cdot 18 = 0.$$

To find the semimajor axis a, begin with the fact that the z-coordinate of the point A is equal to the z-coordinate $z = \frac{6(12 \cdot 20 + 3 \cdot 2)}{14 \cdot 20} = \frac{3(120+3)}{7 \cdot 10} = \frac{9 \cdot 41}{7 \cdot 10}$ of the center C. Setting $z = \frac{9 \cdot 41}{7 \cdot 10}$ in the equation and solving for x, we find, after an arithmetic slog, that $x = \pm \frac{3}{\sqrt{70}}$. It follows that $a = \frac{3}{\sqrt{70}} \approx 0.36$ feet, or about 4.3 inches. Because a and b determine the focal points of the ellipse (Figure 5.53 shows how) and $k = 2a$, the ellipse can now be drawn with precision.

While our mathematical analysis of perspective has focused on horizontal floors, vertical walls and ceilings can be studied in the same way. So we now understand why the circular window of the Chartres Cathedral pictured in Figure 3.24 is represented as an ellipse and why Panini's painting in Plate 21 depicts the circular arches on the side of the nave of St. Peter's as elliptical arcs.

Our story about the architecture of the Renaissance has reached its end. It has been a tale about some of the magnificent structures of Brunelleschi, Alberti, Bramante, Palladio, Michelangelo, della Porta, Bernini, and others. These structures rely on the architectural forms of classical Greece and Rome, but they reflect the confident, rational, and artistic spirit of this age. In the same way that Renaissance artists refined their paintings and relief sculptures by using the principles of perspective, Renaissance architects infused the design of their buildings with the proportion, balance, and transcending order derived from geometry and musical ratios.

Problems and Discussions

The first set of problems deals with mathematical matters that are related to themes developed in the text.

Sebastiano Serlio (1475–1554) was an important architectural theorist of the Italian Renaissance. His *Five Books of Architecture*, published in several installments from 1537 to 1551, explains geometry and perspective drawing, illustrates antique Roman buildings as well as Renaissance works of Bramante and Raphael, sets out the five orders of architecture (see Discussion 5.1), provides designs of centralized churches, contributes many designs of his own, and comments on construction practices and materials. Serlio's treatises influenced the architecture of his age to a much greater extent than his buildings. Arch and column combinations of the sort that Palladio used in the bays of the basilica of Vincenza (see Figure 5.14) were given emphasis by Serlio and are today referred to as *Serlianas*. Figure 5.58a is one of Serlio's designs for an entrance to a building.

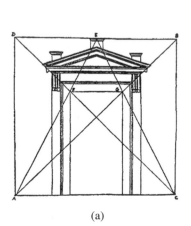

(a)

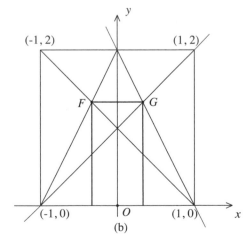

(b)

Figure 5.58

Problem 1. Put Serlio's figure into a coordinate plane as shown in Figure 5.58b. Determine the coordinates of the points F and G. Use this information to find the ratio of the height of the structure to the height of the door, as well as the ratio of the height of the door to its width. Are these proportions consistent with Alberti's principles relating proportion in architecture to harmony in music?

The Pythagoreans used three different averages, or means, of two positive numbers a and b: the arithmetic mean $\frac{a+b}{2}$, the geometric mean $\sqrt{ab}$, and the harmonic mean $\frac{2ab}{a+b}$. For $a = 1$ and $b = 2$, the three means are $\frac{3}{2}$, $\sqrt{2}$, and $\frac{4}{3}$, respectively. The first and last provide two of the consonant musical ratios of the earlier section "Alberti, Music, and Architecture."

Problem 2. Let a and b be positive numbers.

i. Let c be the harmonic mean of a and b. Verify that $\frac{a-c}{a} = \frac{c-b}{b}$.
ii. Let c and d be the arithmetic and harmonic means, respectively, of a and b. Show that the geometric mean of a and b is equal to the geometric mean of c and d.

In the designs of Palladio's buildings, the height h of a room is determined by its width w and length l. For rooms with flat ceilings (flat ceilings were almost entirely limited to upper stories), Palladio chose $h = w$. For the vaulted ceilings of rooms on ground floors, his rule was that h should be equal to one of the three Pythagorean means of w and l, or, in the case of a square room, to $\frac{4}{3}w = \frac{4}{3}l$. Palladio's treatise *I Quattro Libri* provides much evidence for this rule (especially for vaulted rooms, as the treatise provides ground floor plans only). Understandably, the ceilings of the rooms on the ground floor also needed to have the same heights. For this requirement and his rule to be in place simultaneously, Palladio had to coordinate the dimensions of his rooms carefully. In his efforts to achieve this coordination, Palladio tolerated exceptions and was satisfied with approximate solutions.

Problem 3. Recall from the earlier section "Palladio's Villas and Churches" that Palladio's plan for the ground floor of the Palazzo Chiericati specifies the dimensions of the large rooms to be 18 by 18 and 18 by 30 Venetian feet and those of the central hall to be 16 by 54 Venetian feet. All the heights of these vaulted rooms are 24 feet. Explain why this choice for the heights is consistent with Palladio's rule.

Let two positive numbers a and b be given and consider segments of lengths a and b. For each of the three Pythagorean means of a and b, it is possible to construct (with straightedge and compass) a segment that has that mean as its length. This fact allowed architects to lay such lengths out.

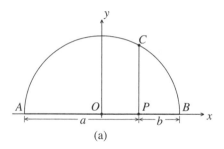
(a)

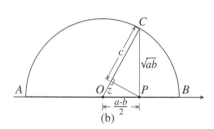
(b)

Figure 5.59

Figure 5.59 shows the essence of the constructions. How they are carried out is explained in Problem 4 below. (Its solution may profit from a review of Chapter 2, "Gods of Geometry.") One of a and b has to be greater than or equal to the other. Assuming that it is a, we take $a \geq b$. Figure 5.59a shows segments of lengths a and b and their respective endpoints A, P, and B.

Problem 4. Explain how to construct the midpoint O of the segment AB. Notice that the segment AO has length equal to the arithmetic mean $\frac{a+b}{2}$. Draw the semicircle with diameter AB and place an xy-coordinate system with origin at O as indicated in Figure 5.59a.

    i. Explain how to construct a segment CP perpendicular to AB. Verify that OP has length $\frac{a-b}{2}$. Use the standard equation of the circle (refer to Chapter 4, "The Coordinate Plane") to show that the length of CP is the geometric mean $\sqrt{ab}$.

    ii. Figure 5.59b shows the segment OC divided into segments of lengths c and z by a perpendicular through P. Use the Pythagorean Theorem twice to solve for $c^2 - z^2 = (c+z)(c-z)$ and then for $c-z$ and $2c$, to show that c is the harmonic mean $\frac{2ab}{a+b}$.

    iii. Explain each link of the inequality $b \leq \frac{2ab}{a+b} \leq \sqrt{ab} \leq \frac{a+b}{2} \leq a$.

Problem 5. The diagrams in Figure 5.60 are taken from Folio 455 of the Codex Atlanticus of Da Vinci's *Notebooks*. Let the radii of the semicircles be two units in length (so those of the smaller circles and semicircles are one unit in length) and determine the total area of the shaded region in each case.

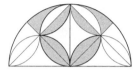

Figure 5.60

Problem 6. In Figure 5.61, C is the center of a circle of radius 1, AD is a diameter, and PB is perpendicular to the diameter. Show that $\alpha = \frac{1}{2}\gamma$. Now let $\gamma = 45°$. Compute the lengths of CB and PB. Use the Pythagorean Theorem to show that $AP = \sqrt{2 + \sqrt{2}}$. Conclude that $\sin 22.5° = \frac{\sqrt{2}}{2(\sqrt{2+\sqrt{2}})}$, $\cos 22.5° = \frac{1}{2}\sqrt{2+\sqrt{2}}$, and $\tan 22.5° = \frac{1}{1+\sqrt{2}}$.

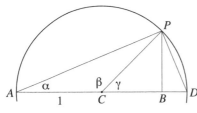

Figure 5.61

The next three problems study the horizontal and vertical sections of the dome of St. Peter's. The fact that the dome has 16 equally spaced ribs tells us that the regular 16-gon is relevant in understanding its geometry.

Problem 7. Figure 5.62a considers a circle of radius 1. Because $16 \times 22.5 = 360$, a regular 16-gon can be inscribed in it by marking off the angle $22.5°$ repeatedly. Let s be the length of the side of this 16-gon. Turn to Figure 5.62b and verify that $s^2 = AB^2 = AC^2 + BC^2 = \sin^2 22.5° + (1 - \cos 22.5°)^2 = 2 - 2 \cos 22.5°$. Use one of the results of Problem 6 to conclude that $s = AB = \sqrt{2 - \sqrt{2 + \sqrt{2}}} \approx 0.3902$. Check that the length of the arc AB is $\frac{\pi}{8} \approx 0.3927$.

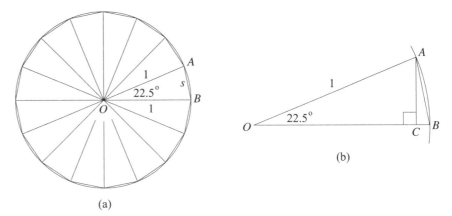

(b)

Figure 5.62   (a)

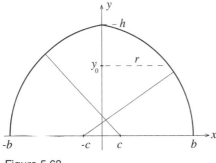

Figure 5.63

Problem 8. Figure 5.63 shows two circular arcs in an xy-coordinate plane. The centers of the two circles that determine them are both on the x-axis, a distance c from the origin. The arcs intersect the x-axis at b and $-b$, respectively. The two arcs determine a Gothic arch. By revolving this arch one revolution around the y-axis a pointed dome is obtained. Use the standard equation for one of the circles to derive the formula $h = \sqrt{b^2 + 2bc}$ for the height h of this dome. Show that the height-to-span ratio of the dome is $\frac{h}{2b} = \frac{1}{2}\sqrt{1 + \frac{2c}{b}}$. All the horizontal cross sections of the dome are circles. Determine the radius r of the circle y_0 units above the base of the dome in terms of b, c, and y_0.

Problem 9. Figure 5.64 shows a section of the dome of St. Peter's from the springing upward. It has the same proportions as the section of Figure 5.37 and confirms that the inner surface of the dome has the shape described in Problem 8 with $\frac{c}{b} \approx \frac{1}{6}$. Use one of the conclusions of Problem 8 to show that the height-to-span ratio of the inside of the dome of St. Peter's is approximately 0.58 and that the same ratio for a hemispherical dome is 0.5.

Discussion 5.1. About the Classical Orders. In architecture, the word *order* refers to one of several carefully proportioned arrangements of a

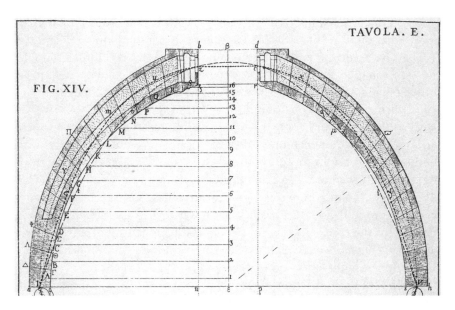

Figure 5.64. Detail from Table E in Giovanni Poleni, *Memorie istoriche della gran cupola el Tempio Vaticano e de'danni di essa, e de'ristoramenti loro, divise in libri cinque*, Padova, 1748. Marquand Library of Art and Architecture, Princeton University Library

column and the related elements. In *The Ten Books of Architecture*, Vitruvius comments on Greek texts (now lost) and singles out three orders: the Doric (the sturdiest, said to be based on the proportions of a man), Ionic (lighter in character to reflect the proportions of a woman), and Corinthian (the most slender and ornate, suggesting the form and proportions of a young woman). The Tuscan and Composite orders are of Roman origin. They modify and combine elements from the earlier orders of the Greeks. In keeping with the rational spirit of this age, the orders were expressed in terms of precise numerical ratios in the Renaissance. This was one of the purposes of Palladio's *I Quattro Libri*. Palladio's approach was influenced by the earlier treatise *Regola delle Cinque Ordini d'Architettura*, or *Canon of the Five Orders*, of Vignola (1507–1573). Vignola, an architect named after the Italian town from which he came, followed Michelangelo (and preceded della Porta) as capomaestro of St. Peter's. He contributed the two small cupolas that flank the dome. (See Figure 5.44.)

Vignola's treatise is an illustrated manual explaining how the orders are to be set out. For example, Vignola fixed the height of a circular column in terms of the diameter of the cross section D at its base as follows: Doric $8D$, Ionic $9D$, Corinthian $10D$, Tuscan $7D$, and Composite $10D$. The structural elements of a column include the base from which the column springs and the capital that tops it. The pedestal is the block on which the base rests and the entablature is the horizontal element that a column supports. Vignola set the proportional relationship between the heights of the pedestal, the column (including the base and the capital), and the entablature at 4 to 12 to 3. He derived these proportions from classical examples. For instance, his Doric order was

Figure 5.65. An Ionic column in the British Museum, London. Photo courtesy Wayne Boucher, © Cambridge 2000

based on the proportions of the Theater of Marcellus in Rome and his Corinthian order on those of Hadrian's Pantheon. (Refer to Figures 2.36 and 2.42.)

The characteristic and distinguishing features of the Ionic column are the scroll-shaped formations that decorate its capital. They are known as Ionic volutes. An example from the British Museum is depicted in Figure 5.65. All methods of constructing the spiraling curve of the Ionic volute (with straightedge and compass) build it as an organized sequence of circular arcs with different centers and radii. Vignola's treatise sets out one such method, based on a procedure of the Greeks and Romans. Before discussing it, we will turn to a construction of the Dutchman Nicolaus Goldmann (1611–1665). Goldmann wrote practical manuals for architects and taught mathematics and architecture in the Dutch town of Leyden. Goldmann's method for the construction of the Ionic volute appeared in the Amsterdam edition of Vitruvius in 1649. It was introduced to the English world of architecture in the eighteenth century.

Goldmann's construction of the volute begins with the eye, namely with the circle at the center. The construction is illustrated in Plates 22 and 23, and it is explained below. Start by drawing the eye as a circle with radius r and center C. The radius r as well as the location of C depend on the diameter of the column and will be determined later. Place the blue points 1 and 4 on the vertical axis so that both are a distance $\frac{r}{6}$ from C. Add the blue points 2 and 3 so that the four blue points form a square of side length $\frac{r}{3}$. (The numbers attached to the points as well as the color coding will become relevant shortly.) Place the green points 5 and 8 on the vertical axis so that each is a distance $\frac{r}{3}$ from C. Choose the green points 6 and 7 so that the four green points form a square of side length $\frac{2r}{3}$. Finally, place the red points 9 and 12 on the vertical axis at a distance of $\frac{r}{2}$ from C and complete them with the red points 10 and 11 to a square of side length r. An important part of the construction is now complete. The points 1 to 12 are the centers of circular arcs that will together comprise the volute. (While it will not be made explicit, this construction of the volute can be carried out with straightedge and compass.)

Now take a compass. Place the sharp point at 1 and stretch it up along the vertical axis to the point where the vertical axis and the circle intersect. From there, draw a circular arc of exactly one quarter-circle in the counterclockwise direction and stop at the point V_1. The color of the arc matches that of its blue center. Its radius is $R_1 = r + \frac{r}{6} = \frac{7r}{6}$. Now place the sharp point at 2, stretch the compass to V_1, draw an exact quarter-circle, again counterclockwise (and again in blue), and stop at V_2. Note that the radius of the second arc is $R_2 = R_1 + \frac{r}{3} = \frac{9r}{6}$. Next, put the sharp point at 3, stretch the compass to V_2, and go for an exact quarter-circle to V_3. The radius is $R_3 = R_2 + \frac{r}{3} = \frac{11r}{6}$. Plate 23 shows how this pattern continues. At each step, the color of the quarter-circle is the same as that of the center point used to draw it. The circular arc from V_3 to V_4 with center the point 4 has radius $R_4 = R_3 + \frac{r}{3} = \frac{13r}{6}$. This completes the blue section of the volute. The first green arc is centered at the point 5 and has radius $R_5 = R_4 + \frac{r}{2} = \frac{16r}{6}$. The second, third, and fourth green quarter-circles are centered at the points 6, 7, and 8. Notice that their radii increase by $\frac{2r}{3}$ at each step. In particular $R_8 = R_5 + 3 \cdot \frac{2r}{3} = \frac{28r}{6}$. The first red quarter-circle from V_8 to V_9 is centered at 9 and has radius $R_9 = R_8 + \frac{5r}{6} = \frac{33r}{6}$. The second, third, and fourth red quarter-circles are centered at 10, 11, and 12. Their radii increase by r at each step. Therefore, $R_{12} = R_9 + 3r = \frac{51r}{6}$. Having arrived at the point V_{12}, the volute is complete.

A final question remains. Precisely where should the center C of the volute be in relation to the column and what should the size of r be? The answers are determined by the diameter D of the Ionic column at its base. Horizontally, the center C needs to be at a distance of $\frac{1}{2}D$ from the central axis of the column. Vertically, the requirement is that C is $\frac{1}{4}D$ below the element that the volute supports. Because the distance from C to V_{12} is $\frac{r}{2} + R_{12}$, this means that

$$\frac{1}{4}D = \frac{r}{2} + R_{12} = \frac{r}{2} + \frac{51r}{6} = \frac{54r}{6} = 9r.$$

Therefore $r = \frac{1}{36}D$.

Goldmann's construction of the Ionic volute is best undertaken by starting at the eye and proceeding outward, and this is the approach taken here. It is also possible to go from the outside in, namely by starting with a quarter-circle from V_{12} to V_{11} centered at the point 12, and descending to the eye from there.

Problem 10. Follow the method of Goldmann just described to construct—carefully with straightedge and compass—an Ionic volute for a column that has diameter $D = 18$ inches at its base.

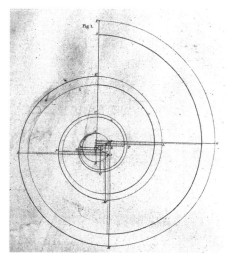

The volute of Figure 5.66 was constructed with Goldmann's method (as the diagram at the center of the figure confirms). A second, inner spiral runs inside and essentially parallel to the first. It is obtained in the same way as the first spiral by changing the centers of the circular arcs and their radii slightly in a systematic way.

A method of Vignola extends a Greek-Roman approach for constructing an Ionic volute. Much like Goldmann's volute, the construction spirals around the eye three times. The basic idea is the same as that of Goldmann's method. The only difference is the position of the centers for the circular arcs. The twelve centering points are shown in Figure 5.67. The points lie within the so-called Renaissance diamond and are labeled from 1 to 12 in the same way as those of Goldmann's method of Plate 22. The circular arc from A to B is centered at 1. The one from B to C is centered at 2. Their radii are shown as dashed lines. Following this pattern completes the volute. The inner spiral is constructed in the same way. The small markers in the figure near the points labeled 1 to 12 provide the positions of the centers for its twelve arcs. The first starts at A' stops at B'. Its radius is another dashed line. The volute in Figure 5.68 is attributed to Palladio. The Renaissance

Figure 5.66. From Sir William Chanders, *A Treatise on Civil Architecture: in which the Principles of the Art are laid down, and Illustrated by a great number of plates, accurately designed, and elegantly engraved by the best hands.* J. Dixwell, London, 1768 (2nd ed.), p. 25

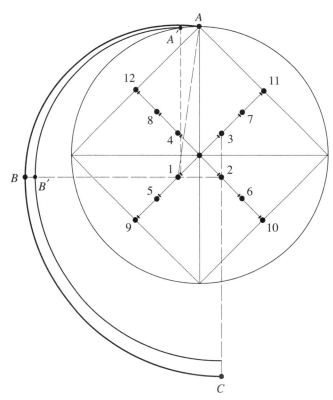

Figure 5.67. The Renaissance diamond

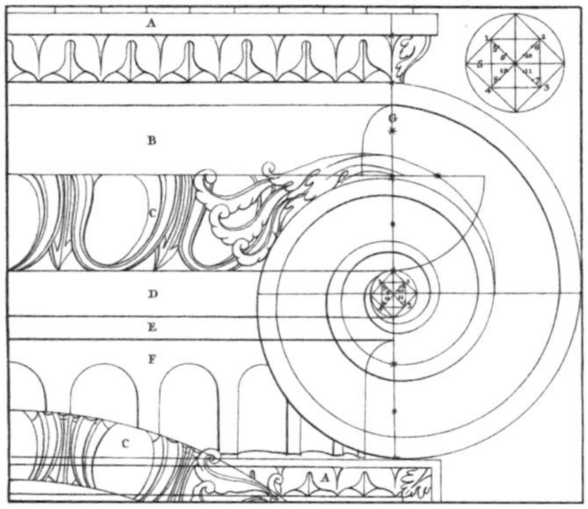

Figure 5.68. An Ionic volute of Palladio

diamond, visible both in the upper left corner and inside the eye of the scroll, confirms that he used the method just described.

Discussion 5.2. The Obelisk of Saint Peter's Square. An Egyptian obelisk stood near Old St. Peter's Basilica for centuries. No sooner had Sixtus V—the great builder among popes—ascended to the papal throne, he announced his intention to have the obelisk moved to the center of the square in front of the new St. Peter's. The obelisk, made of a single piece of red granite, is over 80 feet high and weighs over 700,000 pounds. Moving it to its new location 260 yards away would be a monumental effort of engineering.

The year was 1586. Five hundred mathematicians, engineers, and others came to present proposals about how best to move the obelisk. The strategy of the Roman architect Domenico Fontana won approval. (This is the same Domenico Fontana who would assist Giacomo della Porta with the construction of the dome of St. Peter's two years later.) Fontana had the obelisk encased in wooden planks strapped with iron bands. An elaborate system of scaffolding, ropes, windlasses, and pulleys was devised and put in place. Figure 5.69 shows some of these structures. A team of 900 laborers and 75 horses was on standby to supply muscle power. With a great throng of people watching, the obelisk was lifted off its stone base just high enough to allow a massive carriage (built

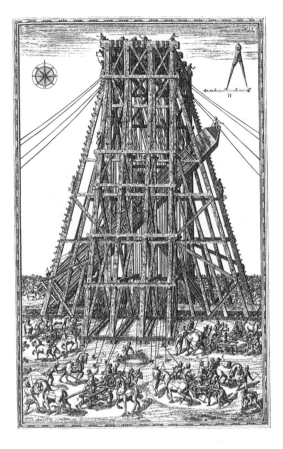

Figure 5.69. Natale Bonifacio, illustration for Domenico Fontana's manuscript *Della Transportatione dell' Obelisco Vaticano*, 1590. Marquand Library of Art and Archaeology, Princeton University Library

specifically for the purpose) to be placed under it. A week later the obelisk was lowered and placed onto the carriage horizontally. After the summer's heat passed, the carriage was pulled and rolled on a special track to the new location. Figure 5.70 depicts the scene. An even greater crowd gathered to watch the obelisk raised into position. The pope issued a strict order for silence, and the ropes, windlasses, pulleys, men, and horses sprang into action once more. When it seemed that one of the ropes was beginning to fail under the strain the shout "water on the ropes!" rang out. This instruction, issued by a sailor who knew his ropes, apparently saved the day. The obelisk was lowered into place successfully. The disobedience was forgiven.

The feat of moving the obelisk 260 yards from the one position to the other was hailed at the time as a triumph of technology. This is an astonishing assessment when one considers the history of the obelisk: It was quarried, moved, and raised for an Egyptian pharaoh before 1000 B.C. The Roman emperor Augustus had it transported to Alexandria, Egypt, between 30 and 20 B.C. A few decades later, the obelisk was shipped across the Mediterranean and taken to Rome. This is amazing when one considers that all that was available to cut such huge obelisks from a single stone and to move them were

Figure 5.70. Natale Bonifacio, illustration for Domenico Fontana's manuscript *Della Transportatione dell' Obelisco Vaticano*, 1590. Marquand Library of Art and Archaeology, Princeton University Library

tools, such as spikes, hammers, ramps, levers, and ropes, powered by men and animals. Obelisks were probably quarried from granite deposits by inserting wooden wedges that, soaked with water, would expand to split the granite into blocks. Egyptian relief sculptures tell us that obelisks were transported on great barges. More than likely, they were placed on their foundations by hauling them up earthen mounds, tilting, and sliding them into position.

Next is a set of problems dealing with perspective. In the context of Alberti's floor in the "Brunelleschi and Perspective" section, we encountered sets of parallel lines that converge to a so-called vanishing point when drawn in perspective. (Lines that are horizontal from the vantage point of the artist are the exception. There is no point of convergence for them.) Sets of parallel lines often define the basic form of a scene or object that we look out on. It follows that there are often several vanishing points in a picture. Accordingly, one speaks of one-, two-, or three-point perspective. The depiction of Alberti's floor in Figure 5.49b with its focus on vertical and diagonal lines is an example of a drawing in two-point perspective.

Problem 11. Study Raphael's *School of Athens* in Plate 20 and Panini's *Interior of St. Peter's* in Plate 21. Identify the location of the vanishing point or points in each.

Figure 5.71. Piero della Francesca (attributed), *Perspective View of an Ideal City*, c. 1470. Galleria Nazionale delle Marche, Urbino

Problem 12. The famous *Perspective View of an Ideal City* of Figure 5.71 is often attributed to Piero della Francesca (from about 1420 to 1492). Study it and locate the vanishing point. Della Francesca was one of the great artists of the early Italian Renaissance. He probably studied the buildings of Filippo Brunelleschi and the paintings of Masaccio in Florence. More than likely he was familiar with Leon Battista Alberti's writings on painting and perspective. He was also one of the leading mathematicians of his time and wrote books that developed arithmetic, algebra, and geometry, as well as his own insights about solid geometry and perspective. Some of his works later found their way into other treatises, notably Luca Pacioli's *De Divina Proportione*.

Problem 13. Albrecht Dürer (1471–1528), a renowned German painter and graphic artist and a contemporary of Leonardo da Vinci, incorporated mathematical considerations in his designs. Figure 5.72 shows Dürer's woodcut *Man Drawing a Lute* from 1525. Write a paragraph that relates the scene that Dürer depicts to Figure 5.47.

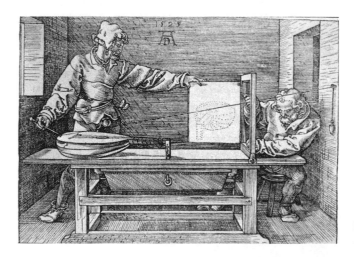

Figure 5.72. Albrecht Dürer, *Man Drawing a Lute*, 1525. Metropolitan Museum of Art

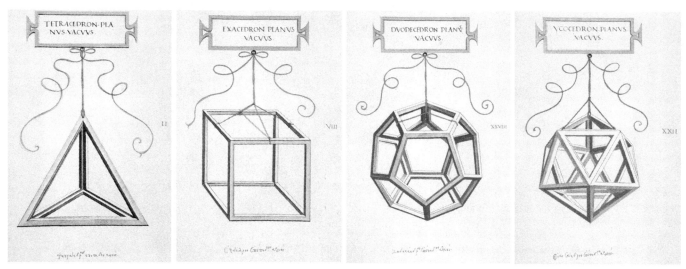

Figure 5.73. Leonardo da Vinci woodcuts. From Luca Pacioli, *Divina Proportione*, Venice, 1509, Biblioteca Ambrosiana, Milan, ms 170 sup, plates II, VIII, XXVIII, and XXII. Marquand Library of Art and Archaeology, Princeton University Library

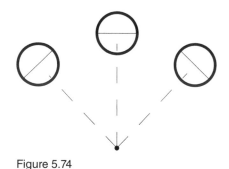

Figure 5.74

Problem 14. Study the perspective drawings in Figure 5.73 of the tetrahedron (4 sides), the cube (6 sides), the dodecahedron (12 sides), and the icosahedron (20 sides). (The octahedron is omitted.) They are attributed to Leonardo da Vinci. Whoever the artist, there seems little doubt that he made a model of each of these Platonic solids and used it as a basis for his drawing.

Problem 15. Think of the three identical black circles with their diameters that are shown in Figure 5.74 to be painted on a horizontal floor. The black dot represents the eye of an observer who stands on the same floor and looks out on the three circles. The three circles are at the same distance from the observer. Draw what the observer of the three circles and their diameters would see.

A set of problems follows that explores aspects of coordinate geometry (in two and three dimensions) presented in the section "Brunelleschi and Perspective."

Problem 16. The two points $P_1 = (3,5)$ and $P_2 = (-2,7)$ determine a line L. Take P_1 first and then P_2, and find two different pairs of parametric equations for L by using the difference $P_1 - P_2$ to compute the coefficients of t. Repeat this twice more to determine four different pairs parametric equations for L.

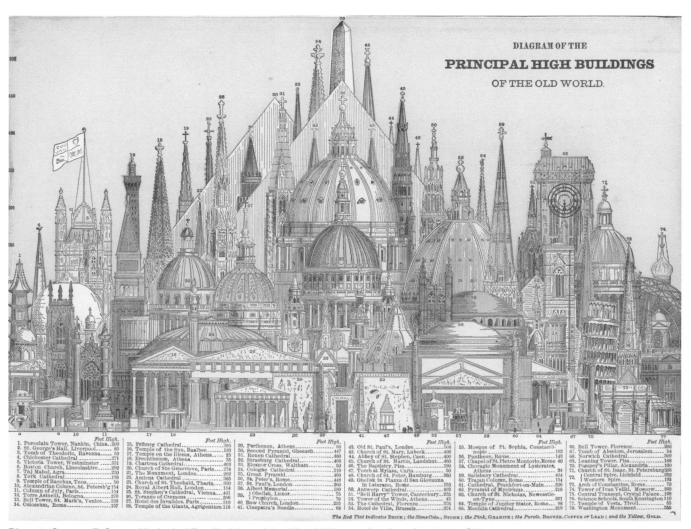

DIAGRAM OF THE
PRINCIPAL HIGH BUILDINGS
OF THE OLD WORLD.

	Feet High.		Feet High.		Feet High.		Feet High.		Feet High.		Feet High.
1. Porcelain Tower, Nankin, China..	200	15. Friburg Cathedral..	385	29. Parthenon, Athens..	66	42. Old St. Paul's, London..	508	55. Mosque of St. Sophia, Constantinople..	182	66. Bell Tower, Florence..	260
2. St. George's Hall, Liverpool..	85	16. Temple of the Sun, Baalbec..	120	30. Second Pyramid, Gheezeh..	447	43. Church of St. Mary, Lubeck..	400	56. Pantheon, Rome..	143	67. Tomb of Absolom, Jerusalem..	54
3. Tomb of Theodoric, Ravenna..	50	17. Temple on the Ilisus, Athens..	25	31. Rouen Cathedral..	460	44. Abbey of St. Stephen, Caen..	400	57. Chapel of St. Pietro Montorio, Rome	40	68. Norwich Cathedral..	309
4. Chichester Cathedral..	271	18. Erechtheium, Athens..	35	32. Strasburg Cathedral..	468	45. Church of St. Martin, Landshut..	460	58. Choragic Monument of Lysicrates,		69. Leaning Tower, Pisa..	188
5. Victoria Tower, Westminster..	331	19. Chartrea Cathedral..	403	33. Eleanor Cross, Waltham..	50	46. The Baptistry, Pisa..	190	Athens..	34	70. Pompey's Pillar, Alexandria..	100
6. Boston Church, Lincolnshire..	292	20. Church of Ste Genevieve, Paris..	274	34. Cologne Cathedral..	510	47. Tomb at Mylasa, Caria..	50	59. Salisbury Cathedral..	404	71. Church of St. Isaac, St. Petersburg	336
7. Taj Mahal, Agra..	220	21. The Monument, London..	202	35. Great Pyramid..	460	48. Church of St. Peter, Hamburg..	380	60. Trajan Column, Rome..	134	72. { Central Spire, Lichfield..	252
8. York Cathedral..	198	22. Amiens Cathedral..	383	36. St. Peter's, Rome..	448	49. Obelisk in Piazza di San Giovanna		61. Cathedral, Frankfort-on-Main..	326	{ Western Spire..	192
9. Temple of Bacchus, Teos..	50	23. Church of St. Theobald, Tharin..	320	37. St. Paul's, London..	360	in Laterano, Rome..	153	62. Pyramid of Mycerinus..	218	73. Arch of Constantine, Rome..	70
10. Alexandrian Column, St. Petersb'g	154	24. Royal Albert Hall, London..	154	38. Albert Memorial..	180	50. Antwerp Cathedral..	403	63. Church of St. Nicholas, Newcastle-		74. Tower of Ivan Veliki, Moscow..	260
11. Column of July, Paris..	154	25. St. Stephen's Cathedral, Vienna..	441	39. { Obelisk, Luxor..	75	51. "Bell Harry" Tower, Canterbury..	235	on-Tyne..	201	75. Central Transept, Crystal Palace..	198
12. Torre Asinelli, Bologna..	270	26. Torazzo of Cremona..	396	{ Propylon..	79	52. Tower of the Winds, Athens..	45	64. Temple of Jupiter Stator, Rome..	98	76. Science Schools, South Kensington	110
13. Bell Tower, St. Mark's, Venice..	323	27. Hotel des Invalides, Paris..	310	40. Bow Church, London..	235	53. The Cathedral, Florence..	376	65. Mechlin Cathedral..	319	77. Temple of Vesta, Tivoli..	55
14. Coloseum, Rome..	157	28. Temple of the Giants, Agrigentum	116	41. Cleopatra's Needle..	68	54. Hotel de Ville, Brussels..	374			78. Washington Monument..	555

The Red Tint indicates BRICK ; *the Stone Colo.,* STONE ; *the Pink,* GRANITE ; *the Purple,* BRONZE, COPPER *or* LEAD ; *and the Yellow,* GOLD.

Plate 1. George F. Cram, *Unrivaled Family Atlas of the World*. Lithograph color print, 1884, Chicago.
Note that the explanation of the color scheme at the bottom of the page lacks accuracy.

Plate 2. Prehistoric drawing on the walls of the Chauvet Cave, southern France. Photo by HTO

Plate 3. Hypostyle Hall of the Great Temple of Amun, Karnak, Egypt. Lithograph by Louis Haghe, 1842–1849, from a painting by David Roberts, 1838–1839

Plate 4. The Acropolis of Athens. Photo by ccarlstead.

Plate 5. Giovanni Paolo Panini (Italian, 1691–1765), *Interior of the Pantheon, Rome*, c. 1747. Oil on canvas, 127.0 x 97.8 cm. The Cleveland Museum of Art. Purchase from the J. H. Wade Fund 1974.39

Plate 6. A thirteenth century mosaic of Christ the Savior from the Hagia Sophia. PavleMarjanovic/Shutterstock, © Shutterstock

Plate 7. Gaspare Fossati, *The Hagia Sophia as Mosque*. From *Aya Sofia, Constantinople, as recently restored by order of H.M. the sultan Abdul Medjid, from the original drawings by Chevalier Gaspard Fossati*, lithograph by Louis Haghe, P. & D. Colnaghi & Co., London, 1852. The Marquand Library of Art and Archaeology, Princeton University Library, Albert M. Friend '15 Bequest

Plate 8. Rows of double arches define the vast prayer hall of the Great Mosque of Córdoba. Photo by Timor Espallargas

Plate 9. Three leafed arches and a horseshoe arch of the mihrab of the Great Mosque of Córdoba

Plate 10. The Giralda in Seville. Photo by GrahamColmTalk

Plate 11. The Romanesque Cathedral of Speyer, Germany. Photo courtesy of Karl Hoffmann

Plate 12. The Basilica Sainte Madeleine with its nave, bays, groin vaults, clerestory windows, and apse, in Vézelay, France. Photo by Vassil

Plate 13. The Cathedral of Notre Dame, Chartres. View from the south

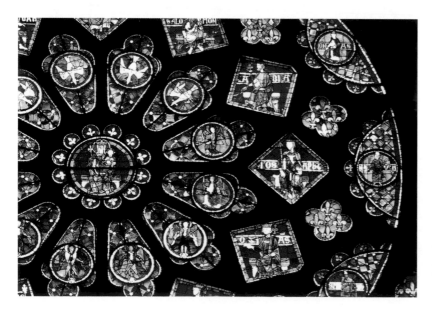

Plate 14. Detail of the rose window on the north side of the Cathedral of Notre Dame, Chartres. The illuminated interior of the rose window depicted in Figure 3.24. It is dedicated to the Virgin and has a diameter of 10.5 meters. Photo by MOSSOT

Plate 15. God as Geometer. Frontispiece of Bible Moralisée Vienna, Codex Vindobo-
nensis 2554 (French, c. 1250), Österreichische Nationalbibliothek, Vienna. Marquand
Library of Art and Archaeology, Princeton University Library

Plate 16. Facade and domes of San Marco, Venice. Photo by Andreas Volkmer

Plate 17. Geometric designs from the Alhambra palace. Far left and top right: photo by Jebulon. Top middle: photo by Jebulon, © GFD and Creative Commons Attribution. Bottom right: photo by R. S. Tan. Bottom middle: photo by Dharvey, © GFD and Creative Commons Attribution

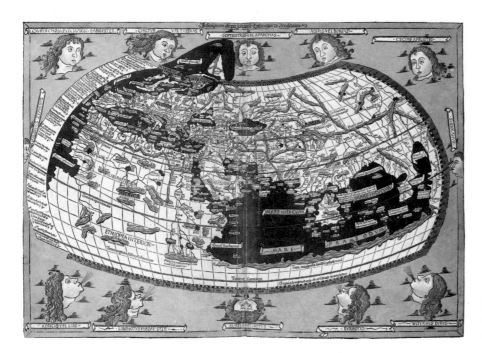

Plate 18. Ptolemy's world map from *Geographike*, Lienhart Holle, Ulm, 1482, engraved by Johannes Schnitzer

Plate 19. The Cathedral of Florence with its dome and its nineteenth-century marble exterior. Photo by Benjamin Sattin

Plate 20. Raphael's *School of Athens,* 1508–1511, Vatican

Plate 21. Giovanni Paolo Panini, *Interior of St. Peter's, Rome*, 1731. Saint Louis Art Museum

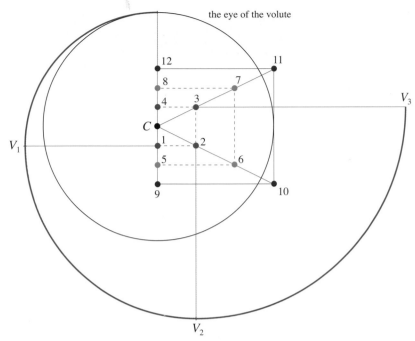

the eye of the volute

Plate 22. Goldmann's construction of the Ionic volute, starting at the eye

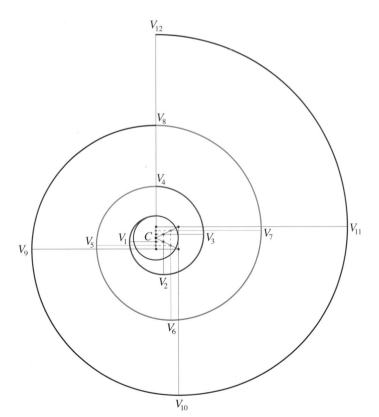

Plate 23. Goldmann's construction of the volute, completing the spiral

Plate 24. Shell and podium structure of the Sydney Opera complex. Photo by Matthew Field

Problem 17. A line in the xy-plane is given by the parametric equations $x = -3 + 5t$ and $y = 4 + 2t$. Check that the points $(-3, 4)$ and $(7, 8)$ are on the line. Show that the ratio $\frac{2}{5}$ of the coefficients of t is equal to the slope of the line. Are the points $(0, 5.2)$ and $(5, 7.2)$ on the line? What about $(1, 5.5)$?

Problem 18. Find parametric equations of the line through $(1, 2)$ and $(-3, -4)$ and of the line through $(2, 1)$ and $(3, 6)$. Use them to find the point at which the two lines intersect.

Problem 19. Consider the line L determined by the two points $P_1 = (1, 2, 3)$ and $P_2 = (4, 6, 8)$. Write down a set of parametric equation for L.

  i. Consider the plane $x - 2y - 3z = 4$ and determine the point of intersection of L and this plane.
 ii. Consider the sphere $x^2 + y^2 + z^2 = r^2$. For which r does L pass through the sphere?

Problem 20. The intersection of any two planes that are not parallel is a line. Determine the parametric equations for the line that is the intersection of the planes $3x - 4y + 5z = 2$ and $2x + y - 3z = 4$.

Problem 21. Find equations for two planes that both contain the line given by $x = 2 + 3t$, $y = -1 + 2t$, $z = 1 + t$. [Hint: $\frac{2}{3}(x - 2) - \frac{1}{2}(y + 1) - (z - 1) = 0$.]

Problem 22. Repeat the perspective drawing of the tile floor of Figure 5.49b by taking a different e and d.

Problem 23. Modify an argument in the section "Brunelleschi and Perspective" to show that any line with nonzero slope m in the xy-plane of the floor converges to the point $(\frac{d}{m}, e)$ on the horizon line when represented on the xz-plane of the canvas.

Discussion 5.3. More about Perspective. The explanation of perspective drawing in "Brunelleschi and Perspective" within an xyz-coordinate system can be adapted to apply to any scene or object and not just Alberti's horizontal floor. However, the point P in Figure 5.47 must be taken more generally as $P = (x_0, y_0, z_0)$ so that it can also represent locations above and below the floor (and not just on it). Make this change in the problem that follows, but keep everything else in Figure 5.47 the same.

Problem 24. After a review of lines and their parametric equations, write down parametric equations for the line determined by $E = (0, -d, e)$ and $P = (x_0, y_0, z_0)$. Then show that the point of intersection Q of this line with the plane of the canvas is given by $Q = (x_1, 0, z_1)$ where $x_1 = \frac{dx_0}{d + y_0}$ and $z_1 = \frac{dz_0 + ey_0}{d + y_0}$.

In Problems 25 to 29 the values for e, d, and h are taken to be $e = 8$, $d = 2$, and $h = 22$ feet. The unit of length for the xy-plane of the floor in Figure 5.47 is the foot and the tiles of Alberti's floor are understood to be 1 foot squares. For the xz-plane of the canvas, the unit of length is the inch.

Problem 25. Turn to Figure 5.49. Show that the points Q_1, Q_2, Q_3, Q_4, and V are given by $Q_1 = (-3, 88)$, $Q_2 = (3, 88)$, $Q_3 = (2\frac{2}{5}, 89\frac{3}{5})$, $Q_4 = (-2\frac{2}{5}, 89\frac{3}{5})$, and $V = (0, 96)$ (all coordinates in inches).

Figure 5.75a depicts Alberti's 6 by 6 arrangement of tiles with 1 unit representing 1 foot. Since $h = 22$, the leading edge of the floor is 22 feet from the x-axis. Figure 5.75b represents the floor drawn in perspective on the canvas with 1 unit representing 1 inch. It makes use of the conclusions of Problem 25. The vertical dashed segments in Figures 5.75a and 5.75b indicate that the placement of the two diagrams above the x-axis is not to scale. The solution of each of the Problems 26 to 30 requires a copy of Figure 5.75. When solving these problems, don't forget to convert from feet to inches before transferring information from the situation of the floor of Figure 5.75a to that of the canvas in Figure 5.75b.

Problem 26. Consider the point $P = (-2, 25)$ on the plane of the floor in Figure 5.75a. Place the perspective image Q of the point P into Figure 5.75b. Next consider the two lines $x = 2$ and $y = x + 24$ in the plane of the floor. Draw them into Figure 5.75a. Place the perspective images of the two lines into Figure 5.75b. Where do these two lines meet the horizon line $z = 96$?

Problem 27. Consider the line $y = 8x + 22$ in the plane of the floor in Figure 5.75a. Where does its perspective image cross the line $z = 88$ in Figure 5.75b? Use the result of Problem 23 to draw the perspective image of the line into Figure 5.75b.

Figure 5.75. (a) Alberti's floor at a scale of 1 unit = 1 foot; (b) Alberti's floor drawn in perspective on the canvas at a scale of 1 unit = 1 inch

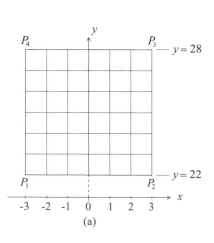

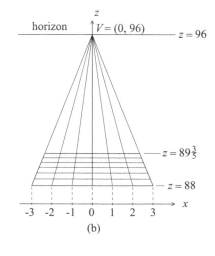

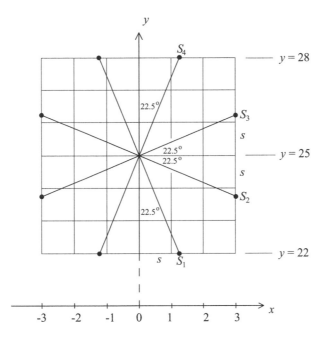

Figure 5.76. An octagon on Alberti's floor at a scale of 1 unit = 1 foot

Problem 28. Figure 5.76 depicts Alberti's arrangement of tiles of Figure 5.75a along with four line segments. Each of the line segments goes through the center of the arrangement and makes an angle of 22.5° with either the x- or y-axis. It follows by similar triangles that the eight endpoints of the segments determine a regular octagon.

    i. Use the conclusion $\tan 22.5° = \frac{1}{1+\sqrt{2}}$ from Problem 6 to show that the distance s in the figure is $s = \frac{3}{1+\sqrt{2}} \approx 1.24264$.

    ii. Determine the coordinates of the points S_1, S_2, S_3, and S_4 in terms of s. Place these four points and then the entire octagon into Figure 5.75a using the approximation $s \approx 1\frac{1}{4}$.

    iii. Let T_1 T_2, and T_3 be the images of the points S_1, S_2, and S_3 on the canvas of Figure 5.75b. Use $\frac{3}{1+\sqrt{2}} \approx 1.24$ to show that $T_1 \approx (1.24, 88)$ and that the z coordinates of T_2 and T_3 are approximately 88.55 and 89.20 respectively. Place the points T_1, T_2, and T_3 carefully into Figure 5.75b.

    iv. Place the image T_4 of S_4 into Figure 5.75b by making use of the vanishing point. Then complete the points T_1, T_2, T_3, and T_4 to a perspective drawing of the regular octagon.

Problem 29. Return to the 6 by 6 tile floor of Figure 5.47 and extend it upward to a 6 by 6 by 6 cube. Refer to Figure 5.75a and the points P_1, P_2, P_3, P_4 and designate by P_5, P_6, P_7, P_8 the upper corners of the cube that lie respectively above them. Find the coordinates of these points. Then find parametric equations for the lines that the two edges P_5P_8 and P_6P_7 determine. Show that when the perspective images of these edges on the canvas are extended, they meet at the vanishing point V. [Hint: Start by applying the conclusion of Problem 24 to typical points on each of the two lines.]

Problem 30. Use the conclusions of Problem 29 to make a perspective drawing of the cube. Start with the perspective drawing of the base as given in Figure 5.75b and draw in the four upper vertices to complete it.

We turn next to study examples of quadratic equations and conic sections and then consider problems involving perspective.

Problem 31. The graphs of the equations $y = \frac{1}{2}x^2$, $3x^2 + 4y^2 = 6$, and $3x^2 - 4y^2 = 12$, are, respectively, a parabola, an ellipse, and a hyperbola. Study Figures 5.52, 5.53, and 5.54, and use this information to determine the locations of the focus and the directrix of the parabola, the semiminor and semimajor axes of the ellipse, and the equations of the two lines that guide the shape of the hyperbola. Sketch the three graphs.

Problem 32. Draw the graphs of the equations $\frac{x^2}{6^2} + \frac{y^2}{4^2} = 1$ and $\frac{(x-2)^2}{6^2} + \frac{(y-4)^2}{4^2} = 1$. [Hint: Both graphs are ellipses. First draw the boxes that determine them.]

Problem 33. By factoring the left side of the equation $x^2 + 4xy + 4y^2 + 6x + 12y + 9 = 0$ in two steps show that the equation is degenerate and that its graph is a line.

Problem 34. The graph of the equation $x^2 + 4y^2 - 6x + 8y + 9 = 0$ is a conic section. Apply the $B^2 - 4AC$ criterion to show that it is an ellipse. By completing squares, rewrite the equation in such a way that you can identify the center of the ellipse as well as its semimajor and semiminor axes. Sketch the graph of the ellipse.

Problem 35. Let $Ax^2 + Bxy + Cy^2 + Dx + Ey + D = 0$ be the equation of a conic section. It is a fact that under a translation or a rotation of the conic section, the equation of the conic section changes but the term $B^2 - 4AC$ remains the same. On the other hand, two very different-looking conic sections can have the same $B^2 - 4AC$. Why do parabolas provide examples of this? Turning to ellipses, write down the equation of a very flat ellipse that has the same $B^2 - 4AC$ as the circle $x^2 + y^2 - 1 = 0$.

Problem 36. The hyperbolas $x^2 - y^2 = 1$ and $\frac{x^2}{2} - \frac{y^2}{2} = 1$ have the same focal axis and are shaped by the same pair of intersecting lines. Determine the pair of lines and sketch the graphs of the two hyperbolas in the same coordinate plane.

Problem 37. Consider the graph of the equation $xy - 1 = 0$, or, equivalently, $y = \frac{1}{x}$. Why does the theory of the preceding section imply that its graph is a hyperbola? By considering the graph of $y = \frac{1}{x}$, determine the pair of intersecting lines that shape this hyperbola. Determine the focal axis of the hyperbola. This hyperbola can be transformed into one of the two hyperbolas of Problem 36 by a rotation around the origin. By how many degrees (and in what direction) does it have to be rotated and which of the two hyperbolas of Problem 36 is the result?

Return to Alberti's tile floor and its image on the canvas as both are depicted in Figure 5.75. The solutions of Problems 38 and 39 require familiarity with Basic Facts 2 and 3 as well as the derivation of the equation of the ellipse carried out in the preceding section. Consider the nine circles on Alberti's floor shown in Figure 5.77a. Figure 5.77b shows their images in perspective on the canvas.

Problem 38. Pick one of the three circles depicted in Figure 5.77a that has its center on the y-axis and write down an equation for this circle.

   i. Determine an equation for the ellipse of Figure 5.77b that is the perspective image of the circle you picked. Do this for a general e, d, and h, rather than the specific values $e = 8$, $d = 2$, and $h = 22$. Express the equation in the form $Ax^2 + Bxz + Cz^2 + Dx + Ez + F = 0$.

   ii. Show that $B^2 - 4AC = -4e^2((d + y_{\text{cen}})^2 - 1)$, where $(0, y_{\text{cen}})$ is the center of the circle.

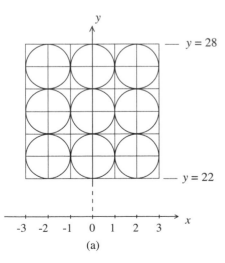

(a)

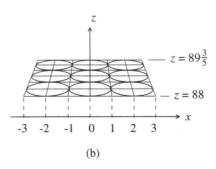

(b)

Figure 5.77. (a) Circles of radius 1 on Alberti's floor at a scale of 1 unit = 1 foot; (b) circles drawn in perspective as ellipses at a scale of 1 unit = 1 inch

iii. In the situation of Figure 5.77, $d = 2$ and $y_{cen} = 22 + 3 = 25$, so that $B^2 - 4AC < 0$ and the graph of the equation of (i) is an ellipse. Would this be so for *every* combination of d and h? Discuss the various possibilities.

Problem 39. Pick one of the circles in Figure 5.77a that lies on the left or right of center and write down an equation for it.

i. Determine an equation for the ellipse of Figure 5.77b that is the perspective image of this circle. Do this for a general e, d, and h, rather than the specific values $e = 8$, $d = 2$, and $h = 22$. Express the equation in the form $Ax^2 + Bxz + Cz^2 + Dx + Ez + F = 0$.

ii. If you answered Problem 15 correctly, you'd expect that this ellipse is rotated so that its focal axis is *not* horizontal. Does your answer to (i) confirm this?

The drawing of Figure 4.48, especially the depiction of the cylinder, made use of insights gained from Problem 39.

6

A New Architecture:
Materials, Structural Analysis, Computers, and Design

An Industrial Revolution begins in the middle of the eighteenth century, develops quickly in the nineteenth, and ushers in a new age. The machines that the steam engine and the utility of iron make possible dramatically change the scale and range of human activity. Powerful steamships and locomotives are able to transport large quantities of cargo and people. Machine-powered foundries, mills, and other factories employing large numbers of workers produce goods made of metals, fabrics, artificial fibers, and plastics economically and in great quantities. The hoists, lifts, and cranes deployed in building projects and the vehicles that bring the materials to the site are no longer driven by humans and pack animals, but by engines powered by oil-based fuels and electricity. The larger scale of production, construction, and transportation require an infrastructure of roads, railways, bridges, buildings, and other facilities. This fuels the use of cast iron, wrought iron, reinforced masonry, and reinforced concrete, and the need for experts on materials, structures, and mechanics. Each of these fields of expertise develops into its own discipline of engineering. Building on the science of force and motion and the mathematics developed by Archimedes, Galileo, and Newton, these disciplines study loads, thrusts, stresses, and displacements in building materials and structural elements. And they subject these materials and elements to experiment and testing. The engineer is the new professional who develops the infrastructure for the new age.

The Europe of the eighteenth and nineteenth centuries is dominated by a number of monarchies, Austria, England, France, and Russia among them. The architecture of this time continues to be influenced by the tastes and demands of the ruling order and is to a large extent an architecture of churches, palaces, mansions, government buildings, banks, libraries, museums, and opera houses. For the most part, it is an architecture that continues along the classical arc. It develops Baroque, a classical style embellished with rich overlays of decorative elements, and relies on revivals of Gothic, Renaissance, and Palladian traditions. The new materials of construction and disciplines of engineering do have an impact, but it is an impact that remains largely behind the facades. Both become visible with the construction of the Eiffel Tower in Paris in 1887–1889. This elegant symbol of the Industrial

Age is made of curving iron trusses in a design that is shaped by structural engineering. Nationalist aspirations lead to democratic movements and wars of independence, both successful and not. These currents begin to impact art and architecture in the last years of the nineteenth century. The split of a group of artists and architects from the artistic establishment of Vienna under the slogan "for every age its art, and for every art its freedom" is symbolic of this spirit. At the same time, from Chicago on the other side of the Atlantic comes the emphatic dictum that in architecture "form should follow function." The combination of the spirit of free architectural expression and the demands of function on a structure have critically influenced—often in tension—architectural design ever since.

The rule imposed by the empires persists until 1914 when the conflicts between them erupt into a world war. The use of the machinery and technology of the industrial age makes it a war of unparalleled destruction and devastation. This inferno brings an end to Europe's old imperial structure. A period of instability follows as newly created countries and new democratic institutions attempt to establish themselves. Several influential architects of the first part of the twentieth century—Frank Lloyd Wright, Le Corbusier (born Charles Édouard Jeanneret, he took a version of his grandfather's name Le Corbésier), Walter Gropius, and Ludwig Mies van der Rohe among them—pursue architecture as an expression of a new order. It should aim to bring the living and working spaces of people into line with the aesthetics, technology, and culture of the times. Architecture is "not the container, but the space within" and a house is "a machine for living in." Rectangular buildings supported by steel skeletons, clad by exterior glass, incorporating cantilevered concrete slabs, and featuring flat roofs come to represent a new International Style. The new building materials and the invention of elevators make it possible to build living and working spaces in skyscrapers at heights that far exceed those of the tallest Gothic structures. Computers, especially high-power computers, and their interfaces with design and manufacture accelerate the developments just described. Architects of the second half of the twentieth century and the beginning of the twenty-first—including Eero Saarinen, Jørn Utzon, Frank Gehry, Santiago Calatrava, and Norman Foster—take advantage of the new materials and technologies to create completely novel architectural forms.

The purpose of this chapter is to tell the story of the new materials (reinforced masonry, cast iron, reinforced and prestressed concrete, and finally titanium) and the emerging discipline of structural engineering (the testing of materials, mathematical analysis, and modern computer technology). In so doing, it aims to illustrate the expanded scope of architectural design and the range of the structures that could be built. It goes without saying that these are complex issues and that this chapter can only pull on a few of the important threads.

The first section explores some of the great domed buildings of the eighteenth and nineteenth centuries. The buildings are St. Paul's Cathedral in London, the Panthéon in Paris, St. Isaac's Cathedral in Saint Petersburg, and the United States Capitol Building in Washington. The exteriors of their domes are very similar and strongly influenced by the designs of Bramante's Tempietto and Michelangelo's St. Peter's. But the supporting elements underneath the domes tell an evolving story of the use of iron-reinforced masonry and cast iron and progress in structural engineering.

The second and third sections take a detailed look at important contributions to structural engineering made by English, Italian, and French engineers and scientists in the eighteenth century. This is an account of basic applied geometry and physics. In an effort to determine when an arch or a dome is structurally sound, it investigates lines of thrust and bending moments, and it studies the impact of tension, shear, and friction. The conclusions influenced the design of the domes mentioned above and enabled the evaluation of their stability.

The fourth section of the chapter tells the story of the Sydney Opera House. Built from 1957 to 1973, its construction exhibits a very interesting tension between the demands of the creative spirit of the architect, the insistence of the engineer on staying on a buildable course, and the economic advantage of standardized production. Prestressed concrete and spherical geometry are both essential and an early version of computer analysis plays an important role. The concluding fifth section is an essay that discusses the impact on architecture of computers and the advances in computer-aided design and manufacture. This technology enables the execution of architectural designs that are completely disconnected from the conventions of the past. We will see that the Guggenheim Museum in Bilbao has a flowing, abstract shape. Freed from the constraints imposed by traditional building materials and methods of construction, its architecture *is* modern art.

Evolving Structures: Domes from St. Paul's in London to the Capitol in Washington.

The exteriors of Saint Paul's Cathedral in London (built from 1675 to 1710), the Panthéon in Paris (built from 1757 to 1790), Saint Isaac's Cathedral in Saint Petersburg (built from 1818 to 1857), and the Capitol Building in Washington (built from 1856 to 1868) are all classical. The facade of each has one or more porticos similar to that of the Roman Pantheon and makes important use of Corinthian columns and pilasters. The dominant and most striking feature of each is its dome. As Figure 6.1 shows, the domes are very similar in design. A solid cylindrical base supports a circular arrangement of columns called a peristyle. Behind it is the dome's drum. The peristyle

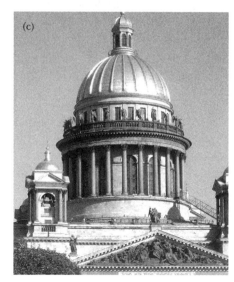

Figure 6.1. The domes of (a) St. Paul's in London (1705–1708) photo by Bernard Gagnon, (b) the Panthéon in Paris (1775–1781) photo by Siren-Com, (c) St. Isaac's in Saint Petersburg (1840–1842), (d) the Capitol Building in Washington (1856–1864), photo by Túrelio

supports a circular platform that is bordered by a railing called a balustrade. From it, recessed and punctured by windows, rises an attic. It is the base of a ribbed hemispherical dome that is crowned by a tall lantern. Figure 5.30 shows that this design is influenced—directly or indirectly—by Bramante's Tempietto. In each of the four structures, the elements just described are sized and proportioned differently, so that the aesthetic impact of the four domes is different. Table 6.1 tells us that the dimensions of the domes also vary considerably. (Note that none matches the size of the dome of St. Peter's.) However, the most important difference among the four domes is

Table 6.1. Dimensions in feet of the four domes and those of the dome of St. Peter's

	St. Paul's	Paris Panthéon	St. Isaac's	U.S. Capitol	St. Peter's
Diameter, exterior of peristyle	137	111	110	125	187
Diameter, exterior of drum	110	81	87	105	159
Diameter, interior of drum (at base)	112	69	74	98	138
Height, interior of dome	58	34	42	98	92

the supporting structure underneath. What is "under the hood" is the focus of this section.

The Great Fire of London in 1666 devastated two-thirds of the city and reduced St. Paul's Cathedral to a ruin. The king of England appointed Christopher Wren (1632–1723) to oversee the reconstruction of the city and to rebuild St. Paul's. Wren was a mathematical scientist with scholarly interests in geometry, astronomy, navigation, and surveying, as well as the instruments relevant to these disciplines. Along with his contemporaries Isaac Newton (who was reshaping science by developing calculus and applying it to his theory of the motion of the planets) and Robert Hooke (a powerful scientist with interests broadly similar to his own), Wren was a member of London's Royal Society, a prominent organization devoted to science and its promotion. Architecture had long been accepted as part of the mathematical sciences and Wren had become interested in architecture—aesthetics, proportion, and the strength of building components—long before his appointment. During the two years before the Great Fire, Wren visited Paris. He observed the construction of several large buildings and had discussions with their architects. Parts of the Louvre—the residence of the Sun King Louis XIV and his court—were being reconstructed at the time. Wren wrote that

> The Louvre for a while was my daily Object, where no less than a thousand Hands are constantly employed in the Works . . . which altogether make a School of Architecture, the best probably, at this Day in Europe.

The construction of the summer palace at Versailles outside Paris also began at around this time. It would provide a safer and more lavish location for Louis and his entourage. Wren met Bernini, who, having just about completed his work on details of the Baroque interior of St. Peter's and the great colonnade of its square, had come to Paris to contribute several designs for the new eastern facade of the Louvre (none of which would be adopted).

The construction of the new St. Paul's began in 1675. It would be the largest building project of the time. A number of aspects of the construction go back to the Middle Ages. The walls and piers are built with rubble (fragments of stone and masonry) and mortar cores that are encased by facings of massive cut stone, or ashlar. Wren called this "rubbish-stone" construction and knew that it deforms under big loads. But economic considerations forced him to build the piers and walls of the new cathedral in this way. St. Paul's external walls are thick to reduce the need for excessive exterior buttresses. The masonry vaults were built with elaborate timber centering that reached up from the floor. Over the vaults are roofs with triangular, oak timber trusses (much like those depicted in Figure 3.13) that are supported independently by the walls of the nave and aisles. The tie-beams are single pieces of timber. Those over the nave are 42 feet long.

As a new landmark of the city of London, as well as a symbol of both the royal house and the Church of England, the new cathedral needed to be a large and impressive structure. Wren's concept called for a church in traditional basilica form topped by a dome that would soar high over the city. The design of the dome would be a major concern throughout the construction of the church. It progressed through several versions and revisions. Wren was fully aware of the outward thrust that vaults, domes, and the loads they support exert on a structure. He knew about the cracks along the meridians of the dome of St. Peter's. He studied plans and sections of the dome and vaults of the Hagia Sophia. He was informed about the timber-supported, lofty, bulging outer shells that were added to the originally flat domes of San Marco in Venice. He also knew about the novel design of some of the domes of Paris. There is no evidence that either Wren or Hooke, his consultant and collaborator during the construction of St. Paul's, was able to estimate the magnitudes of the loads and thrusts in large structures with any accuracy. However (as we will see later in "Hanging Chains and Rising Domes"), they did know what structural shapes were best able to resist such loads and thrusts.

The dome of St. Paul's cathedral has an outer, middle, and inner shell. See Figure 6.2. The shell in the middle is the essential structural component. It is in the shape of a cone that rises from its circular base at the level of the balustrade to support the massive lantern. It is almost completely hidden from view. Made of brick with walls only 18 inches thick, it gains stability from the compression that the lantern provides. The hoop stresses that it is subject to are contained by four iron chains built into the brick at various heights. The visible, nearly hemispherical outer dome is supported by a framework of timber that rests on the brick cone. Figure 6.3 shows the outer surface of the brick cone and some of the timber. Rounded sheets of lead cover the exterior of the outer shell. The inner shell is a structurally independent brick hemisphere that supports no loads other than its own weight. It rests on the inner drum and is braced by a single iron chain around its base. The frescoes on its interior surface and the light that streams into it through the 24 large windows between the columns of the peristyle make it the aesthetic focal

Figure 6.2. Section through Wren's triple-shell dome of St. Paul's. From Jacques Heyman, *The Stone Skeleton: Structural Engineering of Masonry Architecture*, Cambridge University Press, 1995, fig. 8.9

point of the interior of the cathedral. The windows and painted surface at the top of the interior of the brick cone are visible through the oculus of the inner shell. A second oculus at the top of the cone creates the illusion that the inner shell is the lower part of a structure that extends continuously to the open, decorative cylinder inside the lantern. The interior of the church is in English Baroque, not opulent, but in understated, classical lines.

Comparisons with St. Peter's in Rome are instructive. The construction of St. Peter's had required 120 years (25 more if the interior and the colonnade are included) and 12 architects. St. Paul's was completed in 35 years under the direction of a single lead architect. The exterior of the dome of St. Peter's continues the surging verticality of its facade, so that it seems to soar. The horizontal gap between the balustrade and the dome of St. Paul's dissolves the upward vertical emphasis of the columns of the peristyle. The dome of St. Paul's appears to float. (Compare Figures 5.44 and 6.1a.) Two spires, symmetrically placed, complete the facade of St. Paul's. Similar towers had been planned for St. Peter's, but the unstable subsoil had put an end to Bernini's construction of them. Table 6.1 tells us that the dome of St. Peter's is much larger than the dome of St. Paul's. The weights of the two domes (each around 100 million pounds) are roughly the same. Recall that the dome of St. Peter's had to be retrofitted with five iron chains at various elevations to contain the excessive hoop stresses that had caused extensive cracking. No such revisions have been required for St. Paul's. Wren's triple dome with its conical inner shell is structurally stronger.

Figure 6.3. Inner timber structure resting on the brick cone in support of the outer dome. Conway Library, The Courtauld Institute of Art, London. Photo by A. F. Kersting

The church of Ste-Geneviève in Paris is a Neoclassical structure built from 1756 to 1790. Renamed Panthéon, it is today tomb and place of honor for many of France's great intellectuals and literary figures. It has the plan of a traditional basilica and a portico similar to that of the Roman Pantheon. Like St. Paul's, the dome over its crossing has three shells, all three made of masonry. Nothing extraordinary so far. But now comes the interesting part of the story. Its French architect Jacques Germain Soufflot wanted to build a church that combined a classical Greek geometry with the structural lightness of a Gothic cathedral. This idea was revolutionary, in fact it was thought to be contradictory. Soufflot's design did not conform to the traditions and practices of the time. The masonry of the shells of the dome was to be thinner, the piers supporting them more slender, and the windows of the nave much larger. Figure 6.4 shows sections of St. Peter's, St. Paul's, and Ste-Geneviève on the same scale. A comparison of the medium and darkly shaded areas of the figure tells us that Wren's dome is supported by eight piers and Soufflot's by only four. It also shows that Soufflot's vaults and walls are lighter and that sets of slender, free-standing columns play a structural role.

How did Soufflot intend to pull off the execution of a light structure with a large dome? By reinforcing the masonry of the structure with wrought iron

St. Paul's

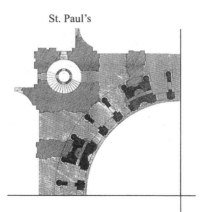

St. Peter's

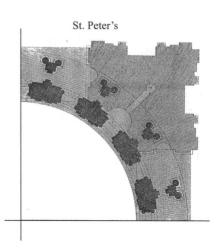

Ste-Geneviève

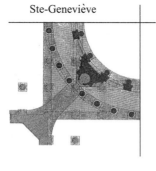

50 feet

Figure 6.4. The three plans are identical in scale. The darkest tones indicate structural elements of the dome at the level of the drum; medium tones indicate supports at floor level; light tones indicate bridging arches and pendentives below the drum. From J. Rondelet, *Memoire historique sur le dôme du Panthéon Française*, Paris 1797. Marquand Library of Art and Archaeology, Princeton University Library

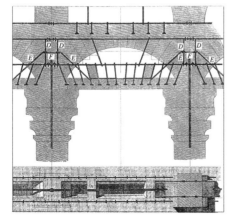

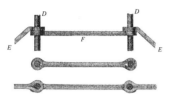

Figure 6.5. From J. Rondelet, *Traité théorique et practique de l'art de bâtir*, Paris 1827–1832. Marquand Library of Art and Archaeology, Princeton University Library, purchase: Elizabeth Foundation

clamps and ties. This would add the tensile strength required to counter the loads effectively. Figure 6.5 shows the system of clamps and ties embedded within the stone structure of the portico. The diagrams at the bottom enlarge some components of the system. The Romans had used bronze to strengthen the portico of the Pantheon, St. Peter's dome had been braced by chains of iron, and Wren had equipped St. Paul's with reinforcing iron ties, but Soufflot's extensive and systematic use of iron was unprecedented.

Not surprisingly, the established order criticized the design. In response, Soufflot and his collaborators, including Jean-Baptiste Rondelet (who would take over as architect in 1770 after Soufflot's death), carried out comprehensive compression tests on stone samples and made the proposed structure the subject of mathematical analysis. (The section "Analyzing Structures: Statics and Materials" below describes what this entailed.) These studies concluded that the piers and reinforced supporting structure would be strong enough to sustain the weight of the dome. For the first time in the recorded history of architecture methods of structural engineering had been applied to the design of a building.

The construction proceeded. By the time the church was finished in 1790 the French Revolution had broken out and soon thereafter Ste-Geneviève became the Panthéon. When the masonry of the piers later developed cracks, new controversies flared up. The cracks were studied and compared against cracks induced by stress tests. New calculations of the thrusts were undertaken. Yet again it was determined that the building was sound. The problem had been the quality of the construction. The mortar beds of the piers varied in thickness in such a way that the dome was supported primarily by the exterior stone casings of the piers. In response, the piers were made more massive early in the nineteenth century. The use of the building alternated from religious to secular over the next hundred years. As the Panthéon, it now stands as monument for great French women and men, among them Voltaire (philosopher, dramatist, and essayist), Rousseau (philosopher and author), Victor Hugo (novelist, poet, and dramatist), Émile Zola (novelist), Marie Curie (physicist), and Louis Braille (teacher of the blind).

Soufflot's structure with its extensive reliance on iron reinforcement had taken masonry construction beyond its natural limits. Nothing comparable would be built until reinforced concrete came into use about a century later. Modern methods of structural engineering have shown what Soufflot and his colleagues could not have known. Reenforcing iron clamps and ties in masonry can produce stress concentrations in the masonry that can result in fracture over time. Other long-term effects such as slow chemical changes, changes in the subsoil, and the actions of heat and wind on masonry structures are complex and their consequences difficult to predict. The Panthéon currently exhibits some problematic deformations. The most significant of these are in the four great arches, each about 100 feet in span, that carry the drum from which the columns of the peristyle rise.

The territory of the Russia of Tsar Peter the Great included the northern parts of eastern Europe and Asia. Its capital city Saint Petersburg would soon have half a million inhabitants and become the fourth largest city in Europe behind London, Paris, and Constantinople (by that time known as Istanbul). Its St. Isaac's Russian Orthodox Cathedral was in a deteriorating state, and rebuilding the church dedicated to their patron saint became important to the tsars of the Romanov dynasty. After Russia's victory over Napoleon in 1812, the time had come. The new St. Isaac's was to be both a splendid house of worship and a symbol for the newly powerful country and its capital. The restoration of the French monarchy after Napoleon's defeat prompted the architect Ricard de Montferrand (1786–1858) to leave France and to seek his fortunes in Russia. Montferrand's architectural drawings came to the attention of the tsar, who was impressed enough to put the inexperienced

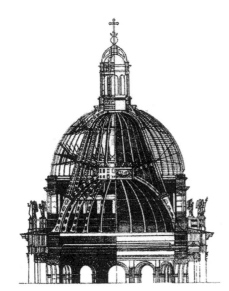

Figure 6.6. A section of the dome of St. Isaac's Cathedral. Ricard de Montferrand, *Église cathédral de Saint Isaac*, Paris, 1845

young architect in charge of the project. Work began in 1818. At that time there was a kind of "ferromania" (*ferrum* is Latin for iron) in Europe. Iron was being produced in ample quantities, engineering had advanced, and large buildings framed or covered with iron had become common. During the beginning of the construction of the new church a large cast iron Gothic hall was completed in Saint Petersburg to house the archives of the general staff of the Imperial Russian Army. Montferrand himself had earlier served as assistant in the building of an iron-framed dome for a commercial building in Paris. Therefore, Monteferrant's choice to provide the church with a dome carried by a cast iron frame was by then not a revolutionary idea. He developed a triple-shell design, much like that of Wren's dome of St. Paul's. Compare Figures 6.2 and 6.6 to see the similarity of the two geometries and the central role that the inner cone plays in both. Figure 6.7 shows the curving iron frames and supports of the dome of St. Isaac's. Its weight is an advantage. At 5 million pounds it is much lighter than the dome of St. Paul's.

St. Isaac's is the fourth largest domed church in the world. It can accommodate 14,000 worshippers. Its great golden dome dominates the skyline of the city. The exterior of St. Isaac's, with its pediments and porticos, is executed in a classical style. Its rich interior has a Byzantine aspect. It is lavishly decorated with paintings, mosaics, friezes, sculptures, marbles, and semiprecious stones. St. Isaac's has been both praised and criticized. Some historians refer to its brilliant architect and call it the greatest classical monument in Russia. Others consider its flaws, refer to the enormous expense, and call it a "sham structure of cast iron."

The cornerstone of the United States Capitol Building was laid in 1793 by George Washington himself. It was completed in 1828, but by 1850 it had already become too small to accommodate the increasing number of legislators from the newly admitted states. In response, the Senate authorized a

Figure 6.7. The inner structure of the dome of St. Isaac's Cathedral. From Ricard de Montferrand, *Église cathédral de Saint Isaac*, Paris, 1845

substantial extension of the building. Its new wings would more than double the width of the facade. Architect Thomas Walter (1804–1887) was appointed to supervise the construction. Work on the wings was interrupted soon after it began by a fire that destroyed the old Library of Congress housed in the western part of the building. Most of the volumes, many irreplaceable, were lost. The sad event had a powerful impact on both Walter and the Capitol. He would build all the roof structures, decorative ceilings, paneling, and trim within the new wings with fireproof cast iron. Never before had iron been employed so extensively in an important public building. The choice of iron also made it possible to add richly molded decoration inexpensively.

It soon became apparent—even to the politicians—that a new larger dome would be needed to give the extended building the proportions and balance that good design required. In preparing his design Walter studied the domes of St. Peter's, St. Paul's, and the Panthéon in Paris. During an earlier brief visit to Europe he had a firsthand look at all three. Walter chose Wren's dome as his principal model. He took the dimensions for the diameters of the drum and the peristyle, as well as the vertical proportions of the peristyle and the hemisphere above it from St. Paul's. But Walter's composition of the cylindrical attic just above the peristyle differs from that of St. Paul's. Its tall slender windows extend the rising verticality of the peristyle much more effectively than the small, square windows of St. Paul's. Another new feature of Walter's design is the curving band of coffers that link the pilasters between the slender windows of the attic to the ribs of the hemisphere above them. A comparison of Figures 6.1a and 6.1d confirms the greater elegance of the dome of Washington's Capitol.

When it came time to choose the building materials for the dome, Walter turned to cast iron. Several factors informed his choice. He had already used cast iron extensively in the new wings of the Capitol in his effort to make them fireproof. Building with cast iron was much cheaper than building with masonry. Then there was the matter of thrust. The existing base could not have absorbed the horizontal thrusts of a masonry dome of the required size without first being extensively reinforced down to its foundations. This was not an option within the existing central section of the Capitol and its open cylindrical Rotunda. A cast iron dome, however, would be lighter and could sit on the existing drum not unlike a lid on a pot. Finally, there was the precedent of St. Isaac's. Walter had architectural sketches that provided the configuration of its dome and the iron cone at its center.

The final design for the support structure of the dome of the Capitol was neither Montferrand's cone nor a cone of Walter's design, but a curving array of ribs proposed and developed by Walter's German assistant August Schoenborn (1827–1902). A look at Figure 6.8 gives a sense of the differences between Walter's cone and Schoenborn's ribs. Schoenborn's ribs rise as open trusses and fuse with the drum into a single structural unit. It is evident that

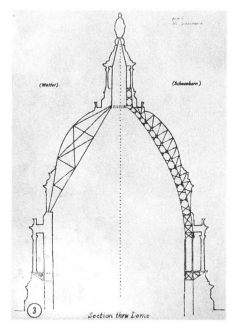

Figure 6.8. A preliminary design and the final design for the rib structure of dome of the Capitol. Walter's original design is on the left. Schoenberg's final design is on the right. From T. Bannister, *The Genealogy of the Dome of the United States Capitol*, 1855

the trussed ribs and the shorter strut structure holding up the dome's outer hemisphere are structurally sounder.

The construction of the dome began in 1856. A temporary roof was placed over the Rotunda and the old dome was removed. The drum was made higher (and parts of it replaced) with the addition of 5 million pounds of new brick and mortar. A single narrow trussed timber tower was erected at the center with a platform that supported a crane. The 80-foot mast and 80-foot beam of the crane and the steam engine that powered it raised all the iron components of the structure. A circle of large triangular brackets was affixed to the outside of the drum. Two such brackets support each of the 27-foot-high columns of the peristyle. Figure 6.9a shows a bracket, a column, and the supporting vertical extension of the drum. The peristyle has 36 columns in a circle of 124 feet in diameter. Figure 6.9b shows a cast iron configuration below a bracket. It is a part of the substructure for a curtain that extends the drum outward by 10 feet. This curtain and the outward position of the peristyle give the lower section of the dome the wider proportions that balanced the extended, winged facade of the new Capitol.

After the lower section of the dome was completed, Schoenborn's 36 arching ribs began to rise. In 1861, most construction was suspended by the

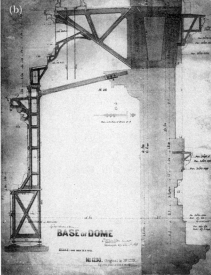

Figure 6.9. (a) Vertical extension of the drum and a column of the peristyle; (b) top of the drum with its outward cast iron extension. From *Glenn Brown's History of the United States Capitol*, House Document No. 108-240, Plates 196 and 197

Civil War and the Capitol was used briefly as military barracks, hospital, and bakery. But work resumed in 1862. The horizontal trusses that encircle and connect the ribs at various elevations were put in. Arrays of cables under carefully adjusted tension secured the ribs further. The cast iron components of the outer shell of the dome and the struts that hold them were fixed into place. Near the top, the ribs were merged three at a time and extended to support the dome's lantern. When the bronze Statue of Freedom was hoisted into place at the top of the lantern, the exterior of the dome was done. Work turned to its interior. A cast iron inner dome with ornate octagonal coffering was attached to the rib structure. It has a circular oculus of 65 feet in diameter at the top. Finally, a canopy in the shape of a section of a sphere was suspended above the oculus to be visible through it. Figure 6.10 shows these elements within the completed upper part of the interior of the dome. In 1866 the dome of the United States Capitol Building, with its exterior height of 288 feet (as viewed from the east side, not from the much lower mall side) and its 9 million-pound cast iron structure, was complete. (The canopy's fresco venerating a godlike George Washington was not finished until much later.)

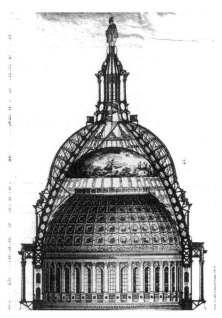

Figure 6.10. Ribs, cast iron structure, top of inner dome, and canopy. From *Glenn Brown's History of the United States Capitol*, House Document No. 108-240, detail from Plate 186

Thomas Walter's stately white dome has been a shining symbol of American democracy ever since. But it has its critics. They see a lack of consonance between its form and the materials of its construction and say that a dome with such an elegant classical geometry should be built entirely with classical materials—stone and brick—and certainly not with cast iron. Much more to the point, however, is the observation that a building has to be constructed with materials that measure up to the demands of its design. Classical or not, it has to be executed with materials that are able to adequately absorb the loads that are generated. We have seen in this text that the masonry of large and famous domes has sustained their loads only with difficulty (and after being reinforced with iron chains). How has the dome of the Capitol Building fared in this regard? A detailed inspection in 1933 established that the dome of the Capitol is structurally unproblematic. Inspections begun in the 1990s have revealed cracks, breaks, and peeling paint on the decorative surfaces and coffers of the inner shell of the dome. However, a major structural review that involved two- and three-dimensional computer models confirmed that the dome continues to be in excellent condition.

Hanging Chains and Rising Domes

A dome with a triple-shell design had been built in Paris before Wren finalized his design for the dome of St. Paul's. In particular, the idea that a middle dome should hold up a timber frame in support of an outer dome and that decorative elements on the inside of the middle dome should be visible through an oculus of an inner dome was not new. But Wren's decision to give this middle dome the shape of a cone was as unprecedented as it was unusual. Of all possible shapes, why a cone?

The scientist Robert Hooke was Wren's collaborator on the St. Paul's project. He would play a significant role. The primary challenge that faced Wren and Hooke was the stability of the dome and in particular its ability to support the massive, nearly 2 million-pound lantern. At around the time the construction of St. Paul's began, Hooke had an insight of fundamental importance. It was the relevance of the hanging chain to the question of the stability of arches and domes. He formulated the following principle: "As hangs the flexible line, so but inverted will stand the rigid arch." With this idea in mind, Hooke took a light, flexible chain mesh and attached an arrangement of weights to it. Holding the mesh in place and letting it hang, he had a bowl-shaped form in front of him that was fixed in place by gravity and the tension in the chain mesh. Imagining this shape turned upside down, Hook had the model of a dome. In this mental picture, the weights on the mesh were now the loads on the dome. Hook regarded this geometry to be ideal for the support of the given arrangement of loads.

Hooke's idea applied to the design of the dome of St. Paul's is depicted in Figure 6.11. The two outward-slanting forces in Figure 6.11a hold up the two ends of a chain. The arrows at the bottom represent weights that model the downward push of the loads, especially of the lantern. In Figure 6.11b the shape is inverted and regarded to be rigid. The arrows representing the loads are now at the top. The slanting lines are the cross section of a model of a cone-shaped dome that supports these loads in an ideal way. The slanting forces at the bottom, now pointing inward, represent the push of the dome's base. The fact that the wall of the cone is relatively steep means that the horizontal components of the downward forces generated by the loads are relatively small. They do exist, of course, and are contained by the four iron chains built into the brick cone at various heights. There is little doubt that Hooke's penetrating analogy between "flexible lines and rigid arches" led Wren to base his design of the dome of St. Paul's on the cone-shaped inner shell that Figure 6.2 depicts.

Recall from Chapter 5, "Bernini's Baroque Basilica," that hoop stress in the dome of St. Peter's caused serious cracking in the inner shell soon after it was finished. During the next century, the cracks worsened and by the middle of the eighteenth century, alarm bells rang in the Vatican. During the years 1742 and 1743, the pope convened several committees and commissions of architects, master masons, and mathematicians to assess the stability of the dome. Completely different opinions emerged. One view was that the collapse of the dome was imminent and that the structure was in urgent need of major modifications. But according to another, the cracks were not critical and the dome was stable. To resolve the matter, the pope appointed Giovanni Poleni (1683–1755), a famous mathematician and structural engineer (and

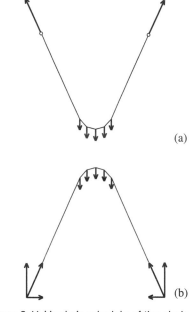

(a)

(b)

Figure 6.11. Hooke's principle of the chain

marquis!) from Padua, to assess the state of the dome. Poleni, a member of London's Royal Society (his membership overlapped with Newton's and Wren's), analyzed the shells and the cracking pattern thoroughly and concluded that there was no immediate danger of failure.

A key aspect of Poleni's analysis was the application of Hooke's insight about the connection between the shape of a hanging chain and the soundness of a dome's structure. Poleni conceived the dome to be divided into 50 identical tapered slices. Refer to Figure 6.12. He regarded two opposite slices to be paired into an arch and thought of the dome as an arrangement of 25 such arches, each supporting one–twenty-fifth of the weight of the lantern. Making use of the drawing of the cross section of the dome shown in Figure 6.13, Poleni then made a model of the typical arch and the load it supports by taking a flexible string and attaching 32 weights along it. Each of the weights was proportional to the estimated weight of a corresponding section of the arch and lantern. The string is shown in the lower half Figure 6.13. The circles represent the carefully positioned weights. The larger the circle, the greater the weight. The decrease in size of the descending circles corresponds to the taper in the arch and the large circle at the bottom to the load of the lantern. Poleni observed, and a careful look at his figure confirms, that the inverted shape of his loaded string fits within the inner and outer

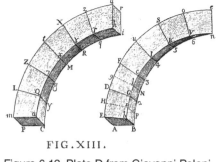

FIG. XIII.

Figure 6.12. Plate D from Giovanni Poleni, *Memorie istoriche della gran cupola el Tempio Vaticano e de' danni in di essa, e de' ristoramenti loro, divise in libri cinque*, Padova, 1748. Marquand Library of Art and Archaeology, Princeton University Library

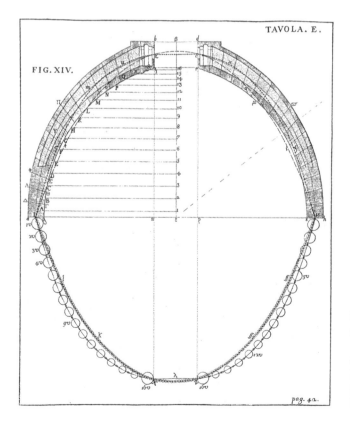

Figure 6.13. Plate E from Giovanni Poleni, *Memorie istoriche della gran cupola el Tempio Vaticano e de' danni in di essa, e de' ristoramenti loro, divise in libri cinque*, Padova, 1748. Marquand Library of Art and Archaeology, Princeton University Library

FIG XVIII!

Figure 6.14. Plate F from Giovanni Poleni, *Memorie istoriche della gran cupola el Tempio Vaticano e de' danni in di essa, e de' ristoramenti loro, divise in libri cinque*, Padova, 1748. Marquand Library of Art and Archaeology, Princeton University Library

cross section of his diagram of the dome. This was the decisive feature of his model that persuaded Poleni that the dome was safe. (A careful look at the figure also shows that an inverted chain without additional loads falls below the inner cross section in part.) Given the prospect of further deterioration, however, he recommended that five more iron chains (in addition to the original three) be placed around the dome of St. Peter's to secure it.

Work on the dome began immediately. Elaborate scaffolding was erected on the inside and outside of the dome. The cracks were filled. Iron bars for the links of the chains were forged and tested. More than a dozen tons of iron were needed for the five chains. Each link is an 18-foot-long curving rod of high-quality iron with eyelets at the ends. The links were placed into grooves in the masonry. Figure 6.14 is a study of a link. It shows the iron wedges that were hammered into the eyelets to assemble and tighten the chains. Completed chains were cemented into the masonry. The dashed lines labeled A, B, C, D, and E in Figure 5.37 show where the five chains encircle the dome and its 16 vertical ribs. The restoration of the dome was finished in 1748. With the restoration of the dome of St. Peter's the prestige of the papacy was restored as well. A grateful pope rewarded Poleni with a golden box, medals of silver and gold, and a pension. The honors and gifts came quickly, but it would take more than two centuries before Poleni's loaded string analysis of the dome of St. Peter's was validated.

The story of the validation begins with a study undertaken by the French scientist and Jesuit priest Pierre Varignon (1654–1722). Published in 1725 in the book *New Mechanics or Statics*, it provides a graphical approach for determining the shape of a hanging string that has weights attached to it. Varignon's construction requires close attention, but it involves little more than the basic properties of vectors already discussed in Chapter 2.

Consider a string, a cord, or a chain and suppose that it is completely flexible and does not lengthen when stretched. Figure 6.15 shows a string held in place at its two ends at the points A and B by the forces R_1 at A and R_2 at B. The segment AB that A and B determine is horizontal. Weights W_1, W_2, and W_3 are attached to the string at the points P_1, P_2, and P_3 at the indicated distances d_1, d_2, d_3, and d_4. (More weights may be attached, but three are enough to illustrate what follows.) The weight of the string is understood to be negligible compared to any of the attached weights. The weights are allowed to hang freely, and the string and weights are assumed to be in equilibrium. Nothing moves. Both a unit of length and a unit of force are given and a vector of length x represents a force of magnitude x. The same (or a different) unit of length can be used to draw to scale the configuration that has been described. It is the point of the discussion that follows to assume that the forces R_1 and R_2, the weights W_1, W_2, and W_3, as well as the distances d_1, d_2, d_3, and d_4 are known, and to show how this information determines the exact shape and hence also the length of the string.

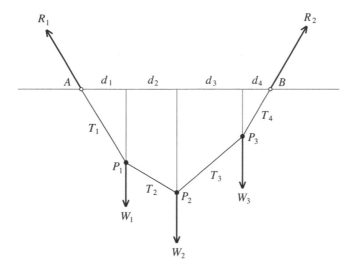

Figure 6.15

The forces with which the four segments of the string pull are denoted in Figure 6.15 by T_1, T_2, T_3, and T_4, respectively. The magnitude of the force with which a string pulls at a point is the tension in the string at that point. Because nothing moves, the tension in each segment of the string is the same at all points along it. In particular, the downward pull of the first segment of the string at A is equal in magnitude to the upward pull of this segment at P_1. A similar thing is true for each of the other three segments of the string. However, the tension in one segment may differ from that in another.

How does the given information determine the path of the string? And what precisely is this path? Because the system is in equilibrium, the first segment of the string pulls down at A in the direction precisely opposite to R_1 with a tension equal to the magnitude of R_1. So the first segment of the string stretches from A in the direction shown in Figure 6.16a until it reaches the point P_1 that the distance d_1 determines. To have equilibrium at P_1, the pull of the second segment of the string at P_1 must counter the resultant of T_1 pulling up and the weight W_1 pulling down. It follows that T_2 is determined by the force diagram of Figure 6.16b. The second segment of the string stretches from P_1 in the direction of T_2 until it reaches the point P_2 that the distance d_2 determines. For equilibrium at P_2, the pull of the third segment of the string at P_2 needs to balance the resultant of T_2 pulling up and the weight W_2 pulling down. So T_3 is determined by the force diagram of Figure 6.16c. Figure 6.17a shows the third segment of the string stretching from the point P_2 in the direction of T_3 to the point P_3 that the distance d_3 specifies. Connecting the points P_3 and B completes the path of the string. The force diagram in Figure 6.17b shows that the resultant of T_3 and W_3 pulls with the same magnitude as R_2 but in the opposite direction. This is consistent with the equilibrium of the system at B.

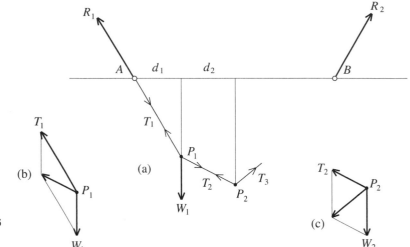

Figure 6.16

As advertised, the forces R_1 and R_2, the weights W_1, W_2, and W_3, and the distances d_1, d_2, d_3, and d_4 have determined the precise path of the string. The polygon $AP_1P_2P_3B$ is called the funicular polygon of the string for the set of forces R_1, R_2, W_1, W_2, and W_3. (In Latin, *funiculus* = string.) In general, a funicular polygon is the form that a string takes on under an assigned pattern of forces.

The tensions in and the directions of the four segments of the string can be determined quickly by collecting the essential information into a diagram. Figures 6.18a and 6.18b show the horizontal and vertical components H_1 and V_1 of R_1 and the horizontal and vertical components H_2 and V_2 of R_2. Because the system is in equilibrium, the total downward pull is equal to the total upward pull, and it follows that the magnitudes of the vector sums $W_1 + W_2 + W_3$ and $V_1 + V_2$ are equal. A study of Figure 6.17a provides the following information. Equilibrium at A implies that the magnitude of H_1 is equal to the magnitude of the horizontal component of T_1. Equilibrium at P_1 implies that the magnitudes of the horizontal components of T_1 and T_2 are equal. Analogous things are true at the points P_2 and P_3. Finally, equilibrium at B means that the magnitudes of H_2 and the horizontal component of T_4 are equal. By combining this information, we find that the magnitudes of H_1 and H_2 are equal.

In Figure 6.18c both sets of vectors W_1, W_2, W_3 and V_1, V_2 are placed end to end as shown. The triangular configuration in Figure 6.18c is determined as follows. Its vertical base is aligned with the weight vectors as shown. The length of the vertical base is equal to the magnitude of the vector sum $W_1 + W_2 + W_3$ or, equivalently, the magnitude of the sum $V_1 + V_2$. The horizontal height of the triangle is determined by a segment placed as shown. Its position is determined by the tip of the vector V_2 and its length is equal

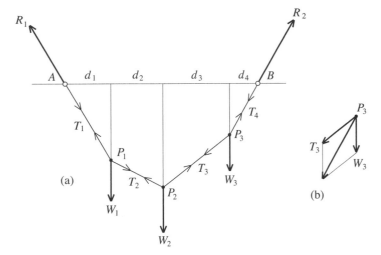

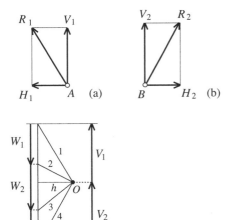

Figure 6.17

to the common magnitude h of the vectors H_1 and H_2. The point O is its right endpoint. The segments labeled 1, 2, 3, and 4 are placed as shown in the figure. Comparing, in sequence, Figure 6.18a with the way segment 1 is placed, Figure 6.16b with the way segment 2 is placed, Figure 6.16c with the way segment 3 is placed, and Figure 6.17b with the way segment 4 is placed tells us that the segments 1, 2, 3, and 4 determine (with their lengths) the respective magnitudes of the vectors T_1, T_2, T_3, and T_4, and (with their directions) the positions of the four segments of the string in Figure 6.17a. The diagram of Figure 6.18c is known as the force polygon for the given pattern of forces. Given that the distances d_1, d_2, d_3, and d_4 are known, the segments 1, 2, 3, and 4 of the force polygon quickly determine the shape of the string of Figure 6.17a.

Keep the loads W_1, W_2, W_3 and the distances d_1, d_2, d_3, d_4 as they are. Keep the vertical components V_1, V_2 of the forces R_1 and R_2 the same, but change—and this is the *only change*—the common magnitude h of the horizontal components of R_1 and R_2. Figure 6.19a shows a force polygon with an increased

Figure 6.18

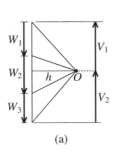

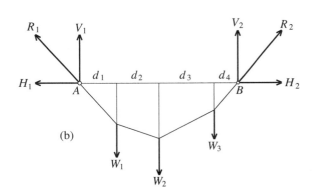

Figure 6.19

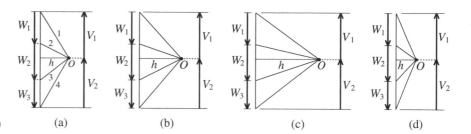

Figure 6.20   (a)   (b)   (c)   (d)

h. It is obtained by stretching the force polygon of Figure 6.18c horizontally. The new force polygon gives rise to a new pattern of string segments. The new shape is quickly derived from the force polygon and it is shown in Figure 6.19b. It should not be surprising that the configuration of the string in Figure 6.19b is flatter than that in Figure 6.17a. (Your experience should tell you that if a string is fixed at one end and pulled on the other it will sag less if pulled with a greater horizontal force.)

Figure 6.20 shows four force polygons. Those of Figure 6.20a and 6.20b correspond to the string configurations of Figures 6.17a and 6.19b, respectively. The force polygon in Figure 6.20c has a relatively large *h* and the *h* of the force polygon in Figure 6.20d is relatively small. The diagonals within the triangles show how the string segments rise and fall in each case.

The analysis just undertaken for the hanging string also applies to an abstract version of the rising arch. Flip the diagram of Figure 6.17a upside down. The loads W_1, W_2, and W_3 and the distances between remain as they are. The vertical components V_1 and V_2 of the forces R_1 and R_2 also remain the same, but their horizontal components H_1 and H_2 reverse direction. A thin rigid rod, or strut, takes the place of each of the four segments of the string. The result is shown in Figure 6.21a. The four rods are hinged where they meet (and at *A* and *B*). In Figure 6.17a, the forces on the four segments of the string are tensile forces acting along the segments of string. In Figure 6.21a they are compressions acting along the rods. By flipping the force polygon of Figure 6.18c, the force polygon in Figure 6.21b is obtained. It provides

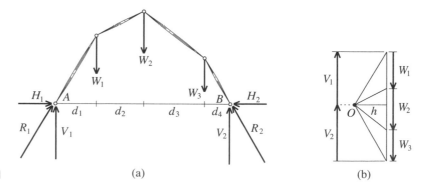

Figure 6.21   (a)   (b)

the magnitudes of the compressions as well as the directions of each of the four rods. The system of strings and forces of Figure 6.17a is in stable equilibrium. Pulling on or pushing against a weight or string will deform the configuration, but it will eventually return to its original position after the push or pull is stopped. The system of rods and forces of Figure 6.21a is also in equilibrium. But it is *not stable.* If subjected to a push or pull, the rods will rotate at the hinges and the system will collapse. In a real arch, the thickness of the voussoirs (recall that these are the wedges of the arch) prevent such a rotation and provide stability.

Figure 6.22 shows an arch under loads $W_1, W_2, \ldots$. The loads include the weights of the voussoirs. Let A and B be points at the base of the arch at the edges of the two lowest voussoirs. The two forces V_1 and V_2 are the vertical components of the forces that act at A and B to hold up the arch at the base. The stability of the arch requires that $W_1 + W_2 + \cdots = V_1 + V_2$. Let H_1 and H_2 be horizontal forces of the same magnitude acting at A and B both in an inward direction. The vector sums of V_1 and H_1 and V_2 and H_2 are the forces that support the arch at points A and B, respectively. Using this information in the abstract and proceeding exactly as in Figure 6.21b provides a force polygon. This in turn gives rise to a funicular polygon with base AB of the sort shown in Figure 6.22.

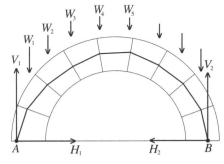

Figure 6.22

Varignon's work was published years before Poleni's investigation of the stability of the dome of St. Peter's and Poleni must certainly have been familiar with it. While it does provide quantitative models of an arch, it does not validate Poleni's loaded string analysis. For this validation, we need to fast forward all the way to 1966 and a powerful insight by Jacques Heyman (who would later become professor and head of the Department of Engineering at the University of Cambridge). Heyman assumed that the masonry with which the arch is made is unlimited in resisting compression, that it has no tensile strength, and that sliding failure of the arch does not occur. It is a fact that masonry is strong in resisting compression and relatively weak in resisting tension. In addition, the compression that an arch is under enhances the cohesion between its voussoirs and provides additional resistance to sliding failure. (Figure 6.29b illustrates sliding failure. Figure 6.29a depicts the more common failure due to hinging.) Therefore, the three assumptions are aligned with the properties of masonry arches. With these assumptions in place, Heyman establishes

The Safe Theorem. Suppose that an arch is subject to loads $W_1, W_2, \ldots$, and that the vertical components V_1 and V_2 of the supporting forces at its base balance them. Assume that the masonry of the arch (i) is unlimited in resisting compression, (ii) has no tensile strength, and that (iii) sliding failure of the arch does not occur. Then the existence of *at least one* funicular polygon (based on the given loads on the arch) that falls completely within the inner and outer boundaries of the arch, implies that the arch is safe.

Let's return to Poleni's diagram of Figure 6.13. The arrangement of circles and the hanging string through them model the weights of components of the arch that the cross section represents. The inversion of this hanging string—Poleni sketches it as a dashed curve—provides a funicular polygon for the arch that falls within the boundaries of the cross section. So the Safe Theorem applies and the arch is safe. So all of Poleni's 25 arches are safe and hence the dome is safe. More than 200 years after it was undertaken, Heyman's theorem gave validation to Poleni's study of the dome of St. Peter's. (By the way, the proof of the theorem is as much beyond the scope of this book as it was beyond the grasp of Poleni.) The Safe Theorem is astonishing. The funicular polygon in question must be derived from the loads that the arch is subject to, but *it might not at all reflect the way the forces within the masonry of the arch actually behave*! So the path of Poleni's inverted loaded string might have little to do with the line of thrust—a theoretical line that represents the path of the resultants of the forces acting through the masonry and mortar—of the arch.

As astonishing as Heyman's Safe Theorem is and as much insight into the behavior of masonry arches it provides, it is a statement about an idealized situation. After all, masonry does have some tensile strength and its ability to resist compression is limited. Furthermore, as the arch of Figure 2.61 demonstrates, sliding failure does occur. The fact is that an in-depth study of arches and domes (and indeed any architectural structure) needs to take the strength of the materials and the related static behavior between the components into account.

Analyzing Structures: Statics and Materials

Statics is the discipline of structural engineering that studies physical systems that are in motionless equilibrium. The forces that act on and within such a system are in a state of balance. None of its components moves in relation to any other. (In Greek, *statikos* means standing or causing to stand.) The related topic strength of materials analyzes building materials under loads, the stresses in the components of a structure, as well as their response and their susceptibility to failure.

In Chapter 4, "Remarkable Curves and Remarkable Maps," we encountered Archimedes (285–212 B.C.) as a mathematical genius. We will now meet him again as the grandfather of the field of statics. It was noted in Chapter 4, "A Line of Numbers," that the Flemish mathematician Simon Stevin (1548–1620) promoted the use of the decimal number system in Europe. The fact is that he was also one of the first to represent force as a vector and to recognize the parallelogram law of forces. Galileo Galilei (1564–1642) analyzed motion by splitting acceleration vectors into components, studied the strength of materials, and reflected about the stability of structures. After another 50 years, Isaac Newton (1642–1727) formulated the basic laws of force and motion and used them to analyze problems of statics as well as dynamics (the field to which his definitive explanation of the motion of the objects in the solar system belongs). Chapter 2 already made fundamental use of Newton's laws. The First Principle of Structural Architecture is a direct consequence of Force = Mass × Acceleration and its corollary that a nonzero force acting on a particle at rest *will* move the particle. And the stability of the beam and column shown in Figure 2.19a is an illustration of the principle that every action has an equal and opposite reaction.

Let's have a look at the basic of concepts of physics on which the studies of statics and strength of materials rely. Consider a material object. The weight of the object is simply the magnitude of the force of gravity acting on it. The weight of an object varies with its location. Its weight is one thing at sea level, another at the top of Mount Everest, and it would be very different if it were to be weighed on the Moon. The concept of mass, on the other hand, is fundamental. The object's mass m is determined by Newton's equation $F = mg$, where F is the object's weight and g is the gravitational acceleration (on the surface of the Earth $g \approx 32$ ft/sec^2, in the units feet and seconds). The constant g and hence the weight F of the object change from location to location. The mass m is the same no matter where the object is. But at any given location, g is fixed and mass and weight are proportional to each other.

When the actions of forces on an object are analyzed, it can often be *assumed that the forces act on a single point of the object*, its center of mass. This important concept not only simplifies such analyses, it makes them possible. We'll specify the center of mass of an object "experimentally" by suspending it from strings. For a one- or two-dimensional object, the center of mass is that point C on the object such that it is in a state of balance when suspended on a string that is attached at C. This is illustrated in Figures 6.23a and 6.23b. To pinpoint the center of mass of a three-dimensional object, suspend the object on two strings that are attached at two different points on the surface. The intersection of the object with the vertical plane that the two strings determine is shown in Figure 6.23c. Now pick a point P on the surface of the object that is not in this plane and suspend the object on a string attached at P. The center of mass of the object is the point of intersection C of the extension of the line of this string with the plane identified earlier. Refer to Figure 6.23d. If the object is made of a completely homogeneous material, then the center of mass is the geometric center, or the centroid, of the object.

A force can have a rotational effect on an object. Think, for example, of the rotational effect that the weight of an Olympic diver standing at the free end of a diving platform produces at the interface between the platform and the supporting vertical wall. The fulcrum and the lever tell us how to

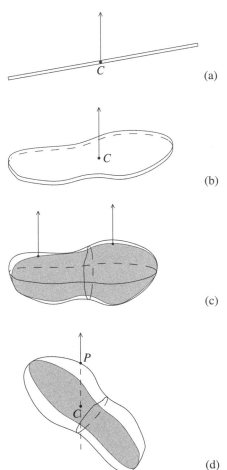

(a)

(b)

(c)

(d)

Figure 6.23

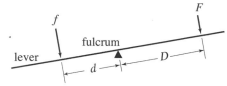

Figure 6.24

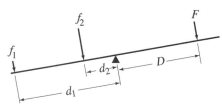

Figure 6.25

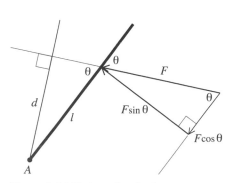

Figure 6.26. Varignon's criterion

quantify such rotational effects. A lever is a thin, rigid rod (like the one in Figure 6.24). The lever does not bend or break when forces are applied to it. The fulcrum is a fixed point around which the lever is free to rotate without slippage, friction, and obstruction. Suppose a force of magnitude f acts perpendicularly to the lever at a distance d from the fulcrum. We will see that the capacity of the force to produce a rotation about the fulcrum is measured by the product $f \cdot d$. Let a second force with magnitude F act perpendicularly to the lever on the other side of the fulcrum at a distance D from the fulcrum. Notice that the two forces are in competition. The first is trying to rotate the lever in a counterclockwise direction and the second in a clockwise direction. Archimedes's law of the lever informs us that if fd and FD are equal, then the lever will not rotate, so that the system is in balance. (It is understood that no other forces, e.g. gravitational, are acting on the lever.)

Consider two forces f_1 and f_2 that act—*one at a time*—perpendicularly to the lever on the same side of the fulcrum as f at the respective distances d_1 and d_2 from the fulcrum. Refer to Figure 6.25. If $f_1 d_1 = FD$ and $f_2 d_2 = FD$, then by Archimedes's law, each one of these two forces will balance the system. It follows that the forces f_1 and f_2 have the same capacity to effect a rotation of the lever about the fulcrum, precisely if the products

$$f_1 d_1 = f_2 d_2$$

are equal.

Pushing against a door gives a sense of what we have discussed. For instance, if you push perpendicularly against the plane of a door with a force of 5 pounds at a distance of 3 feet from the vertical line determined by the hinges, then the rotational effect is equal to $5 \times 3 = 15$ pound-feet. If you push with the same force at a distance of 2 inches, or $\frac{1}{6}$ of a foot, from that line, then the rotational effect is $5 \times \frac{1}{6} = \frac{5}{6}$ pound-feet, more than 15 times smaller than before. This is why much more effort is required to push a door shut near the hinges than near the handle.

So far we have considered only forces that act perpendicularly to a lever. We will now consider a force of magnitude F pushing at an angle θ against a rigid beam that is fixed at the point A. As depicted in Figure 6.26, the force acts at a distance of l units from A. Separate the force into its components perpendicular to the beam and along the beam. We learned in Chapter 2, "Dealing with Forces," that the magnitudes of these components are $F \sin\theta$ and $F \cos\theta$, respectively. The perpendicular component produces a rotational effect of $(F \sin\theta)l$ around A. The component $F \cos\theta$ puts the beam under compression or tension (compression in the case illustrated in the figure) but does not have the capacity to rotate the beam. A look at Figure 6.26 shows that $d = l \sin\theta$. Therefore, $(F \sin\theta)l$ is equal to $F \cdot d$. So if a force of magnitude F is acting on a structure, then its capacity to effect a rotation around a point A of the structure is equal to $F \cdot d$, where d is the perpendicular distance from A to the line of action of the force. The quantity

$$F \cdot d$$

is called the moment of force about the point A. The equality of $(F\sin\theta)l$ and $F \cdot d$ for the moment of force was discovered by the same Pierre Varignon who analyzed strings under loads.

Note an important fact (that Archimedes had already made ingenious use of in his computations of areas and volumes): in the computation of the moment that the weight of an object generates around a point, the entire weight of the object can be assumed to be located at its center of mass.

Given the discussion above, we now add a *Second Principle of Structural Architecture* to the first such principle set out in Chapter 2. It says that *for a structure to be stable, then for every point of the structure the combined moments of force about the point must be zero.* If this is not so, the structure will rotate about the point. As in the case of the first principle, all forces must be considered, including internal reactions, compressions, and tensions.

Charles Augustin Coulomb (1736–1807), a French physicist and member of the French military corps of engineers, investigated the strengths of building materials and the statics of basic structures in the latter part of the eighteenth century. If Archimedes is the grandfather of structural engineering, then Coulomb is the father of this discipline. Much of his important work is contained in his *Essay on Problems of Statics* of 1773.

Coulomb starts his *Essay* by describing the way he tested materials. Here is the essence of what he does. He takes a beam of the material of cross-sectional area A and embeds it firmly and horizontally into place. He applies a horizontal force T along the axis of the beam. He increases the magnitude of this tensile force until the beam breaks. The magnitude of T just before breakage is the ultimate tensile stress of the material of the beam. Coulomb finds that it is equal to $\tau \times A$, where τ is a constant that depends on the particular material. He then applies a vertical force S at the point the beam is embedded and increases the magnitude of this shearing force until the beam breaks. The magnitude of S just before breakage is the ultimate shear stress of the material of the beam. As in the earlier case, Coulomb discovers that it is equal to $\sigma \times A$, where σ is a constant that depends on the material. For a stone found near Bordeaux, Coulomb's tests showed that $\tau = 215$ pounds per square inch and $\sigma = 220$ pounds per square inch. For a well-made brick from the Provence, he measured that τ was between 280 and 300 pounds per square inch. For some masonry materials with mortar, τ was 50 pounds per square inch, but there was a variation by a factor of 2 or 3 depending on the quality of the mortar that was used.

Coulomb also considers friction. Figure 6.28 shows a block being pushed along a dry flat plane by a force just barely strong enough to move the block. So the magnitude of the force is essentially equal to the magnitude F of the opposing frictional force. Coulomb observes that F depends only on the weight N of the block (and not on the area of the block's base) and that $F = \mu \times N$,

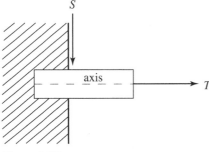

Figure 6.27

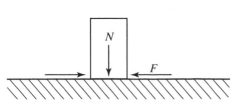

Figure 6.28

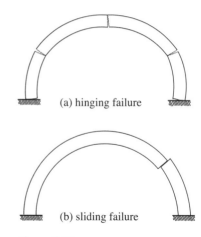

(a) hinging failure

(b) sliding failure

Figure 6.29

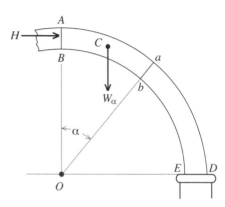

Figure 6.30

where μ is a constant. The constant, known as the coefficient of static friction, depends on the nature of the surfaces of both the plane and the block.

Two primary types of failure, hinging failure and sliding failure, had been observed in masonry arches for some time. They are illustrated in Figure 6.29. The formation of hinges illustrated in Figure 6.29a is more common. Such a gap is caused by an imbalance of the moments of force about the point at which the hinge opens. Excessive shearing forces can produce the kind of slippage shown in Figure 6.29b.

Coulomb's analysis of the stability of a masonry arch considers both types of failure. As shown in Figure 6.30, he starts his study by letting *ABED* represent one half of an arch. (The arch is circular in the figure, but this assumption is not needed.) The point *A* is its highest point, the section *AB* is vertical, and *ED* is the horizontal base. The point *O* is the intersection of the vertical line through *AB* and the horizontal line through *ED*. The horizontal thrust at *AB* is designated by *H*. For an angle α at *O* with $0° \leq \alpha \leq 90°$, *ab* is the cross-section of the arch that α determines. Let A_α be the cross-sectional area of the arch at *ab* and let W_α be the weight of the segment *ABba* of the arch. The point *C* is the center of mass of the segment. The constants of shear and tension of the masonry materials along *ab* as well as the coefficient of static friction may depend on the location of *ab* and hence on α. We will label them respectively by σ_α, τ_α, and μ_α.

Coulomb assumes that the segment *ABba* of the arch is solid, so that a break is possible only along *ab*. His analysis makes use of the basic properties of vectors developed in "Dealing with Forces" in Chapter 2. In particular, Coulomb relies on the information contained in the diagrams of Figure 6.31. They show that the components of *H* and W_α along the cross section *ab* are $H\sin\alpha$ and $W_\alpha\cos\alpha$, respectively, and that those perpendicular to *ab* are $H\cos\alpha$ and $W_\alpha\sin\alpha$, respectively. It follows from the diagrams that the combined effect of *H* and W_α along the cross section *ab* is

$$W_\alpha\cos\alpha - H\sin\alpha$$

and that the combined effect of *H* and W_α perpendicular to *ab* is

$$H\cos\alpha + W_\alpha\sin\alpha.$$

Preventing Sliding Failure. Suppose first that $W_\alpha\cos\alpha \geq H\sin\alpha$ (or equivalently that $\frac{W_\alpha}{H} \geq \tan\alpha$). It follows from a comparison of the two force triangles of Figure 6.31 that *H* and W_α combine to produce a downward force of magnitude

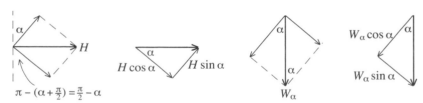

Figure 6.31 $\pi - (\alpha + \frac{\pi}{2}) = \frac{\pi}{2} - \alpha$

$W_\alpha \cos\alpha - H \sin\alpha$ on *ABba* along the cross section *ab*. This downward force is counteracted by the resistance $\sigma_\alpha A_\alpha$ of the arch to shear. The fact that the arch is under compression increases this resistance. Coulomb accounts for this by adding the frictional force $\mu_\alpha(H \cos\alpha + W_\alpha \sin\alpha)$ produced by the compression $H \cos\alpha + W_\alpha \sin\alpha$ to the resistance to shear along *ab*. Coulomb can now argue that the downward slippage of the arch along *ab* is prevented if

$$W_\alpha \cos\alpha - H \sin\alpha \le \mu_\alpha(H \cos\alpha + W_\alpha \sin\alpha) + \sigma_\alpha A_\alpha.$$

This inequality is equivalent to $W_\alpha \cos\alpha - \mu_\alpha W_\alpha \sin\alpha - \sigma_\alpha A_\alpha \le H \sin\alpha + \mu_\alpha H \cos\alpha$, so that there is no downward slippage along *ab* if

$$\frac{W_\alpha \cos\alpha - \mu_\alpha W_\alpha \sin\alpha - \sigma_\alpha A_\alpha}{\sin\alpha + \mu_\alpha \cos\alpha} \le H. \tag{i}$$

Let α range over $0° \le \alpha \le 90°$, consider the term on the left of the inequality, and let F_0 be the maximum value that arises. Coulomb can conclude that if

$$F_0 \le H,$$

then (**i**) holds for all α, so that there will be no downward sliding failure at any cross section of the arch.

Suppose next that $H \sin\alpha \ge W_\alpha \cos\alpha$ (or that $\frac{W_\alpha}{H} \le \tan\alpha$). In this case, H and W_α combine to push the segment *ABba* upward along *ab*. After reasoning as above, Coulomb can assert that the upward displacement of the arch along *ab* is avoided if

$$H \sin\alpha - W_\alpha \cos\alpha \le \mu_\alpha(H \cos\alpha + W_\alpha \sin\alpha) + \sigma_\alpha A_\alpha.$$

A little algebra (and the fact that only the case $\sin\alpha - \mu_\alpha \cos\alpha > 0$ is relevant) tells him that the upward displacement along *ab* is prevented if

$$H \le \frac{W_\alpha \cos\alpha + \mu_\alpha W_\alpha \sin\alpha + \sigma_\alpha A_\alpha}{\sin\alpha - \mu_\alpha \cos\alpha}. \tag{ii}$$

Let α range over $0° \le \alpha \le 90°$ and let F_1 be the minimum value of the term to the right of this inequality. It follows that if

$$H \le F_1,$$

then (**ii**) holds for all α. In this case there will be no upward displacement of the arch along any cross section. In summary, Coulomb concludes that if the horizontal force H at *AB* satisfies

$$F_0 \le H \le F_1,$$

then the arch will not experience sliding failure anywhere.

Coulomb next turns to the hinging failure illustrated in Figure 6.29a. Such hinge formation is brought about by an imbalance of the moments of force acting on the arch.

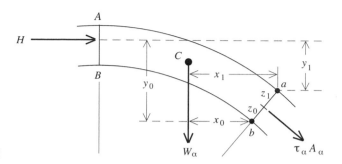

Figure 6.32

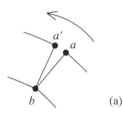

(a)

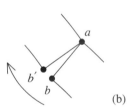

(b)

Figure 6.33

Preventing Hinging Failure. Coulomb's analysis relies on the computation of the moments of the forces on the segment *ABba* first around the point *b* and then around the point *a*. Figure 6.32 adds notational detail to Figure 6.30. Let the distances between the lines of action of the forces W_α and H to the point *b* be x_0 and y_0, respectively. Notice that W_α will tend to rotate the segment *ABba* counterclockwise around *b* and that H will tend to rotate it clockwise around *b*. For a separation of the sort shown in Figure 6.33a to develop, the moment $W_\alpha x_0$ has to prevail over the moment Hy_0. But if the difference between these moments is counteracted by the moment of the tensile force $\tau_\alpha A_\alpha$ around *b*, then the counterclockwise rotation of *ABba* around *b* is prevented. (Note in this regard that the moments of friction and shearing along *ab* are both zero.) Therefore, such a rotation is prevented if $W_\alpha x_0 - Hy_0 \leq (\tau_\alpha A_\alpha)z_0$. This is so if $W_\alpha x_0 - Hy_0 \leq 0$, or, equivalently, if $W_\alpha x_0 \leq Hy_0$ or $\frac{W_\alpha x_0}{y_0} \leq H$. Let G_0 be the maximal value achieved by $\frac{W_\alpha x_0}{y_0}$ as α ranges over $0° \leq \alpha \leq 90°$. If

$$G_0 \leq H,$$

then $\frac{W_\alpha x_0}{y_0} \leq G_0 \leq H$ for all angles α, and Coulomb concludes that there will be no hinging failure at any point along the interior edge of the arch.

With an identical analysis of the moments generated by the forces on the segment *ABba* around the point *a*, Coulomb shows that if G_1 is the minimum value of $\frac{W_\alpha x_1}{y_1}$ (for α between $0°$ and $90°$) and

$$H \leq G_1,$$

then there will be no hinging failure at any point on the exterior edge of the arch. Therefore, if

$$G_0 \leq H \leq G_1,$$

then there will be no such rotational failure at any point around either the inner or outer edge of the arch.

Of Coulomb's two stability conditions $F_0 \leq H \leq F_1$ and $G_0 \leq H \leq G_1$, the second is more relevant because resistance to shear (even at the interface between two voussoirs) is usually large enough to prevent sliding failure.

With the focus on $G_0 \leq H \leq G_1$, Coulomb's *Essay* goes on to discuss the determination of the values G_0 and G_1. Coulomb says that this is always easier by trial and error than by "exact methods" (meaning calculus). Given that the geometry and the materials of the arch are understood, he suggests that the start be made by calculating $\frac{W_\alpha x_0}{y_0}$ for $\alpha = 45°$, for example. If the repetition of this calculation for say, $\alpha = 40°$ produces a larger result, then "it will be certain that the rupture point of the arch is between the keystone and the first joint." By repeating this calculation and moving in the direction of the keystone in the process, G_0 "will easily be found." He goes on to say that G_1 can be determined in the same way. "The Calculus of Moments and Centers of Mass" in Chapter 7 will develop information about G_0 and G_1 that calculus provides. Some of the problems at the end of that chapter will explore Coulomb's analysis including his assertions about G_0 and G_1.

Coulomb considered both external forces on the arch (gravity and the push of one component on another) as well as internal forces (tensile, compression, and shearing forces) within the arch, and he analyzed the arch by balancing these forces as well as the moments that they generate. The estimates of the horizontal force H and the weight of half the arch provide information about the structural requirements on the pier supporting the arch, so that the design of the pier can proceed. By dividing a dome into arches (the way Poleni did for St. Peter's), Coulomb's theory can also be applied to domes. Theories such as Coulomb's were used to demonstrate the stability of the dome of the Panthéon in Paris and to inform the construction of St. Isaac's in Saint Petersburg.

Major advances in the disciplines of strength of materials and statics continued in the eighteenth and nineteenth centuries. The emergence of cast iron, steel, and reinforced and prestressed concrete as materials with which large structures were later built, along with the computer technology that assessed them, led the analysis of structures off into new directions in the twentieth and twenty-first centuries.

The Sydney Opera House

In January 1957, the young Danish architect Jørn Utzon (1918–2008) won a design competition for an opera and concert hall complex to be built on a dramatic piece of land that juts out into Sydney Harbor. Utzon, the son of a director of a shipyard and designer of yachts, submitted a design with an arrangement of high vaulted roofs that looked like a cluster of sailboats under full sail. Figure 6.34 shows one of his sketches. The panel of experts that assessed the competing designs described Utzon's entry as follows:

> The drawings submitted for this scheme are simple to the point of being diagrammatic. Nevertheless, as we returned again and again to the

Figure 6.34. A sketch of the sail-like vaults in Utzon's competition entry. Collection of Powerhouse Museum, Sydney. © Jan Utzon

study of these drawings, we are convinced that they present a concept of an opera house which is capable of being one of the great buildings of the world.

And one of the great buildings of the world it would become. But the realization of the architect's vision was a monumental challenge. The plans called for two large auditoriums, a restaurant, and the necessary support structures. There was to be a major hall for musical events of large scale such as grand opera and orchestral concerts, and a smaller minor hall for theatrical performances, recitals, chamber music, and lectures. On the recommendation of the panel, Ove Arup (1895–1988), a Dane with British citizenship, was named as the lead structural engineer for the project. Arup's firm had completed a large concrete roof for a factory in Wales in the 1950s. It was built in the form of a square arrangement of nine identical concrete shells, each arching over a 62 by 82 foot rectangular space. As the largest roof of this kind at that time, it received much acclaim. The offices of Utzon and Arup formed the nucleus of the design team for the project. They were later joined by consultants with expertise in acoustics, theater design, and mechanical and electrical engineering.

The analogy to sailing ships extended to the building phases of the project, both architecturally and functionally. There is the substructure "below the deck," the essential "mast and sail structure," and, finally, there is everything else "above the deck." Accordingly, the project split into three basic stages: the podium, the curving shells of the roof vaults, and the rest, including the interiors of the major and minor halls, especially the acoustics and seating schemes.

The Podium. This concrete substructure was to extend over a large rectangular area of 380 by 610 feet, cover almost the entire site, and reach from the foundations to the level of the seating in the auditoriums. It was to include some performing spaces (a chamber music hall, experimental theater, rehearsal room), entrance lobbies, a kitchen for the restaurant, box offices, storage and dressing rooms, a cafeteria, administrative areas, meeting rooms, paint and carpentry shops, and electrical and telecommunications facilities. The podium needed to be built in such a way that each of the performance venues, especially the two principal auditoriums, were isolated from sounds and vibrations generated elsewhere in the complex. It was to include a large roofed circulation space for cars, buses, and trucks, known as the concourse. The top of the flat roof of the concourse was to be a large open platform, reached by a wide and grand staircase, that provided the main approach to the auditoriums for opera- and concertgoers. Quite clearly, building the podium of the Sydney Opera House was a large and complicated undertaking.

The construction of the podium began in March 1959. From the perspective of this text, the most interesting aspect of the construction is the roof of the concourse. Given the expanse of this roof, Utzon's plans called for an array of columns to support it. Arup's team was able to do away with these columns to create a concourse area with a clear space of about 160 feet by 312 feet, a clear space the size of a football field! In the context of the discussion in Chapter 2 (the section "Dealing with Forces") about the limitation of column and beam construction (due to the tensile forces that are generated in the beam), this is extraordinary. How was Arup's team able to achieve such a large clear space? With the installation of beams that made use of prestressed concrete and combined a "folded-slab" design with an ingenious variable cross-sectional geometry.

First a word about prestressed concrete. As was already pointed out (in Chapter 2, "The Roman Arch"), unreinforced concrete is strong in resisting compression, but much weaker in resisting tension. The ratio between the compressive strength and tensile strength of concrete is roughly 12 to 1. Reinforcing concrete beams by embedding iron or steel rods and bars strengthens them, but not enough for very demanding applications. The process of prestressing or, more accurately, precompressing concrete relies on the ability of concrete to resist compression. Consider a concrete beam that is cast from a form or mold. Include as part of the mold a carefully, lengthwise-positioned set of metal ducts. After the concrete has been poured and has hardened, steel cables with great tensile strength are threaded through the ducts. These cables are pulled tight (with hydraulic jacks) and anchored—under great tension—firmly into place at the ends of the beam. The finished beam is under great compression from the pull of the cables. The remaining space in the ducts is pressure grouted with a special mortar. This fixes the cables into place within the beam and prevents corrosion. If this is done correctly—and here is the point—then the beam will always be under compression, no matter what loads and tensile forces it is subjected to! Prestressed concrete, now common in structures that are

Figure 6.35. Diagram of the cross section of a concourse beam. The *x*-axis and the points with coordinates 0, *x*, *b*, and 2*b*, were added to the original figure and play a role in the analysis of the geometry of the beam. 1. Upper part of beam 2. Prestressing cables 3. Concourse beam 4. Prestressing cables 5. Sub-structure 6. Tie-beam 7. Ground level 8. Prestressing cables 9. Sliding bearing 10. Jacking point 11. Cross-head for tie-beam. From Yuzo Mikami, *Utzon's Sphere: Sydney Opera House—How It Was Designed and Built*, photographed by Osamu Murai, Shokokusha, Tokyo, 2001, fig. 5.11. Marquand Library of Art and Archaeology, Princeton University Library

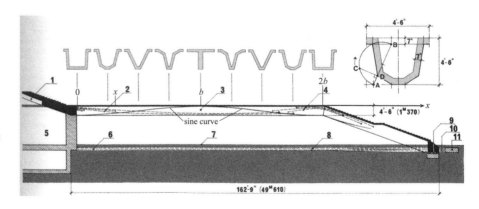

subject to large loads, such as bridges and viaducts, was just coming into wider use at the time of the opera project.

Arup's engineers created the large clear space of the concourse by building its flat roof with a parallel arrangement of 52 prestressed concrete beams, each 6 feet wide and about 160 feet long. Figure 6.35 shows the cross section of a concourse beam along its length. The slanted part supports the grand staircase. The diagram shows that the prestressing cables of the beam at midspan (the point labeled *b*) run along the bottom of the beam. This is a response to the fact that the tension on a beam at midspan is greatest at the bottom. Refer to the illustration provided by Figure 2.19b. This figure also shows that the left and right vertical edges of the beam lean inward. In reference to the concourse beam this effect explains why the prestressing cables run near the top at the points labeled 0 and 2b (thus preventing the tension that the beam would be subjected to there). The prestressed tie-beam in the foundation is designed to counteract the outward thrust that the beam generates.

There is more to the story of the concourse beams. It is common knowledge that cardboard is strengthened when it is made with corrugated paper, paper with parallel ridges and grooves. Arup's engineers applied this idea in the design of the beam. So that the beam can sustain the prestressing adequately and be structurally most effective, the corrugation is designed to vary along the length of the beam. From one support to midspan to the other support, the cross section of the beam changes from U-shaped to V- shaped to T-shaped, and back again to V and U. The sequence of Us, Vs, and T in Figure 6.35 illustrates the pattern. But how should a continuous beam with such a variable pattern of cross sections be designed? How should its cross section transition from one shape to the next? An interesting geometric construction provides the answers.

Place an *x*-axis through the middle of the beam at the top along its length. As can be seen from Figure 6.35, the beam stretches horizontally from *x* = 0, to $x = b$ at the beam's midspan (b is approximately 54 feet), to $x = 2b$ (before it slants downward). Let x be any coordinate between 0 and $2b$. The critical question is this: With the cross sections at $x = 0$ (the U), $x = b$ (the T), and $x = 2b$ (the U again) given, what should the cross section of the beam be at a typical x *between* 0 and $2b$? As a first step in the answer, Arup's engineers consider a circle with circumference equal to $2b$. The radius R of this circle satisfies $2\pi R = 2b$, so that $R = \frac{b}{\pi}$. A part of the circle is depicted in Figure 6.36. The angle θ is chosen in such a way that the curved boundary of the wedge that θ determines has length x. Since the part of the circumference of a circle cut out by a central angle is proportional to the angle, it follows that $\frac{\theta}{x} = \frac{180°}{b}$ and hence that $\theta = (\frac{x}{b} \cdot 180)°$.

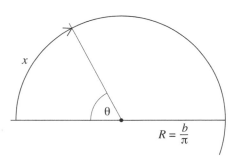

Figure 6.36

To begin the next step, complete the x-axis to an xyz-coordinate system as shown in Figure 6.37. The foot is the unit of length for each axis. Now turn to Figure 6.38. It is an elaboration of the diagram depicted in the upper right corner of Figure 6.35 placed into the yz-plane. The thickness of the concrete slab is $c = 7$ inches or $c = \frac{7}{12} \approx 0.583$ foot. The points A and F are determined by the left edge of the U-shaped cross section that the beam has at the support. (The U over the 0 in Figure 6.35.) The segment FA is not quite vertical, as A is 1 inch closer to the z-axis than F. The points B and G are determined by the left vertical edge of the T-shaped cross section of the beam at midspan. (The T over the b in Figure 6.35.) The points A, F, B, and G in the diagram

Figure 6.37

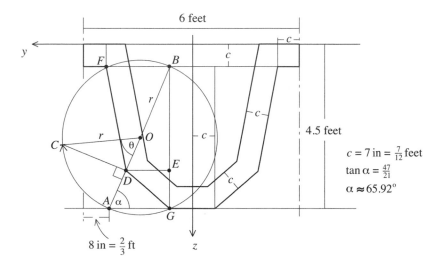

6 feet

$c = 7\,\text{in} = \frac{7}{12}\,\text{feet}$

$\tan \alpha = \frac{47}{21}$

$\alpha \approx 65.92°$

4.5 feet

8 in $= \frac{2}{3}$ ft

Figure 6.38

are fixed, as is the circle with diameter AB and center O. A careful look at the data supplied in Figure 6.38 tells us that $AG = 36 - 8 - 7 = 21$ inches, or $\frac{21}{12} = 1.75$ feet, and $BG = 54 - 7 = 47$ inches, or $\frac{47}{12} \approx 3.92$ feet. So $\tan \alpha = \frac{BG}{AG} = \frac{47}{21} \approx 2.24$ and (by the touch of the inverse tan button of a calculator) $\alpha \approx 65.92°$. By the Pythagorean Theorem, the radius of the circle is $r = \frac{1}{2} AB = \frac{1}{2} \sqrt{AG^2 + BG^2} = \frac{1}{2} \sqrt{21^2 + 47^2} = \frac{1}{2} \sqrt{2650} \approx 25.74$ inches, or $\frac{\sqrt{2650}}{24} \approx 2.14$ feet.

Now take the angle θ of Figure 6.36 and place it into Figure 6.38 by choosing the point C on the circle so that $\angle AOC = \theta$. Let D be the point on the diameter AB determined by the perpendicular from C and let E be the intersection of the horizontal through D with BG. Because F and G are fixed, D determines the outer edge of the diagram of Figure 6.38 (along with the fact that the diagram is symmetric about the z-axis). The inner edge of the diagram is determined by the outer edge and the thickness c of the slab. The diagram that D determines *is the design* of the cross section of the beam at x. Observe that as x increases from 0 to b, θ varies from 0° to 180°. In the process, the point D moves from A to B, the left outer edge FDG of the design changes from FAG to FBG, and the design morphs from U-shaped to T-shaped. For some angle between 0° and 180°, FDG is a straight segment and the design is V-shaped. As x increases from b to $2b$, θ increases from 180° to 360°. In the process, the point D moves from B back to A, and hence the design changes from T-shaped, back to V-shaped, and U-shaped. So as x moves from 0 to $2b$, the design changes exactly as depicted by the sequence of cross sections of Figure 6.35. Arup's engineers had achieved exactly what was required: the design of a beam with a cross section that varies smoothly along its horizontal length from U-shaped to V-shaped to T-shaped and back again.

To conclude this discussion, we'll compute the y- and z-coordinates of the point D. A look at ΔOCD tells us that $\cos \theta = \frac{OD}{r}$. So $OD = r \cos \theta$ and $BD = r + r \cos \theta = r(1 + \cos \theta)$. From ΔBDE, we see that $\cos \alpha = \frac{DE}{BD}$ and $\sin \alpha = \frac{BE}{BD}$. Therefore, $DE = BD \cdot \cos \alpha = (r \cos \alpha)(1 + \cos \theta)$ and $BE = BD \cdot \sin \alpha = (r \sin \alpha)(1 + \cos \theta)$. It follows that the y and z coordinates of the point D are given by

$$y = (r \cos \alpha)(1 + \cos \theta) + c = (r \cos \alpha)\left(1 + \cos\left(\frac{x}{b} \cdot 180\right)^°\right) + c$$

and

$$z = (r \sin \alpha)(1 + \cos \theta) + c = (r \sin \alpha)\left(1 + \cos\left(\frac{x}{b} \cdot 180\right)^°\right) + c.$$

These equations provide a precise mathematical description of the design of Arup's prestressed concourse beam. The graph of $z = r \sin \alpha (1 + \cos (\frac{x}{b} \cdot 180)^°) + c$ is depicted in Figure 6.35 as the solid curve (labeled "sine curve") that flows through the cross section of the beam.

Arup's powerful array of concrete beams did more than create the large clear space for the concourse. From below, each curving beam looked like the underside of the hull of a boat. The flowing surfaces of the ceiling of the concourse provided a visual effect that was consistent with the geometric accents set by Utzon's sail-shaped design of the vaults of the opera complex.

The Roof Vaults. The path from Utzon's imaginative design—with its cluster of billowing, sail-like roofs—to its realization was extremely difficult. What should the explicit geometric definition of these freely flowing sculptural forms be? With what combination of materials and methods of construction should these completely new vaulted roofs be built? It would take from 1957 to 1963—years of exploration, analysis, disagreements, and hard work—to answer these questions.

Roof structures of such magnitude and complexity cannot be built without an explicit geometry that can be expressed mathematically. Without such a mathematical model it is not possible to calculate the loads, stresses, and rotational forces that the vaults would be subjected to, and to estimate the impact of wind and temperature changes on their stability. Without an explicitly defined geometry (such as the spherical geometry that Figure 4.26 provides for the dome of the Hagia Sophia), the necessary computations and computer analyses could not be undertaken, and the construction of the unprecedented structure could not proceed. Parabolas (or more accurately paraboloids, given the three-dimensional aspect of the vaults) were Utzon's first choice for the profiles of the vaults. See Figure 6.39. At a later point, ellipses (or again more accurately, ellipsoids) were considered. For reasons that we will explore shortly, neither of these geometries provided a buildable option.

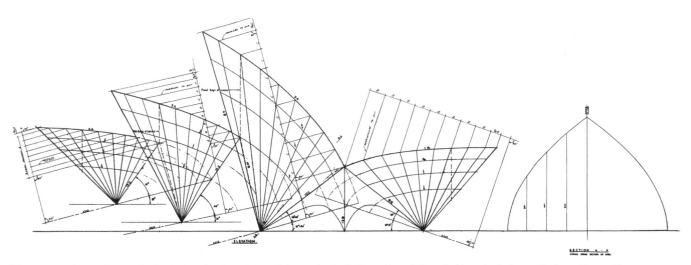

Figure 6.39. An early parabolic design for the shells of the major hall. From Yuzo Mikami, *Utzon's Sphere: Sydney Opera House— How It Was Designed and Built*, photographs by Osamu Murai, Shokokusha, Tokyo, 2001, fig. 4.10. Marquand Library of Art and Archaeology, Princeton University Library. Courtesy of Arup

It had been Utzon's initial thought that his vaults would be built as thin, concrete membrane structures. Such eggshell-like concrete roofs were common at the time. This is how Arup's firm had executed—with concrete shells only 3 inches thick—the roof of the factory in Wales. For such a membrane structure to be stable, two basic conditions need to be satisfied. Its shell needs have a geometry such that at every height the outward and downward thrusts generated by the weight of the shell are countered (almost) perfectly by the push of the shell from below. (This geometry will be studied in "The Shape of an Ideal Arch" in Chapter 7, where it will be seen to be the inverted geometry of the hanging chain.) In addition, the shell needs to be surrounded by strong and rigid edge beams that contain the outward thrusts at its perimeter. If these two conditions are met, then the weight of the shell is borne by compressions within the shell and the response of the surrounding rigid edge. However, the pointed, steeply rising vaults of Utzon's design were not compatible with either of these requirements. Since a departure from his architectural concept was not an option, the idea to build the roof vaults as concrete membranes was abandoned.

The wide and high open spaces between the sail-like shells were another complicating factor in Utzon's design. These were spaces for smaller shell structures as well as the large window areas that would bring both light and the exciting vistas of Sydney Harbor into the lobby areas of the complex. But the existence of these open spaces meant that the areas of contact between the shells and the podium would have to be narrow and concentrated. Given the various parameters of the design, Arup became convinced that each of these sail-like roof structures could be built only as a sequence of curving ribs—narrow at the bottom and increasingly wide as they rise—that would spring from a common point at the podium and fan outward and upward from there. Each roof vault would consist of two such curving fanlike structures—one the mirror image of the other—rising from opposite sides to meet at a circular ridge at the top. Utzon endorsed this concept enthusiastically: "I don't care how much it costs, I don't care how long it takes to build or what scandals it causes, but this is what I want."

This solved one problem, but the problem with the geometry remained. The large size of the shells meant that they would have to be constructed in sections or components. The demands of economy and time meant that these components would have to be mass produced. A parabolic or elliptical shell would not do because then each rib would curve differently. Refer to Figure 6.39. Was there a geometry that would make it possible to build the curving sail-like structures of the sketch that won the Sydney Opera competition with standardized, identical components? If the answer was no, then it would be impossible to execute Utzon's design and the project would collapse.

Suddenly Utzon had a flash of an idea. Ideally, the surface of the shells should curve in the same way in all directions. But the only surface with this

SOH ELEVATION 1/16 SCALE
FINAL SCHEME
UTZON 1/2 - 62 " THE RESULT OF THE ARCHITECTURAL COMPETITION WAS PUBLISHED 1/2-57

Figure 6.40. The shell sequence of the major hall, sketched by Utzon in February 1962. From Yuzo Mikami, *Utzon's Sphere: Sydney Opera House—How It Was Designed and Built*, photographs by Osamu Murai, Shokokusha, Tokyo, 2001, fig. 9.1. Marquand Library of Art and Archaeology, Princeton University Library. © Jan Utzon

property is a sphere of a given radius. Utzon's flash was the realization that a limitless variety of curving triangles could be drawn on a sphere. So all the shells for the roofs could be designed as curving triangles from the same sphere! This was the idea that saved the project. Both Utzon and Arup finally saw light at the end of their tunnel. On February 1, 1962, Utzon sketched what he called the final scheme. See Figure 6.40. As he observed at the very bottom corner of the sketch (in fine print), it was to the day exactly five years after the official announcement that he had won the competition.

Let's analyze the curving triangles that Utzon envisioned. Consider a sphere with center O and radius r. Place an xyz-coordinate system so that the center of the sphere is at the origin O. Refer to Figure 6.41. Let A, B, and C be three points on the sphere and rotate the sphere so that A lies on the y-axis. The curves through A, B, and A' and through A, C, and A' mark the circles given by intersecting the sphere with the planes determined respectively by the points A, O, A', B and A, O, A', C. Any circle obtained as the intersection of the sphere with a plane *through the origin O* is called a great circle. So the arcs AB and AC both lie on great circles. Now project the points B and C into the xy-plane by pushing parallel to the z-axis. The lines of projection determine a plane through B and C perpendicular to the xy-plane. The intersection of this plane with the sphere is a circle that provides the arc BC that completes the determination of Utzon's triangle ABC.

There are many (infinitely many) possible planes through the points B and C, so that there are (infinitely) many ways to connect B and C with a circular arc that lies on the sphere. Utzon's choice of the circular arc BC that lies in a vertical plane meant that he could design each of the roof vaults by

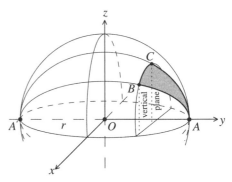

Figure 6.41. Utzon's spherical triangles

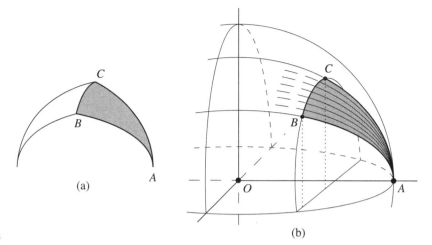

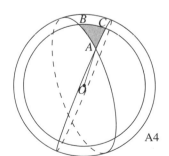

Figure 6.42. Utzon's vaults and Arup's ribs

joining a spherical triangle *ABC* and its mirror image at the circular ridge *BC*. This is illustrated in Figure 6.42a and (more dramatically) in Figure 6.47. Utzon's sphere idea was also compatible with Arup's rib concept. Great circles could divide the spherical triangles into fan-shaped arrangements of ribs, and each rib could be built as a sequence of prestressed concrete segments with edges determined by the same great circles. Refer to the diagram in Figure 6.42b. Utzon and Arup decided on 246 feet as the radius of the sphere from which the design of all the matching pairs of spherical triangles would be derived. In all cases, the angle between any two consecutive ribs at the springing point *A* would be the same 3.65°.

Figure 6.43 depicts the four triangles for the shell sequence of the major hall. Each of these diagrams is a special instance of the diagram of Figure 6.41 with the sphere rotated in such a way that the vertical plane through *B* and *C* is parallel to the plane of the page. Shell A1 provides the vault of the entrance foyer. Shell A2 is the vault over the stage area and in particular the

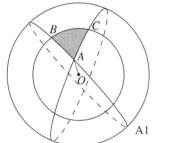

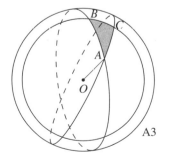

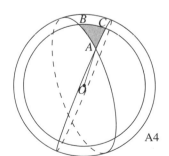

Figure 6.43. The spherical triangles for vaults A1 to A4 of the major hall. From Yuzo Mikami, *Utzon's Sphere: Sydney Opera House—How It Was Designed and Built*, photographs by Osamu Murai, Shokokusha, Tokyo, 2001, fig. 8.10. Marquand Library of Art and Archaeology, Princeton University Library

stage towers that would house the group of large elevators to move stage sets. With its height of about 220 feet, it is the highest vault of the roof complex. Shell A3 is the roof vault over the main auditorium, the seating area, as well as the acoustical ceiling. Shell A4 covers a spacious lounge area. It is closed off by a wall of glass through which opera- and concertgoers could experience a dramatic panoramic view of Sydney Harbour Bridge and the shipping activity of the harbor. Figure 6.44 shows the spherical triangles for shells A1 to A4 as well as the spheres that give rise to them, in position in Utzon's design of the major hall.

A comparison of Figures 6.34 and 6.40 shows that Utzon's single-sphere solution had changed the visual quality of the profile of the roof sequence from one that was loftier and more elegant to one that seemed more rigid and heavy. But Utzon and Arup knew that the idea of the single spherical geometry had been a critical advance. It was now possible to construct the

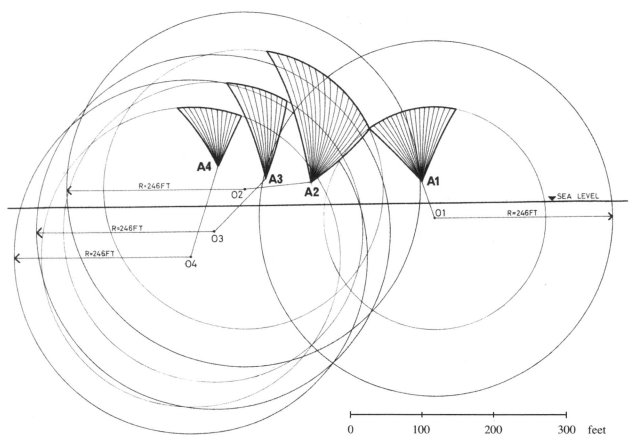

Figure 6.44. Shells A1 to A4 for the major hall in position. From Yuzo Mikami, *Utzon's Sphere: Sydney Opera House—How It Was Designed and Built*, photographs by Osamu Murai, Shokokusha, Tokyo, 2001, fig. 8.11. Marquand Library of Art and Archaeology, Princeton University Library

vaults from mass-produced repeating components. When Arup recalled later that "we did not want to pull the architect down to hell, but we wanted him to pull us up to heaven," it was this breakthrough that he had in mind. Utzon was certainly influenced by Arup's single-minded focus on "how do we build it?" However, the spherical solution had been his.

The teams of Utzon and Arup could now complete the particulars of the design. These particulars are illustrated by Figure 6.45 for shell A2. The segments of the ribs called for prestressed concrete. Their cross sections are designed to vary from a narrow T near the podium, to a narrow solid Y, to a wider, open Y higher up. Rib segment number 7 of Figure 6.45c is an example of an open Y. The sets of three holes in the flanges at the top and bottom are those of the ducts cast into the segment for the prestressing cables. The pattern of segments comprising the ribs is detailed in Figure 6.45b. This pattern of segments flows up in exactly the same way for each of the shells. The arch formed by rib number 12 and its matching pair is shown in Figure 6.45a. It recalls, and is in principle the same, as the Gothic fifth arch of the dome of the Cathedral of Florence (studied in Chapter 4).

The building of the vaults could begin. The various concrete rib segments were prefabricated at the site. A total of 1498 standard and another 280 nonstandard rib segments, each 15 feet long, were cast. There are twelve different types. Of the seven types in the lowest parts of the shells, respectively 280, 280, 260, 196, 174, 110, and 82 were made. The use of so many identical prefabricated components simplified the construction of the roof vaults. Even more critically, it reduced cost and saved time. Figure 6.46 shows the construction of two shells in progress in 1966. Notice the closed rib segments in the lower portions of the shells and the open segments higher up. The image also shows the smaller shells at the transition between the two growing vaults. These side shells were constructed before the main vaults were completed.

The construction of the ribs of the vaults faced a problem. As can be seen from Figures 6.42a and 6.45a, each rib of a triangular shell has a matching rib in the mirror image of the shell on the opposite side. The construction of each matching pair of ribs from its segments was in principle the same as the construction of an arch from its voussoirs. The additional complication was that the ribs lean and expand as they rise. We learned in Chapter 2, "The Roman Arch," that an arch is not stable until it is complete and that a centering structure needs to hold it up until it is. How was this attended to in the construction of the ribs of the vaults? The key was an erection arch made of curving triangular steel trusses. It functioned as follows. A rib segment was lifted by one of the great cranes and placed above the last rib segment already placed. There it would be supported on one side by the previously completed rib and on the other by the erection arch. Just before a segment was lowered into place above the one already in position, a coat of epoxy resin was applied to glue the two matching surfaces together. The erection arch could

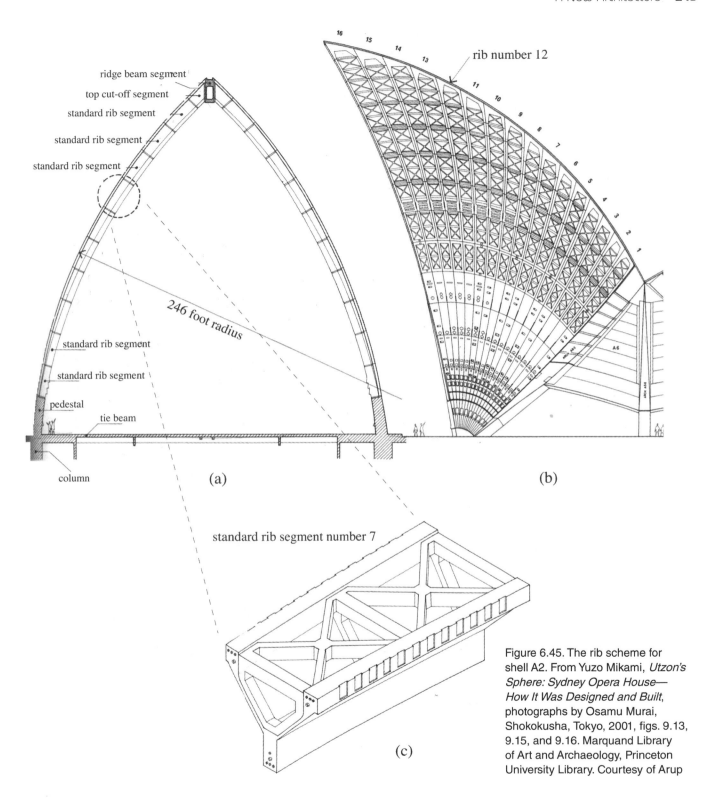

ridge beam segment

top cut-off segment

standard rib segment

standard rib segment

standard rib segment

246 foot radius

standard rib segment

standard rib segment

pedestal

tie beam

column

(a)

rib number 12

(b)

standard rib segment number 7

(c)

Figure 6.45. The rib scheme for shell A2. From Yuzo Mikami, *Utzon's Sphere: Sydney Opera House— How It Was Designed and Built*, photographs by Osamu Murai, Shokokusha, Tokyo, 2001, figs. 9.13, 9.15, and 9.16. Marquand Library of Art and Archaeology, Princeton University Library. Courtesy of Arup

Figure 6.46. Work on the shells in 1966. Photo by Max Dupain. © Max Dupain and Associates Pty. Ltd.

be expanded or contracted as required by the size and lean of the rib being constructed. It can be seen as the dark arch at the boundary of the vault on the left in Figure 6.46. The lowest several segments of the growing rib next to it are already in place. Above them is the gap for the upper segments. In this way, segment by segment and rib by rib, the shells rose to completion.

The remaining challenge was to cover and seal the outer surfaces of the shells. Utzon did this with a sophisticated tile system. He chose two types of square tiles, one matte and one glossy, in slightly different off-white colors. They had to protect the structure from the elements and they needed to be arranged in an attractive pattern. It was decided to cement the tiles in V formation on chevron-shaped concrete slabs. Called lids, these tiled slabs are about 6 inches thick and about 7.5 feet long. Three such lids can be seen in position in Figure 6.46, high on the shell on the right. The lids are reinforced with several layers of steel mesh. They are light so as not to add great weight to the shells. Their backs are sprayed with insulating foam to reduce the thermal effects on the shell underneath. So that they can follow the contours of the shells perfectly, the lids are sections of a sphere of radius slightly larger (by 8.5 inches) than the 246-foot radius of the sphere of the shells. Finally, each lid is locked into place with brackets and bolts that are adjusted to give it the precise position it needed to have on the spherical surface. It goes without saying that much thought as well as mathematics and computer analysis went into the complex design of these tiled lids. The spherical geometry of the lids made it possible to standardize the manufacture of the more than 4000 that were required. When the last lid was lowered into position in January 1967, the roof vaults of the opera complex were finally complete. Their tiles glistened in the sunlight in V-shaped arrays to striking visual effect.

Sailing into a Storm. The stage was set for the construction of the interiors of the major and minor halls. Three aspects required particular attention: the seating, both capacity and arrangement, the acoustics, and the large glass walls that close off the open ends of the vaulted lounge areas. These kinds of matters had been dealt with before. Surely, therefore, the opera project would now sail toward a successful conclusion. Not so fast! Problems over seating capacity, acoustical properties, cost overruns, and the timely completion of the project had been building. Divisive party politics would make them worse.

The costs of the project rose in proportion to the technical challenges and the consequent delays in the construction. The cost estimates that Utzon provided soared from 9.6 million Australian dollars in March 1958, to 18.6 million in August 1961, to 27.5 million in April 1962, and to 50 million in July 1965. (During this time the Australian dollar was worth about 89 U.S. cents.) The lotteries that financed the building were becoming more and more frequent. (The total cost would eventually rise to 102 million Australian dollars when the structure was finished.)

Originally, the major hall was to serve as both opera house and concert hall, each with its own configuration of seats. It was to be the home of Opera Australia as well as the Sydney Symphony Orchestra. But Utzon's design put pressure on the available space. For opera, the small wing spaces for the main stages were problematic, and stage towers with large elevators were included in the design instead. Operatic stage sets would be moved not horizontally but vertically, in the same way that the planes of an aircraft carrier are moved to the deck from the hangars below. For concert hall use, the Australian Broadcasting Commission wanted a seating capacity of at least 2800. It also wanted a stage area big enough not only for the orchestra, but also for a large choir. Expectations about the acoustics added to the problem. The Commission wanted a reverberation time of 2.0 seconds in the middle frequencies. This meant that the major hall needed to have an interior volume of at least 1 million cubic feet.

The Opera Project had been the idea of Australia's Labor Party. But in 1965 the Liberal Party won national office after 24 years of Labor rule. In their campaign the Liberals promised to do something about the escalating costs and the delays in the completion of the opera complex. The new Liberal Minister for Public Works charged Utzon with having supplied only vague information about his design for the acoustical aspects of the interiors and decided to withhold further funding for the construction. On February 28, 1966, when the shells were nearing completion and when the most formidable challenges facing the construction had been met, Utzon was forced to resign. A panel of architects was appointed to replace him. With Arup's firm by their side, they finished the structure. The status of the major hall was changed from a dual-purpose facility to a single-purpose concert hall. The machinery that had already been installed for moving stage sets was removed and the opera house was moved into the minor hall. The glass facades for the lounge areas were completed in a design different from what Utzon had intended. They now descend from the top in the shape of elliptical cylinders and fan out at their base in cone-shaped form. The facility opened on October 20, 1973, with Queen Elizabeth in attendance. Sydney celebrated with Beethoven's Ninth Symphony and its "Ode to Joy."

The Sydney Opera is proof that the age-old qualities of great architecture—suitability to the site, response to light, creation of space, scale and proportion, and use of appropriate materials—are still the crucial qualities that a great building needs to have. It is a contemporary building that can be measured against the achievements of past civilizations. It ensures Utzon's position as one of the most original and important architects of the twentieth century. He rethought the industrial prefabrication of standard elements to produce complex, expressive, curving forms that depart from a functional rectangular order. As Figure 6.47 and Plate 24 confirm, the Sydney Opera is an extraordinary building. It is a large white sculpture that catches and mirrors the sky of its harbor setting with all its varied lights from dawn to dusk and day to day. According to the influential American architect Louis Kahn, "The sun did not know how beautiful its light was, until it was reflected off this building." It has captured the imagination of people the world over and has become a symbol of the city of Sydney. Jørn Utzon received the prestigious Pritzker Prize for Architecture in 2003 and his creation was named a UNESCO World Heritage Site in 2006.

We continue our story of the interconnections between building materials, technology, and design by turning to another iconic structure of the twentieth century, the Guggenheim Museum in Bilbao, Spain.

Computers, CAD, CAM, and the Guggenheim Museum in Bilbao

Let's begin by taking a step back and by engaging in what is now—given what we have learned—a rather obvious reflection about the traditional practice of architecture. Before large and complex buildings can be constructed, architects first need to produce detailed representations of it. Together with engineers they need to provide evidence that a building will be structurally

Figure 6.47. The roof vaults of the major and minor halls and the restaurant. Photo by David Messent. From Anne Watson, ed., *Building a Masterpiece: The Sydney Opera House*, Lund Humphries, 2006, p. 176. Marquand Library of Art and Archaeology

sound, that it can sustain extreme conditions of weather and earthquakes, and that it can do so over time. Architects and engineers need to establish that its interior spaces, the flow of people between them, lighting, heating, and acoustic conditions will measure up to expectations. The purpose of the architectural process is to create representations of different approaches and proposals, to analyze them with respect to the relevant criteria, and, finally, to transform a definitive representation of the proposal that is selected into full-scale, physical reality. The fundamental variables that have to be considered are the function, size, and shape of a building, feasibility, materials, and methods of construction, budgetary considerations, and time to completion and occupancy.

Traditionally, architects have relied on Euclidean geometry to give form to their ideas. They used configurations of points, straight lines, planes, arcs of circles, and sections of spheres drawn on paper with pencil, straightedge, and compass to create their designs. These graphic constructions were transformed into buildings when the tools of the carpenter and the mason gave them full-scale material form. Bricks are rectangular and fit together in a line or at right angles. They grew into flat walls in parallel courses. Saws most easily produce straight cuts and planar surfaces. Ropes—stretched, rotated, and in vertical plumb lines—guided the linear and arching shapes of buildings. Concerns about the stability of a structure—as the Gothic Age and in particular the story of the Cathedral of Milan illustrate (in Chapter 3)—were addressed by basic geometric principles informed and modified by experience gained over time. In sum, architects relied on basic Euclidean tools to draw what they could build and to build what they could draw. The buildings were made of wood, stone, brick and mortar, lifted into place by simple levers, hoists, and cranes powered by humans and pack animals. This is how humanity made its buildings for millennia. This is the story told in the first five chapters of this book.

The industrial revolution, and the technological revolution that followed, transformed this process. It provided new construction materials, including iron, cast iron, steel, and reinforced concrete. It invented cranes and derricks powered by steam, diesel, and electricity. It developed new approaches to assess the soundness of a structure that included the testing of materials and mathematical methods of structural engineering. All this allowed architects to expand the scope of what they could build. "Evolving Structures," earlier in this chapter, described some aspects of this. In the late nineteenth and early twentieth centuries, buildings of greater size and complexity, including skyscrapers with mechanically and electrically serviced interiors (lighting, heating, cooling, and, very importantly, elevators) became common. However, traditional drafting instruments such as the graduated ruler, T-square, parallel bar (for the quick placement of parallel lines), protractor (to subdivide angles), and graph paper (with its grid of rectangular modules) continued to be the tools of the architectural trade. Structural engineers were aware of sophisticated analytical methods, but for the most part questions of stability and performance of a building continued to be addressed by simple formulas, tables, empirical data, the slide rule, and the rule of thumb.

This was the state of things when Jørn Utzon first sketched his designs for the Sydney Opera competition. He drew the shapes and found the rhythms for the curving, sail-like roof structures with pencils of thick and soft lead. His assistants converted his ideas into blueprints and working drawings with traditional drafting instruments, standard techniques of descriptive geometry, and hard, sharp pencils. Utzon's design was not a cluster of boxes topped by hemispherical vaults, it was an unprecedented, massive, freely flowing sculpture. As we saw in the preceding section, the effort of translating his vision into a buildable structure in prestressed concrete and glass was herculean and took years. Utzon and Arup experimented with vault geometries from 1957 to 1962 to try to find a feasible configuration. They explored single-skin concrete shells, double-skin concrete shells, concrete shells carried by steel trussed frames, as well as parabolic and elliptical geometries. Each time a new configuration and geometry was proposed (and before the brilliant insight came to "cut" all the shells from the same sphere and to build them with concrete ribs that fan out from their base), the implications on the stability of the structure—in terms of forces, compressions, and deflections—had to be assessed. The first part of this task was to find a precise mathematical description of the surface of the shells (like the one in Chapter 4 for the spherical dome of the Hagia Sophia). With this in place, the necessary calculations could—in principle—be undertaken.

Now came the hardest part. Arup and his team realized that these calculations would require computer power far exceeding what humans could provide. The first electronic computers had become commercially available in the early 1950s, but computers powerful enough to perform the analysis of the shells were just being developed. The necessary mathematical algorithms were also being formulated at the time. These started with the representation of the stresses or deflections at a single point on a shell by a matrix. (Matrices are rectangular arrays of numbers. Simple examples were used in Discussion 3.1 to characterize permutations and symmetry transformations of designs.) The capture of the stresses and deflections on an entire shell required large systems of matrices. The instructions for the solutions of these systems needed to be expressed in a way that the programmers and their computers could deal with. These programming interfaces and languages were also just being created. At times, some of the required data was derived from conventional drawings of the proposed geometry. The data for a particular analysis would then be punched into thin rolls of paper as a long flow of small holes. After the patterns were checked and double-checked for accuracy, the tapes were fed into and "read" by the computer. The storage capacity of the computers available to Arup was about one millionth of today's personal computer, and it took them tens of hours to complete calculations that today's personal computers perform in tens of seconds. Each time a new shell geometry was considered, the technical analyses from the previous study became obsolete and had to be scrapped. The software was modified, the computer programs were rewritten, and the entire process began afresh.

Arup and his team treated the results that their computers produced with caution. The advice "if you don't know the order of magnitude of the answer [in other words, if you don't have a rough sense of what the answers might be], don't use a computer" guided their approach. Arup's engineers also relied on tests of plastic and wooden models of the shells to gain estimates of the distribution of forces and stresses. The result of one of these is captured by Figure 6.48. Even though the power of available computers was minuscule and the mathematical algorithms only in initial phases of development, the structural assessment of the massive sail-like shells could not have been undertaken without them. The fact is that Utzon's imaginative shells would not have seen the light of day without computer technology.

Today's computer technology has made such analyses routine. Elementary systems of computer-aided design (CAD) are accurate and efficient versions of traditional drafting instruments. They provide basic graphic components such as points, straight lines, circles, and ellipses. By combining and modifying these on computer screens with copy, cut, paste, drag, snap to

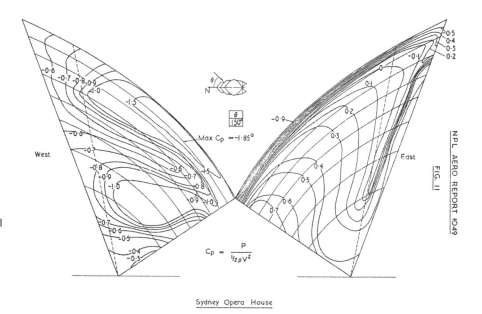

Figure 6.48. The pressure distribution for shell A2 from a wind tunnel test on a model of the shell. From Yuzo Mikami, *Utzon's Sphere: Sydney Opera House—How It Was Designed and Built*, photographed by Osamu Murai, Shokokusha, Tokyo, 2001, p. 312. Marquand Library of Art and Archaeology, Princeton University Library

grid, scaling, and shearing operations, it is simple to produce sophisticated designs. (Such a system, Adobe Illustrator, was used to produce most of the graphics for this book.) In the 1960s and early 1970s computer scientists developed new software for the digital modeling of curved surfaces with systems such as NURBS. The splines the latter makes use of (NURBS is short for nonuniform rational B-splines) are special mathematical functions defined piecewise by polynomials. Such curved-surface CAD systems have become essential tools in automobile, aircraft, and ship design. The entertainment industry uses such software to create cartoon characters and to bring them to life in two- and three-dimensional animations. Ever increasing computer power and ever more sophisticated display technology let designers execute free-form curved surfaces—still or moving—on computer screens with the same or greater ease with which traditional architects execute lines, planes, circles, cylinders, and spheres on paper.

The architect Frank Gehry (1929–) and his team first linked the practice of architecture to this world of digital design in the early 1990s. The monumental fish sculpture (1989–1992), commissioned for the waterfront in Barcelona on the occasion of the 1992 Olympics, was the groundbreaking project. Its flowing, curving surfaces were digitally modeled using a CAD system developed for use in the aircraft industry. The digital model was used for both the development of the design and the analysis of the structure. It also generated the detailed documentation needed for the construction that would have been provided by traditional drawings before. The Barcelona fish set the stage for the successful application of digital curved surface modeling to Gehry's later and larger projects such as the Guggenheim Museum in Bilbao (1991–1997) and the Disney Concert Hall in Los Angeles (1999–2003).

Gehry used the following process both for the design of the historic museum in Bilbao and for its execution. His sketches of curves on paper and shaping of surfaces on models gave free flow to his initial ideas. Many large, freehand physical models were made to explore them. Sophisticated digitizers were used to capture the coordinates of vertices, edges, and other surface elements of the more definitive models. With software such as NURBS it was then possible to devise three-dimensional digital versions of these models that captured the subtleties and nuances of their flowing shapes with great accuracy. Computer-controlled, three-dimensional printers and multi-axis milling machines subsequently produced new physical models. These were compared with the originals and their shapes modified and adjusted until the design team was satisfied with the match. The same strategy is used by designers of automobiles. They too combine freehand sketches made by felt-tipped pens, carefully shaped clay models, and sophisticated computer simulations to create their designs.

At this point, the remarkable advantages of these computer methods become apparent. Recall the difficulties that Arup and his team encountered in their efforts to assess the implications of a given shell geometry on the structural properties of the vaults. For each geometry, the calculations of the compressions, tensions, deflections, and rotational forces required a new large data set, reconfigured versions of the software, and new lengthy computer analyses. All this is easy with today's geometric modeling systems and the fast computers that realize them. Finite Element Methods (FEM) derived from close approximations of surfaces by planar regions are a critical ingredient. The way in which a sphere can be approximated by inscribing the triangular structure of an icosahedron into it (refer to Discussion 4.1 and Figure 4.45e) gives a flavor of what is involved. Increasing the number of points of contact between the linear structure and the surface improves the approximation but increases the size of the system of equations that needs to be solved. The fact that such systems can now be solved speedily with high-performance computers allows FEM to analyze the physical characteristics of complicated structures accurately and efficiently. With FEM, the bends and twists of a structure can be simulated, the distribution of stresses and displacements can be visualized, and the airflow in and around the structure, the effects of thermal conditions, and the acoustical qualities can be studied. With this technology a design can be digitally analyzed, refined, and optimized before it is built.

Once Gehry's design for the Guggenheim Museum of Bilbao was finalized and its performance characteristics understood, the construction began. This was facilitated by computer-aided manufacture (CAM). Just as a laser printer automatically translates and converts a text file into a printed paper output, so CAM fabrication machines can translate three-dimensional digital files into full-scale reality. This can be done with high speed and close to perfect precision. Numerically controlled laser cutters, water-jet cutters, and routers can cut, shape, and transform flat materials into complex shapes with great efficiency. Multi-axis milling machines extend this to the computer-controlled fabrication of three-dimensional components. With such processes, CAM technology erected the large, complex primary steel frame structure of the museum in modular, three meter square sections. The abstract geometry of this frame is specified with complete precision even though it is not symmetric and not repetitive. Between the primary steel structure and the exterior surface of the Guggenheim Museum there are several layers. An inner layer, made of galvanized steel tubes arranged in horizontal ladderlike configurations, establishes the horizontal curvature of the exterior. This layer is connected to the primary structure with joints that were adjustable in all directions. Another layer carries galvanized steel sheathing that is covered with thermal insulation on the inside and a waterproofing asphalt membrane on the outside. It determines the vertical curvature of the exterior. The primary structure and layers are designed to expand and contract to adjust for temperature conditions. Much of the outer surface of the building consists of sheets of titanium. The titanium was chemically treated and cut into flat panels using CAM machinery. The panels were bent and twisted on site, fit to the curving structure, and attached. Only four standard sizes were needed for 80 percent of the titanium-clad areas. The remaining 20 percent required 16 different types of panels. The flowing titanium panels give the museum its complex, sculpted form. (Titanium is costly, so that it was a fortunate coincidence that Russia, the world's largest titanium producer, put huge amounts of the metal on the market right around the time it was needed for the construction. All the titanium that was required was purchased at the dramatically lower price that resulted.) The exterior also has extensive stone surfaces and steel and glass walls. These too were fabricated with the aid of three-dimensional computer-controlled CAM machinery. The assembly of the building materials was facilitated by laser positioning devices guided by three-dimensional coordinate geometry and undertaken by computer-driven robots. In all, 270,000 square feet of titanium, 66,000 square feet of glass, and 1.2 million cubic feet of limestone were used in the construction. Neither the cost of $100 million dollars nor the six years that it took to build were excessive by major museum standards.

When Bilbao's Guggenheim Museum was opened to the public in 1997, it was quickly hailed as one of the world's most spectacular buildings. Figure 6.49 shows what a revolutionary structure it is. The architect Philip Johnson called it "the greatest building of our time." A "Bilbao Effect" has revitalized parts of the city and put Bilbao on the map as a tourist destination.

Jørn Utzon's Sydney Opera and Frank Gehry's Bilbao Guggenheim Museum have both become iconic buildings that have captured the public's imagination. Whereas Utzon had to rely on handcrafted drawings and scale models in his explorations of visual, spatial, and structural effects, Gehry could call on visualization software to produce, almost instantaneously,

Figure 6.49. The Guggenheim Museum in Bilbao (1991–1997). Photo by Ardfern

whatever views and information he needed. The spherical solution that facilitated the completion of the Sydney Opera was elegant and the forms of the finished shells beautiful, but their geometry is stiffer and more rigid than the soaring sails that Utzon had originally imagined. Gehry's design for the Bilbao Guggenheim Museum was a freely flowing sculptural shape, but by that time, the accurate analysis and construction of such shapes was no longer a problem. Architecture, just like modern art, is today able to create, explore, and pursue just about any imaginable form. It is an interesting concluding footnote that the firm Arup and Partners of London, now one of the world's leading engineering firms, pioneered some of the CAD/CAM/FEM technology that has made this possible.

Problems and Discussions

The problems and discussions that follow deal with various mathematical matters that arise from the themes developed in this chapter.

Problem 1. The domes of St. Paul's and of the United States Capitol are similar in size. Study the basic structural configurations of the two domes as these are depicted in the cross sections of Figure 6.2 for St. Paul's and in Figures 6.8 and 6.10 for the Capitol Building. Compare the weights of the two domes and discuss the forces and stresses on the domes and their primary support structures.

It was the objective of the discussion about systems of strings and weights in the section "Hanging Chains and Rising Domes" to assume as given the forces R_1 and R_2, the weights W_1, W_2 and W_3, and the vertical distances between them and to determine the precise configuration of the string segments and the tensions on them. Problems 2 through 7 continue the exploration of such systems. A unit of distance and a unit of weight are given.

Problem 2. Draw (carefully with a straightedge and compass) the funicular polygons that correspond to the force polygons of Figures 6.20c and 6.20d.

Problems 3 and 4 rely on Figure 6.50a. The figure shows a string ACB held in place by the forces R_1 at A, R_2 at B, and the weight W it supports.

Problem 3. Suppose that $d_1 = 4$, $d_2 = 6$ and that $\theta_1 = 45°$. Determine the lengths of the two string segments AC and CB. Given that the magnitude of the horizontal components of R_1 and R_2 are both 75, compute the weight W and the tensions T_1 and T_2.

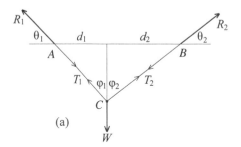

(a)

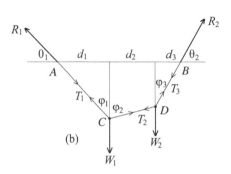

(b)

Figure 6.50

Problem 4. Suppose that $d_1 = 4$, $d_2 = 6$, and that the string segment AC $= 8$. Compute the length of the string ACB and the sines and cosines of the angles θ_1 and θ_2. Suppose that $W = 100$ and determine the magnitudes of the forces R_1 and R_2.

Problems 5 and 6 make use of Figure 6.50b. This figure shows a string $ACDB$ held in place by the forces R_1 at A, R_2 at B, and the weights W_1 and W_2 that it carries.

Problem 5. Suppose that $\theta_1 = 45°$, $\theta_2 = 60°$, $d_1 = 4$, $d_2 = 4$, $d_3 = 2$, $W_1 = 200$, and $W_2 = 150$. Determine the forces R_1 and R_2. Compute the tensions T_1, T_2, and T_3. Draw a careful force polygon for this situation.

Problem 6. Suppose that $d_1 = 4$, $d_2 = 4$, and $d_3 = 2$, and that the points C and D are, respectively, 5 and 4 units below the segment AB. Compute the length of the string $ACDB$. Assume that $W_1 + W_2 = 450$ and compute the tensions T_1, T_2, and T_3 as well as W_1 and W_2.

The diagrams of loaded strings (such as that of Figure 6.15) were based on the assumption that the forces involved and the string that they put under tension are in equilibrium. Any set of weights W_1, W_2, and W_3 and upward vertical forces V_1 and V_2 such that the magnitudes of $W_1 + W_2 + W_3$ and $V_1 + V_2$ are equal, together with horizontal forces H_1 and H_2 of the same magnitude, determine a force polygon. However, if the horizontal distances d_1, d_2, d_3, and d_4 between the points A, B and the weights are also preassigned, it may not be the case that there *exists a string in equilibrium* that satisfies the given force *and* distance requirements. The next problem illustrates this point.

Problem 7. The string and system of forces shown in Figure 6.15 are in equilibrium. The weights W_1, W_2, and W_3 are 100, 150, and 125 pounds, respectively. The magnitudes of the vertical components of R_1 and R_2 are 175 and 200 pounds, respectively, and the magnitude of the horizontal components is 150 pounds. Let a force of $100 \cdot x$ pounds be represented by a vector of length x inches and draw the force polygon for this system using a straightedge and compass. Suppose that the horizontal distances between the points A, B and the weights are, respectively, $d_1 = 1$, $d_2 = 2$, and $d_3 = 1.5$ feet, but that d_4 is (as yet) undetermined. Represent a string of length x feet by a segment of length x inches and draw the system of strings to scale (again with straightedge and compass). What does the distance d_4 have to be? What is the total length of the string?

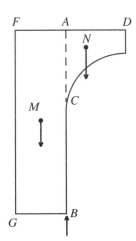

Figure 6.51. From Stephen Wren's *Parentalia*, London, 1750

One of Christopher Wren's studies of an arch includes Figure 6.51. The figure considers the half-arch DAC and its supporting pier $AFGB$. The

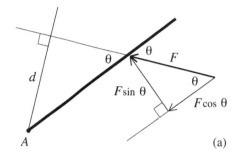

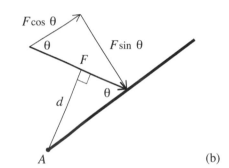

Figure 6.52. Moment diagrams

points N and M are the respective centers of mass of the half-arch and the pier.

Problem 8. At one time Wren thought that the arch is stable if the moment of force produced by the pier around the point B exceeds the moment of force produced by the half-arch around B (and if the same is true for the other half of the arch). Assess this explanation of the stability of an arch.

Problem 9. Figure 6.52 shows two beams acted on by a force. In each case, the beam is fixed at the point A, F is the magnitude of the force, d is the perpendicular distance from A to the line of action of the force, and θ is the angle between this line and the beam. In Figure 6.52a, $F = 800$ pounds, $d = 10$ feet, and $\theta = 50°$. Why is the moment of force of F about the point A equal to 8,000 pound-feet? Resolve the acting force into components perpendicular and parallel to the beam. Discuss the action of each component and compute the moment of force again. Repeat this analysis in the case of Figure 6.52b, where $F = 1200$ pounds, $d = 8$ feet, and $\theta = 60°$.

Problem 10. Review Coulomb's study of hinging in "Analyzing Structures: Statics and Materials." Then focus on the point a and explain why there is no hinging at a, if $Hy_1 - W_\alpha x_1 \le (\tau_\alpha A_\alpha) z_1$. Why is $H \le \frac{W_\alpha x_1}{y_1}$ a sufficient condition for this? Why is it that if $H \le G_1$, then there will be no hinging failure anywhere along the outer boundary of the arch?

The next three problems study the geometry of the concourse beam that Arup designed for the podium of the Sydney Opera. Refer to Figures 6.35 and 6.36 and recall that Figure 6.38 depicts the cross section of the beam for an arbitrary x with $0 \le x \le 2b$. The operative unit of length is the inch in Problem 11 and the foot in Problems 12 and 13. The solutions of these problems rely on facts from sections "The Coordinate Plane" and "Coordinate System in Three Dimensions" of Chapter 4.

Problem 11. Observe from Figure 6.35 that as x moves from 0 to b, the cross section of the beam morphs from U-shape to V-shape to T-shape. The figure suggests that the V-shape occurs for $x = \frac{b}{2}$. Does the precise description of the cross section given by Figure 6.38 confirm this?

    i. Take the inch as the unit of length and use the data of Figure 6.38. Check that $A = (28, 54)$, $G = (7, 54)$, $F = (29, 7)$, and $B = (7, 7)$ in the yz-coordinates of the figure.

    ii. Use the fact that $\cos\alpha = \frac{AG}{2r}$ and $\sin\alpha = \frac{BG}{2r}$ to show that $r\cos\alpha = \frac{21}{2}$ and $r\sin\alpha = \frac{47}{2}$.

    iii. Verify that the point $D = (y, z)$ has coordinates $y = \frac{21}{2}(1 + \cos\theta) + 7$ and $z = \frac{47}{2}(1 + \cos\theta) + 7$.

iv. Show that the slopes of the segments GD and FD are $\frac{-47(1-\cos\theta)}{21(1+\cos\theta)}$ and $\frac{-47(1+\cos\theta)}{23-21\cos\theta}$, respectively.

v. Verify that the angle θ for which the V occurs as cross section of the concourse beam satisfies $\cos\theta = \frac{1}{43}$. Show that this corresponds to $\theta \approx 88.67°$. Conclude that the V occurs at $x \approx 0.493b$ (close to, but not equal to, $\frac{b}{2} = 0.5b$).

Problem 12. The position of D in the yz-plane of Figure 6.38 depends on x, but the y- and z-coordinates of the points A and B are the same for any x. Check that $A = (\frac{28}{12}, \frac{54}{12})$ and $B = (\frac{7}{12}, \frac{7}{12})$ in the yz-plane of the figure and that $z = \frac{47}{12}y - \frac{26}{36}$ is the line that they determine. Recall that $DB = r(1 + \cos\theta)$ and conclude that $AD = r(1 - \cos(\frac{180}{b} \cdot x))$.

Problem 13. Refer to the xyz-coordinate space of Figure 6.37. Problem 12 tells us that the points $A = (0, \frac{28}{12}, \frac{54}{12})$ and $B = (0, \frac{7}{12}, \frac{7}{12})$ lie in the plane P determined by the equation $z = \frac{47}{12}y - \frac{26}{36}$.

i. Why is the point D that determines the geometry of the concourse beam in the plane P for any x with $0 \leq x \leq 2b$?

ii. Set up the following uv-coordinate system for the plane P. The origin is the point $(0, \frac{28}{12}, \frac{54}{12})$ and the unit of length is the foot. The u-axis runs parallel to the x-axis and its positive part points in the same direction as the positive part of the x-axis. The v-axis is perpendicular to the u-axis, lies in the plane $x = 0$ (the yz-coordinate plane of Figure 6.37), and the direction of the positive part of the v-axis is determined by the requirement that the point $B = (0, \frac{7}{12}, \frac{7}{12})$ lies on it. Sketch these two coordinate axes into the xyz-space of Figure 6.37.

iii. Consider the function $v = f(u) = r(1 - \cos(\frac{180}{b} \cdot u))$ with u in the interval $0 \leq u \leq 2b$. Discuss the relevance of the graph of this function to the changing position of the point D in the uv-plane P. Sketch a graph of the function $v = f(u)$ by relying on the points with u-coordinates $0, \frac{b}{4}, \frac{b}{2}, \frac{3b}{4}, b, \frac{5b}{4}, \frac{3b}{2}, \frac{7b}{4}$, and $2b$.

Discussion 6.1. The Flashy Forms of Gaudí. The Spanish architect Antoni Gaudí (1852–1926) built extravagant structures that pushed traditional masonry materials to new limits. His masterpiece is La Sagrada Familia, the Church of the Holy Family, in Barcelona. Its soaring vaults and spires are at once playful, surreal, and inspiring. They are a new interpretation of the Gothic form. Gaudí's designs relied on careful and elaborate studies of complex systems of loaded strings. Figure 6.53 shows one of his sketches and Figure 6.54 depicts one of his models. He photographed the models he created from various angles and used them to design the arching forms of the vaults shown in Figure 6.55. The construction of the huge church has

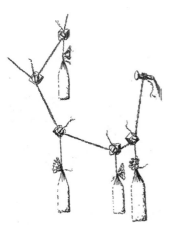

Figure 6.53. One of Gaudí's sketches

Figure 6.54. One of Gaudí's models of a configuration of strings under loads. Photo by Cleftref

Figure 6.55. Some of the vaults of La Sagrada Familia. Photo by Sarika Bedi

proceeded off and on since the 1880s and continues today. As some of the sandstone that Gaudí used is now deteriorating, maintenance and construction go on in parallel, and synthetic sandstone with the same appearance as the original is used for new construction. La Sagrada Familia is scheduled for completion in the year 2030. The thin spires that have been built so far are about 330 feet high but they will be mere foothills for the 560-foot-high central spire that Gaudí's vision calls for. The architectural work of Gaudí is remarkable for its range of forms, textures, and use of colors. The geometries that he devised are complex as well as free and expressive.

Discussion 6.2. Utzon's Triangles and Geodesic Triangles. Recall from "The Sydney Opera House" that all the triangles for the design of Utzon's shells are taken from the sphere of radius 246 feet. Figure 6.56, adapted from Figure 6.45b, shows shell A2 and its rib structure. Two of its circular boundaries have been extended to their point of intersection A where they form the angle α. The respective estimates of 220 feet and 86 feet for the

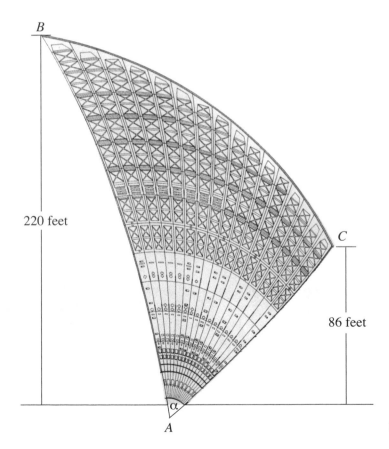

Figure 6.56. The triangle for shell A2

vertical heights of the points B and C of the curving triangle are included in the figure.

Problem 14. Use information provided about the rib structure of the shells to show that the angle α is equal to 58.4°.

If a plane intersects a sphere, then the intersection is a point or a circle. If the plane goes through the center of the sphere, then the circle is called a great circle. The radius of a great circle is the same as that of the sphere.

Problem 15. Consider two distinct points P and Q on a sphere. Explain why there is a great circle on which both P and Q lie. Why is there only one such great circle? Why is the path on the great circle between P and Q the shortest path on the sphere between P and Q? [Hint: The shortest path would have to lie on a plane through P and Q.]

We know (from the discussion describing Figure 6.41) that both of the arcs AB and AC of Figure 6.56 lie on great circles of the sphere of radius 246 feet. Figure 6.57 depicts the two arcs on their great circles along with the angles ϕ_B and ϕ_C that they subtend.

Problem 16. Refer to Figures 6.57a and 6.57b and check that $\sin\phi_B \approx 0.8943$ and $\sin\phi_C \approx 0.3496$. Conclude from this that $\phi_B \approx 63.42°$ and $\phi_C \approx 20.46°$. Use this information to derive the estimates 272 feet and 88 feet for the respective lengths of the arcs AB and AC.

A triangle on a sphere with the property that each of its three sides lies on a great circle is known as a geodesic triangle (the word "geodesic" has the Greek roots *geo* = Earth, *daiesthai* = divide).

Problem 17. Under what assumption is the triangle ABC in Figure 6.41 a spherical triangle? What aspect of Figure 6.43 tells us that none of the triangles A1, A2, A3, and A4 depicted in the figure is a geodesic triangle?

Several basic properties inform our understanding of geodesic triangles. Stating them requires the definition of the radian measure of an angle. This definition is provided by Figure 6.36. Let θ be an angle and place it into a circle of radius R in such a way that its sides meet at the center of the circle. If x is the length of the circular arc that the angle θ cuts out of the circumference, then the radian measure of the angle θ is the ratio $\frac{x}{R}$. (This ratio is the same no matter what radius R is taken for the circle.) Notice, for instance, that the angles 180°, 90°, and 45° have radian measures $\frac{\pi R}{R} = \pi$, $\frac{\frac{1}{2}\pi R}{R} = \frac{\pi}{2}$, and $\frac{\frac{1}{4}\pi R}{R} = \frac{\pi}{4}$, respectively.

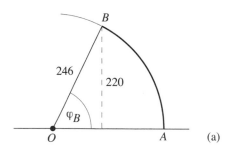

(a)

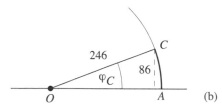

(b)

Figure 6.57. The cross sections of shell A2 along the two great circles

Now turn to Figure 6.58. It depicts a generic geodesic triangle on a sphere with center O and radius R. The angles that the sides of the triangle determine at the vertices are denoted by α, β, and γ, respectively. The lengths of the sides opposite these angles are a, b, and c, respectively. The figure shows the three radii that connect the center O to the three vertices. Observe that the angles at O between two of these radii have radian measures $\frac{a}{R}$, $\frac{b}{R}$, and $\frac{c}{R}$, respectively.

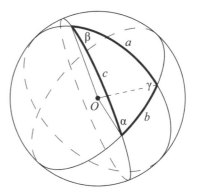

Figure 6.58. A geodesic triangle

The area of a geodesic triangle is given by Girard's Theorem (Albert Girard was a sixteenth century French mathematician and professional lute player) as

$$R^2(\alpha + \beta + \gamma - \pi).$$

This formula has the consequence that the sum of the interior angles of a geodesic triangle always exceeds 180°. A geodesic triangle satisfies the law of sines

$$\frac{\sin\alpha}{\sin\frac{a}{R}} = \frac{\sin\beta}{\sin\frac{b}{R}} = \frac{\sin\gamma}{\sin\frac{c}{R}}$$

and the two laws of cosines

$$\cos\frac{c}{R} = \left(\cos\frac{a}{R}\right)\left(\cos\frac{b}{R}\right) + \left(\sin\frac{a}{R}\right)\left(\sin\frac{b}{R}\right)(\cos\gamma) \quad \text{and}$$

$$\cos\gamma = -(\cos\alpha)(\cos\beta) + (\sin\alpha)(\sin\beta)\left(\cos\frac{c}{R}\right).$$

Refer to Problems 14 and 15 of Chapter 2 and compare these laws with the laws of sines and cosines for a triangle in the plane.

Consider an xyz-coordinate system and refer to the section "Coordinate Systems in Three Dimensions" of Chapter 4. Take the sphere $x^2 + y^2 + z^2 = R^2$. Note that its center is the origin and that its radius is R. Focus on the hemisphere above the xy-coordinate plane. Refer to Figure 4.26, for example. Cut this hemisphere with the xz-coordinate plane as well as the vertical plane given by the equation $y = x$. These two planes determine a geodesic triangle on the hemisphere. Let's denote it by T.

Problem 18. Sketch the geodesic triangle T carefully in xyz-coordinate space.

Problem 19. Refer to the sketch of T and explain why the angles at the two vertices of T that lie in the xy-coordinate plane are each 90° and why the angle at the remaining vertex is 45°. Use the fact that the surface area of a sphere of radius R is $4\pi R^2$ to show that the area of T is $\frac{1}{16}(4\pi R^2) = \frac{1}{4}\pi R^2$. Check that this agrees with the result that Girard's area formula provides.

Problem 20. Refer to the sketch of T and explain why the angle between the two radii that connect O to the vertices of T in the xy-coordinate plane is 45° and why the angle between each of the two remaining pairs of radii is equal to 90°. Organize the available information to show that the triangle T satisfies the law of sines and both laws of cosines.

When the time came to cover the curving triangles of the shells of his vaults with tiles, Utzon needed to know how many tiles he would need. The answer to this question requires an estimate for the surface area of his triangles.

Problem 21. Outline an approach involving geodesic triangles that Utzon could have used to estimate the area of the triangle for shell A2. Once you think you have it right, try to carry out the details. Will your estimate be larger or smaller than the actual area?

A deep theorem from the discipline of differential geometry—a field of mathematics that studies curved surfaces with the methods of advanced calculus—makes it possible to compute the areas of Utzon's triangles exactly. The celebrated work *General investigations of curved surfaces* of 1827 by Carl Friedrich Gauss (1777–1855) is the foundation stone of this discipline. This is the same Gauss whose insights about straightedge and compass constructions are described in Discussion 2.3. Gauss contributed significantly to several other fields, including number theory, statistics, the analysis of functions, geophysics, electrostatics, astronomy, and optics. Along with Archimedes and Newton, he is generally regarded to belong to the list of the top three mathematical scientists ever. The deep result just referred to is the Gauss-Bonnet Theorem. Gauss established a special case and the French mathematician Pierre Bonnett (1819–1892) proved a more general version about twenty years later. Go back to the triangle ABC in Figure 6.41. Let R be the radius of the sphere (instead of r). Let α, β, and γ be the radian measures of the angles at the respective vertices A, B, and C and let d be the (perpendicular) distance from the origin O to the line in the xy-plane that the vertical plane through B and C determines. An application of the Gauss-Bonnet Theorem informs us that

$$\text{Area } \Delta ABC = R^2(\alpha + \beta + \gamma - \pi) - \frac{d \cdot R}{\sqrt{R^2 - d^2}} \cdot \text{length arc } BC$$

Problem 22. The triangle ABC in Figure 6.41 is a spherical triangle in one special case. Show that in this special case the formula for the area of the triangle ABC reduces to Girardi's Theorem.

Basic Calculus and Its Application to the Analysis of Structures

At around the time that Bernini was putting the finishing touches to St. Peter's square in Rome and Wren was beginning to build the new St. Paul's Cathedral, Newton and Leibniz developed calculus, a powerful and widely applicable new mathematics. The circumstances surrounding these two geniuses could not have been more different.

Isaac Newton (1642–1727), an English university student in his early twenties, went back to the family farm when the black plague closed Cambridge University. With extraordinary insights and great powers of concentration he worked there on his own in the years 1665 and 1666. In this short period of time, he formulated the basic laws of the physics of motion, realized that they applied throughout the universe, and developed calculus, the mathematics that allowed him to extract the information that the basic laws provided. He then demonstrated that the parabolic trajectories that Galileo had described and the elliptical orbits of the planets that Kepler had documented are both much more than observed realities—they are mathematical consequences of the fundamental laws of motion. Newton delayed the published version—*The Principia Mathematica*—of this synthesis until 1687 because earlier publications had embroiled him in time-consuming disputes with contemporary scientists. *The Principia Mathematica* is (along with Darwin's *Origin of Species*) the most important scientific volume ever written.

The other genius was Wilhelm Gottfried Leibniz (1646–1716), a German who was in his late twenties when on a diplomatic mission in Paris on behalf of his patron, the duke of a Germanic state. Inspired by some of the the intellectuals of this city, he developed the calculus independently from 1673 to 1676. Leibniz's treatment of the subject was more algebraic and notationally clearer than Newton's more geometric approach. The work of Leibniz had great impact on the development of mathematics. The Jesuit priest Pierre Varignon (1654–1722) (we encountered his contributions to the graphical analysis of structures in Chapter 6, "Hanging Chains and Rising Domes") learned Leibniz's calculus and used it to rework Newton's *Principia*. The two Swiss brothers Jakob Bernoulli (1654–1705) and Johann Bernoulli (1667–1748) also learned calculus from Leibniz's publications and expanded it in new directions. Johann Bernoulli used it to solve the problem of determining the mathematical shape of the hanging chain. A French nobleman, the Marquis de L'Hospital, hired Johann Bernoulli to teach him the new mathematics and then published what he was taught as his own calculus text, *Analysis of the Infinitely Small*, in 1696. It was a calculus text from which the subject could be learned. Another student of the Bernoullis, the Swiss Leonhard Euler (1707–1783), became the most prolific and influential mathematician of the eighteenth century. His work—an incredible 70 volumes in published form—advanced calculus, developed new fields of mathematics, and applied mathematics to study mechanics, artillery, music, and ships (and a number of other subjects). The variational calculus that Euler introduced is one of the key ingredients in the powerful Finite Element Method (FEM) on which modern engineering critically relies. FEM was discussed briefly in "Computers, CAD, CAM, and the Guggenheim Museum in Bilbao" of Chapter 6.

In reference to its application to the analysis of architectural structures, it is important to note that calculus is much more than a computational method that provides solutions to relevant problems. Its central constructions are attuned to and give fundamental insights into the basic concepts that underlie our understanding of such structures. These include volume, weight, force, moment of force, and center of mass. That the methods of calculus inform a number of the structural issues that this text has engaged should therefore not come as a surprise.

The first section of this chapter presents a self-contained review of basic differential and integral calculus including volumes of rotation and lengths of curves. The second section applies calculus to obtain estimates of the volumes and weights of the shells of the domes of the Hagia Sophia and the Roman Pantheon. The third section turns its attention to a careful mathematical analysis of the ideal arch, ideal in the sense that the gravitational forces are perfectly balanced by its reaction to compression. The shape of such an arch is obtained by Hooke's strategy of inverting the shape of a uniformly loaded hanging string. The Gateway Arch in St. Louis is studied and seen to be closely related to the ideal arch. The final section of the chapter investigates centers of mass, moments of force, and analyzes Coulomb's theory of failure in an arch (discussed in Chapter 6, "Analyzing Structures: Statics and Materials") with the methods of calculus.

The Basics of Calculus

The realm of calculus is divided into two parts. There is *differential calculus*, which looks at the slope of a line and is built around the question "how can one think about the slope of a curve?" The essence of *integral calculus* is the organized addition of lots of numbers that are very small. Consider the action of some object within some system. It is often easy to compute the effect of a pointlike particle of the object in the system. Integral calculus asks the question "how can the effects of all the particles be added to tell us what the effect of the entire object is?" Our overview will conclude with a look at the *Fundamental Theorem of Calculus*. This is a miracle that tells us that differential and integral calculus, and in particular the two questions above, are tightly related.

The outline of basic calculus that follows is targeted to the applications that follow. It is intended to be an overview and a review of the subject, but not a text from which calculus can be learned. As we will soon see, calculus lives in the same coordinate plane that was studied in "The Coordinate Plane" in Chapter 4. (We will not pursue the three-dimensional version of calculus or its higher dimensional analogues.) Calculus is organized around the concept of a mathematical function. In fact, expressed briefly, calculus is *the study of mathematical functions*. A function is a rule that assigns to real numbers other real numbers in an explicitly specified way. The rule given by $f(x) = \sqrt{x^2 + x}$ is an algebraic example. The domain of a function is the set of numbers for which the rule is defined (or makes sense). In the example just given, the fact that $\sqrt{x^2 + x}$ is defined only for real numbers x with $x(x + 1) \geq 0$ means that the domain is the set of all real x with $x \geq 0$ or $x \leq -1$. The graph of a function f is the set of all points (a, b) in the coordinate plane with the property that $f(a) = b$. A function is continuous if its graph comes in one connected piece. So the graph of a continuous function has no gaps or breaks in it.

Differential Calculus. The slope, or measure of steepness, of a line has already been discussed and quantified in Chapter 4. Now our discussion turns to the steepness of curves, or, more precisely, the steepness of the graph of a function. Let $y = f(x)$ be a function. The steepness of the graph of $y = f(x)$ will generally vary with x so that it cannot be quantified by a single number. But we will see that the steepness of the graph of f can be quantified by a function. This function is the derivative of f.

At a given point on the graph of f, it makes sense to define the steepness of the graph to be the steepness of the tangent line to the graph at that point. The purpose of the discussion that follows is to make this idea precise. Refer to Figure 7.1. Fix a point x_0 on the x-axis. Let Δx be some positive number

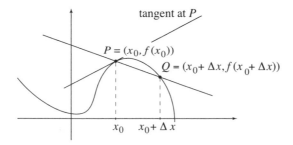

Figure 7.1

and consider the point $x_0 + \Delta x$. (The number Δx can also be negative, but we'll consider the positive case.) Suppose that everything between x_0 and $x_0 + \Delta x$ is in the domain of f. Focus on the points $P = (x_0, f(x_0))$ and $Q = (x_0 + \Delta x, f(x_0 + \Delta x))$ on the graph. In going along the graph from Q to P, the change in the x-coordinate is $\Delta x = (x_0 + \Delta x) - x_0$ and the change in the y-coordinate is $\Delta y = f(x_0 + \Delta x) - f(x_0)$. By an application of a discussion in "The Coordinate Plane" section of Chapter 4, the ratio

$$\frac{\Delta y}{\Delta x} = \frac{f(x_0 + \Delta x) - f(x_0)}{(x_0 + \Delta x) - x_0} = \frac{f(x_0 + \Delta x) - f(x_0)}{\Delta x}$$

of these changes is just the slope of the line through Q and P. Now, while keeping x_0 fixed, push Δx to zero. This keeps P fixed and pushes the point Q to P. In the process—refer to Figure 7.1—the line through Q and P rotates into the tangent line to the graph at the point $P = (x_0, f(x_0))$, and the slope $\frac{\Delta y}{\Delta x}$ of this rotating line closes in on the slope of the tangent. Because it depends on x_0, we'll designate this slope by m_{x_0}.

The process that has produced the slope m_{x_0} of the tangent at P as just described is usually expressed in the shorthand notation of limits as

$$m_{x_0} = \lim_{\Delta x \to 0} \frac{f(x_0 + \Delta x) - f(x_0)}{\Delta x}.$$

The difference $\Delta y = f(x_0 + \Delta x) - f(x_0)$ is the change in the y-coordinate of the graph of the function f that corresponds to the change $\Delta x = (x_0 + \Delta x) - x_0$ in the x-coordinate. So the ratio $\frac{\Delta y}{\Delta x} = \frac{f(x_0 + \Delta x) - f(x_0)}{\Delta x}$ is the average rate of this change as x varies from $x_0 + \Delta x$ to x_0. The limit $\lim_{\Delta x \to 0} \frac{f(x_0 + \Delta x) - f(x_0)}{\Delta x}$ is the rate of change in the y-coordinate *at* x_0.

Let x be a point in the domain of f and do for x what was done above for x_0. Now consider the rule that assigns to x the number m_x. This defines a function called the derivative of f. It is denoted by f'. This function measures the variable steepness of the graph of f. Its rule, illustrated in Figure 7.2, is given by

$$f'(x) = m_x$$

where

$$m_x = \lim_{\Delta x \to 0} \frac{f(x + \Delta x) - f(x)}{\Delta x}$$

is the slope of the tangent to the graph of f at the point $(x, f(x))$. The derivative of a function $y = f(x)$ is also denoted by $\frac{dy}{dx}$ or $\frac{d}{dx} f(x)$.

We say that the function f is differentiable at x, or the derivative of f exists at x, if the limit

$$\lim_{\Delta x \to 0} \frac{f(x + \Delta x) - f(x)}{\Delta x}$$

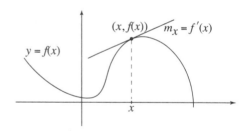

Figure 7.2

exists, or, equivalently, if the rule that defines f' makes sense for x. After thinking about the geometry of the process described above, you will see that if the graph of $f(x)$ is smooth at the point $(x, f(x))$ with a nonvertical tangent, then the function f is differentiable at x. For all the functions that we will consider it is also true that if the function f is differentiable at x, then the graph of $f(x)$ is smooth at the point $(x, f(x))$ with a nonvertical tangent. (It should be noted that functions f with points $(x_0, f(x_0))$ can be constructed such that f is differentiable at x_0, but the graph of f converges to $(x_0, f(x_0))$ from the left and the right in ever smaller sawtoothlike patterns. The graph of such a function—called "pathological" by mathematicians—is not smooth at such a point $(x_0, f(x_0))$.)

Various rules such as the sum, difference, product, quotient, power, and chain rules tell us (often in combination) how to compute derivatives of functions. For example, the power rule says that for any constant exponent r,

$$\text{the derivative of } f(x) = x^r \text{ is } f'(x) = rx^{r-1}.$$

In engineering and the physical sciences, it is often of interest to study rates of increase and decrease of relevant quantities and to estimate their largest or smallest values. Some of the problems that arise can be modeled with functions that vary smoothly. Such functions lend themselves to analysis by calculus.

The derivative of a function provides a strategy for determining the intervals over which the function is increasing or decreasing, as well as the numbers at which the function reaches a maximum or a minimum value. Expressed in terms of the graph of the function these are, respectively, the intervals on the x-axis for which the graph rises from left to right or falls from left to right, and the high or low points of the graph. Consider a function $y = f(x)$. Let c be a number with the property that f is differentiable at c with $f'(c) > 0$. This tells us that the graph of f is smooth at c and has a (nonvertical) tangent line that slopes upward. Because this tangent hugs the graph of f near the point $(c, f(c))$, it follows that the graph rises as it moves through $(c, f(c))$. This is illustrated in Figure 7.3. Completely analogously, if $f'(c) < 0$, then the graph of f falls as it moves through $(c, f(c))$. These considerations also tell us that if $y = f(x)$ has the property that its derivative is zero over a stretch $a \le x \le b$ in its domain, then its graph neither rises nor falls over this stretch. So its graph over this stretch is horizontal. Therefore, $f(x) = C$ with C a constant, for all x with $a \le x \le b$.

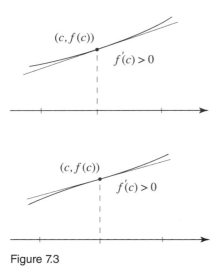

Figure 7.3

An important logical consequence of the observations above is this:

If the graph of f has a high point or a low point at $(c, f(c))$, then either $f'(c) = 0$ or $f'(c)$ does not exist (is not defined) at c.

Numbers c with the property that either $f'(c) = 0$ or $f'(c)$ does not exist are called critical numbers for the function f.

The fact just singled out provides the following strategy for analyzing the behavior of a function f. Compute the derivative f' and determine all the critical numbers of f. Suppose that $c_1 < c_2 < \ldots$ is a complete list of critical numbers. To the left of c_1, f is differentiable throughout, otherwise there would be a critical number to the left of c_1. (There can be no such critical number because c_1 is the first that arises.) Also, either $f'(x) > 0$ (so that f is increasing) for all $x < c_1$ or $f'(x) < 0$ (with f decreasing) for all $x < c_1$, because any increasing/decreasing transition point would supply a critical number to the left of c_1. It follows by the same argument that f is differentiable between any two consecutive critical numbers and either increasing throughout or decreasing throughout the interval between them. In the same way, this is also true to the right of the last critical number. It now remains to sift through all the critical numbers $c_1, c_2, \ldots$, and to determine whether the graph of f has a high point there, a low point there, or if neither of these two options occurs.

Figure 7.4 depicts the graph of a function $y = f(x)$ that illustrates what was just discussed. The numbers $c_1, c_2, \ldots, c_9$ are a complete set of critical points. Over each one of the intervals $x < c_1$, $c_1 < x < c_2$, $c_2 < x < c_3$, $\ldots$, $c_9 < x$ the graph of the function is smooth and the function is either increasing or decreasing throughout. The numbers c_1 and c_2 are critical because the graph comes to sharp points there (so f is not differentiable at either c_1 or c_2). The numbers c_4 and c_6 are critical because the gaps in the graph mean that no tangents can be placed (so that f is not differentiable at either c_4 or c_6). The number c_8 is critical because its tangent there is vertical (so f is not differentiable at c_8). At the numbers c_3, c_5, c_7, and c_9 the graph has horizontal tangents (so the derivative is equal to 0 at all these points). Notice that the graph has high points at c_1, c_3, c_6, and c_9, and low points at c_2 and c_7. There is a "bottomless pit" at c_4. The function reaches maximum values at c_1, c_3, c_6, and c_9, and minimum values at c_2 and c_7.

Finally, there is the question as to *the* largest value and *the* smallest value of a function. For the function depicted above, there is a maximum value—just

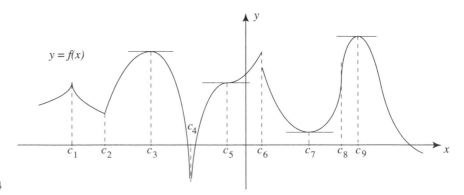

Figure 7.4

check which of $f(c_1)$, $f(c_3)$, $f(c_6)$, or $f(c_9)$ is the largest (a close look shows that $f(c_9)$ wins the prize). However, there is no smallest value, because the pit at c_4 is bottomless. The graph above, when turned upside down provides an example of a function (namely $y = -f(x)$) with a smallest value but no largest value. So a function may or may not have a largest or a smallest value. But there is a situation where *both a largest and a smallest value* are guaranteed. Suppose that the domain of a function $y = f(x)$ contains an interval $a \le x \le b$ such that the function is continuous over the interval. Then $y = f(x)$ with $a \le x \le b$ has both a largest and a smallest value.

Our overview of the calculus of derivatives is complete and we now turn to the calculus of integration. This is the other side of the two-sided coin that calculus is.

Integral Calculus. We know what the area of a rectangle is and how to compute it. In turn, we know how to compute the area of a parallelogram and a triangle. But what about the area of a region in the plane with curving boundaries? What does area mean in such a case? How do you go about assigning a number to it? One way to approach this question is suggested by the rectangular Cartesian coordinate system. Simply fill the region with very fine vertical rectangles. The sum of their areas will closely approximate the area of the region. The thinner the rectangular slices, the better the approximation will be. This approach to area is where the idea of the definite integral begins. Let a and b with $a < b$ be constants. The interval of all real numbers x satisfying $a \le x \le b$ is denoted by $[a, b]$.

Let $y = f(x)$ be a function. We will assume that f is defined for all numbers in $[a, b]$ and that it is continuous over $[a, b]$. So its graph over this interval is a connected curve, a curve with no breaks or gaps in it.

Let n be a positive integer and divide the interval $[a, b]$ into n equal pieces. Notice that each piece has length $\frac{b-a}{n}$. Set $dx = \frac{b-a}{n}$. Put in all the division points between a and b and notice that the distance between consecutive division points is dx. Refer to the number line in Figure 7.5. For a typical division point x (not equal to b) the division point to its immediate right is $x + dx$. For every division point x, starting with $x = a$ and finishing with the last point before b, form the product $f(x) \cdot dx$. Notice that the first product is $f(a)\,dx$ and the last is $f(b - dx)\,dx$. Next, add up *all* these products. With the division points labeled as

$$a = x_0 < x_1 < x_2 < \ldots < x_{n-2} < x_{n-1} < x_n = b,$$

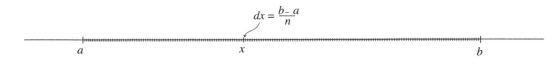

Figure 7.5

this sum is

$$f(x_0)dx + f(x_1)dx + \cdots + f(x_{n-2})dx + f(x_{n-1})\, dx$$
$$= (f(x_0) + f(x_1) + \cdots + f(x_{n-2}) + f(x_{n-1}))dx.$$

Now take n to be huge compared to the length $b - a$ of the interval. For instance, if the length of $[a, b]$ is 5 or 7 or 20, then $n = 1{,}000{,}000$ or $n = 5{,}000{,}000$ is huge. If the length of $[a, b]$ is 1000, then $n = 10^{12}$ (1 trillion) and $n = 10^{14}$ (100 trillion) is huge. With n huge, there are lots of division points and the distance $dx = \frac{b-a}{n}$ between two consecutive division points is very small, very small relative to the length of $[a, b]$. If n is huge and hence dx very small, we will use the notation

$$\int_a^b f(x)dx$$

for the sum just described. The symbol $\int$ is an elongated S that tells you that a "long" sum is being taken. Such a long sum $\int_a^b f(x)dx$ is called the definite integral of $f(x)$ from a to b. For the moment, clear everything from your shelf of facts about the definite integral (the connection with areas and volumes will be discussed shortly) and just *think of $\int_a^b f(x)dx$ as a long sum of small numbers that is obtained by the sequence of steps specified above.*

In actual fact, this "working" definition of the definite integral is not quite correct. The long summation process that was described is only an approximation of the true $\int_a^b f(x)dx$. This is so because no matter how huge the n is that you've chosen, you can always make n larger and form the long sum all over again. This can be repeated again and again. The true $\int_a^b f(x)dx$ is defined to be the limit of this process. *However, let's repeat*: if the number n of subdivisions is huge, then the long sum that was described is essentially equal to the true $\int_a^b f(x)dx$ in the same way that, say, 11.9999999999 is essentially equal to 12.

Sums of the form $\int_a^b f(x)dx$ arise in many different ways and have many different interpretations. They can represent area, volume, or the length of a curve. In physics and engineering, they can represent fundamental concepts such as force, energy, momentum, and moment of force.

We will now see how the definite integral arises in the computation of areas. Suppose that a continuous function f satisfies $f(x) \geq 0$ for all x in $[a, b]$. So the graph of f lies over $[a, b]$ on the x axis. As we've done before, let n be a positive integer and let the points

$$a = x_0 < x_1 < \ldots < x_{n-1} < x_n = b$$

divide the interval $[a, b]$ into n equal pieces each of length $dx = \frac{b-a}{n}$. If x_i is a typical division point, then dx is the distance to the division point to its right. The product $f(x_i) \cdot dx$ is the area of a very thin rectangle of height $f(x_i)$ and base dx. Do this for i ranging from 0 to $n - 1$. The sum of the areas of all the rectangles obtained in this way is

$$f(x_0)dx + f(x_1)dx + f(x_2)dx + \ldots + f(x_{n-2})dx + f(x_{n-1})dx.$$

These rectangles are shown in Figure 7.6. To distinguish one rectangle from the next, they alternate in color between black and gray. The n chosen for the figure is relatively small, so that we can see what is going on. But now suppose that n is huge compared to the distance $b - a$. Then the rectangles fill out the area A under the graph of f and over the interval $[a, b]$. Therefore, the sum above is a tight estimate of A. But this sum is also a tight estimate of $\int_a^b f(x)dx$. After considering the limit process referred to earlier, we can conclude that $A = \int_a^b f(x)dx$.

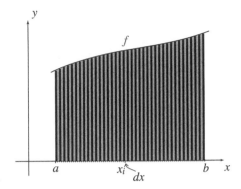

Figure 7.6

Now that we know what the number $\int_a^b f(x)dx$ means, we'll develop a strategy for calculating it. Going through the process of adding a myriad of tiny terms would be laborious at best and impossible at worst. So the question is: Is there an effective way of computing this number?

A continuous function f and an interval $[a, b]$ in the domain of the function are given. Let F be a function that has f as derivative. So $F'(x) = f(x)$ for all x in $[a, b]$. Such a function F is called an antiderivative of f. The definition of the derivative tells us that

$$\lim_{\Delta x \to 0} \frac{F(x + \Delta x) - F(x)}{\Delta x} = f(x).$$

This says that when x is fixed and Δx is pushed to zero, the ratio $\frac{F(x+\Delta x) - F(x)}{\Delta x}$ closes in on $f(x)$. It follows, for a given x and a small dx, that $f(x)$ and $\frac{F(x+dx) - F(x)}{dx}$ are close to each other. In view of the limit, the smaller the dx is, the better the approximation is. Therefore,

$$f(x)dx \approx F(x + dx) - F(x),$$

and the smaller the dx, the better the approximation. With dx small enough, the two terms are essentially equal.

Notice that we are using both Δx and dx. What's the difference? We use Δx to label a quantity that will be pushed to zero, and dx for a tiny quantity that is fixed within the particular discussion.

Let's return to the general situation of a continuous function f with domain $[a, b]$, an antiderivative F of f, and the approximation $f(x)dx \approx F(x + dx) - F(x)$. As required in the definition of the definite integral $\int_a^b f(x)dx$, let n be a huge number and divide the interval $[a, b]$ into n equal segments. Label the division points between a and b by

$$a = x_0 < x_1 < \ldots < x_{n-1} < x_n = b.$$

The distance between any two consecutive points is $dx = \frac{b-a}{n}$. So $x_{i+1} = x_i + dx$. Because the number n is huge, dx is very small. Making use of the approximation just discussed, we get

$$f(x_i)dx \approx F(x_i + dx) - F(x_i) = F(x_{i+1}) - F(x_i).$$

This is valid for $i = 0, 1, \ldots, n-1$. Using this approximation in succession for $i = 0, i = 1, \ldots,$ and finally for $i = n-1$, tells us that the sum of all the terms $f(x_i)dx$ is approximately equal to

$$[F(x_1) - F(a)] + [F(x_2) - F(x_1)] + [F(x_3) - F(x_2)] + \cdots$$
$$+ [F(x_{n-1}) - F(x_{n-2})] + [F(b) - F(x_{n-1})].$$

Notice that the terms $F(x_1) - F(x_1)$, $F(x_2) - F(x_2)$, and so on, and finally $F(x_{n-1}) - F(x_{n-1})$, subtract off in pairs, so that only $F(b) - F(a)$ remains. It follows that $F(b) - F(a)$ is a tight estimate of the sum

$$f(x_0)dx + f(x_1)dx + f(x_2)dx + \cdots + f(x_{n-2})dx + f(x_{n-1})dx.$$

By going to the limit all the approximations above tighten up, and it follows that $\int_a^b f(x)dx$ is equal to $F(b) - F(a)$. This equality is the Fundamental Theorem of Calculus. Let's summarize. Let a continuous function f and an interval $[a, b]$ in the domain of the function be given. The Fundamental Theorem of Calculus tells us that

$$\int_a^b f(x)dx = F(b) - F(a),$$

where F is an antiderivative of f. The Fundamental Theorem of Calculus provides a strategy—in principle—for computing $\int_a^b f(x)dx$. Find any antiderivative F of the function f and form the difference $F(x)\big|_a^b = F(b) - F(a)$. A warning is in order. Finding an explicit antiderivative F of a function f can be a difficult and sometimes impossible task.

We'll conclude this discussion of integral calculus with one last comment. For a given function $y = f(x)$ and an interval $a \leq x \leq b$, the definite integral is a number. This number does not depend on the way the variable of the function is labeled. For example, $\int_1^4 x^2 dx$, $\int_1^4 t^2 dt$, and $\int_1^4 u^2 du$ are all equal to the same thing (namely $\frac{x^3}{3}\big|_1^4 = \frac{t^3}{3}\big|_1^4 = \frac{u^3}{3}\big|_1^4 = \frac{4^3}{3} - \frac{1^3}{3} = 21$). If the upper (or lower) limit is allowed to vary, then the definite integral becomes a function. For example, $\int_a^x f(t)dt$ is a function of x. (So as not to overuse x, we've chosen t as the variable of the function f.) The Fundamental Theorem of Calculus tells us that if F is an antiderivative of f, then $\int_a^x f(t)dt = F(x) - F(a)$. It follows that $\frac{d}{dx}\int_a^x f(t)dt = F'(x) = f(x)$, so that the function $\int_a^x f(t)dt$ is also an antiderivative of f.

Volumes of Rotation and Lengths of Curves. We begin with the fact that the volume of a cylindrical disc is equal to the area of the base times the height. It follows that the disc with height h and circular base of radius r (depicted in Figure 7.7) has volume $\pi r^2 h$.

Let f be a continuous function that satisfies $f(x) \geq 0$ for all x in an interval $[a, b]$. As before, take a huge positive integer n and divide the interval $[a, b]$ into n pieces each of length $dx = \frac{b-a}{n}$. The subdivision points and the graph

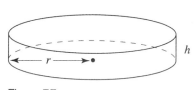

Figure 7.7

of f determine very thin rectangles. A typical one is shown in Figure 7.8. Its left edge is at x, it is dx thick, and its height is $f(x)$. Now rotate the region bounded by the graph, the x-axis, and the lines $x = a$ and $x = b$ one complete revolution about the x-axis. Observe that the volume V of the resulting pear-shaped solid is tightly approximated by adding the volumes of the thin cylindrical discs given by the rotation of all the rectangles that the subdivision points from a to $b - dx$ determine. The typical disc (resulting from the rectangle of the figure) has circular area $\pi \cdot (f(x))^2$ and height dx, so it has volume $\pi \cdot (f(x))^2 \cdot dx$. So V is approximated by our working definition of $\int_a^b \pi f(x)^2 dx$. By going to the limit, it follows that

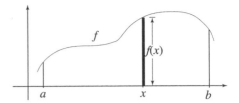

Figure 7.8

$$V = \int_a^b \pi f(x)^2 dx.$$

Refer back to the connection between the integral and area and notice that this definite integral is also equal to the area under the graph of the function $\pi (f(x))^2$ from a to b.

The definite integral can also be applied to the computation of the lengths of curves. This is done as follows. Let f be a continuous function and let $P = (a, c)$ and $Q = (b, d)$ be two points on its graph. See Figure 7.9. Here is the strategy for computing the length L of the curve between the points P and Q. Once more let n be a huge positive integer and divide the interval $[a, b]$ into n equal pieces each of length $dx = \frac{b-a}{n}$. Let x be a typical division point. Again, $x + dx$ is the very next one. Let (x, y) be the point on the graph above x, and use a segment of length dx and the tangent line at (x, y) to construct a right triangle at (x, y). Denote its height by dy. Figure 7.10 shows this triangle "under a microscope." Notice that the slope of the tangent at (x, y) is $\frac{dy}{dx}$. Therefore, $f'(x) = \frac{dy}{dx}$. By the Pythagorean Theorem, the length of the hypotenuse of the triangle is $\sqrt{(dx)^2 + (dy)^2}$. A factoring maneuver tells us that $(dx)^2 + (dy)^2 = [1 + (\frac{dy}{dx})^2](dx)^2$, so that

$$\sqrt{(dx)^2 + (dy)^2} = \sqrt{\left[1 + \left(\frac{dy}{dx}\right)^2\right](dx)^2} = \sqrt{1 + \left(\frac{dy}{dx}\right)^2}\, dx = \sqrt{1 + f'(x)^2}\, dx.$$

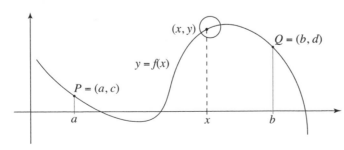

Figure 7.9

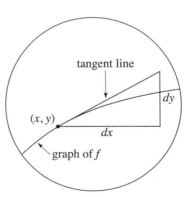

Figure 7.10

Since dx is extremely small, the length of the arc of the graph of $f(x)$ inside the tiny triangle is, for all computational purposes, equal to the hypotenuse of the triangle. The length L of the curve from P to Q is equal to the sum of the lengths of all of these tiny arcs from the point P to the point Q. This is in turn tightly approximated by the long sum of all of the terms $\sqrt{1+(f'(x))^2}\,dx$ as the subdivision points range from a to $b-dx$. Going to the limit, we can conclude that

$$L = \int_a^b \sqrt{1+f'(x)^2}\,dx.$$

With this formula, our survey of the basics of calculus is complete.

Volumes of Spherical Domes

It has been a recurring theme in this book that the outward forces that the weight of a large dome generates are considerable and present a problem for the stability of the structure. The calculus described in the last section provides the mathematical tools that allow us to estimate the weights of the domes of the Hagia Sophia (as discussed in Chapter 3) and the Pantheon in Rome (as described in Chapter 2).

Weighing the Dome of the Hagia Sophia. Figure 7.11 is adapted from Figure 3.3. It shows the cross section of the dome of the Hagia Sophia above the circular gallery of windows around the base of the dome and captures essential information about its shell. We will assume that the inner and outer surfaces of the shell are sections of spheres and focus our attention on the volume of the shell. (Today, these inner and outer surfaces are no longer spherical. The various reconstructions over the centuries have resulted in distortions. See Figure 3.6.) The spheres in question have the same center and radii $R = 52.5$ feet and $r = 50$ feet, respectively. So the thickness of the shell is 2.5 feet. The average weight per cubic foot of the combination of masonry and mortar used in the construction of the dome is 110 pounds per cubic foot. The slanted lines in the figure from the common center of the

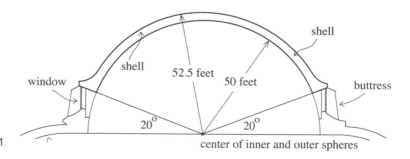

Figure 7.11

inner and outer circles that determine the boundaries of the dome above the row of windows rise at an angle of about 20° with the horizontal. The strategy will be to estimate the volume of the shell in cubic feet and then to multiply this by 110 to get its weight in pounds.

In Figure 7.12 the cross section of the left side of the dome is positioned in an xy-plane. It has been rotated by 90° to put the computation of the volume of the shell into the framework of the method discussed in the paragraph "Volumes of Rotation and Lengths of Curves" of the preceding section. Because $\cos 70° = \frac{a}{r}$, we get $a = r\cos 70° \approx 0.34r = 17$. So we will take $a = 17$ feet. The volume of the shell will be estimated as follows. The volume V_1 obtained by rotating about the x-axis the region below the upper semicircle and above the interval $[a, R]$ is computed first. Then the volume V_2 obtained by rotating about the x-axis the region below the lower semicircle and above the interval $[a, r]$ is computed. The difference $V = V_1 - V_2$ is an estimate of the volume of the shell above the circle of windows of the dome.

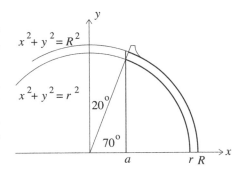

Figure 7.12

The upper half of the outer circle is the graph of the function $f(x) = \sqrt{R^2 - x^2}$ and the upper half of the inner circle is the graph of the function $g(x) = \sqrt{r^2 - x^2}$. Combining what we learned from the discussion of volumes of rotation with the Fundamental Theorem of Calculus, we get

$$
\begin{aligned}
V_1 &= \int_a^R \pi\left(\sqrt{R^2 - x^2}\right)^2 dx = \int_a^R \pi(R^2 - x^2)dx = \pi\left[R^2x - \frac{1}{3}x^3\Big|_a^R\right] \\
&= \pi\left[\left(R^3 - \frac{1}{3}R^3\right) - \left(R^2a - \frac{1}{3}a^3\right)\right] = \pi\left[\frac{2}{3}R^3 - R^2a + \frac{1}{3}a^3\right].
\end{aligned}
$$

In the same way,

$$
\begin{aligned}
V_2 &= \int_a^r \pi\left(\sqrt{r^2 - x^2}\right)^2 dx = \int_a^r \pi(r^2 - x^2)dx = \pi\left[r^2x - \frac{1}{3}x^3\Big|_a^r\right] \\
&= \pi\left[\left(r^3 - \frac{1}{3}r^3\right) - \left(r^2a - \frac{1}{3}a^3\right)\right] = \pi\left[\frac{2}{3}r^3 - r^2a + \frac{1}{3}a^3\right].
\end{aligned}
$$

Therefore,

$$
V = V_1 - V_2 = \pi\left[\frac{2}{3}R^3 - R^2a + \frac{1}{3}a^3\right] - \pi\left[\frac{2}{3}r^3 - r^2a + \frac{1}{3}a^3\right] = \pi\left[\frac{2}{3}(R^3 - r^3) - a(R^2 - r^2)\right].
$$

Putting in the values $R = 52.5$ feet, $r = 50$ feet, and $a = 17$ feet, we get

$$
\begin{aligned}
V &= \pi\left[\frac{2}{3}(52.5^3 - 50^3) - 17(52.5^2 - 50^2)\right] \approx \pi(13{,}135.42 - 4356.25) \\
&\approx 27{,}581 \text{ cubic feet.}
\end{aligned}
$$

The fact that the dimensions are given with an accuracy of only two significant figures means that the result for the volume needs to be rounded to the same level of accuracy. So an estimate for the volume of the shell of the dome is $V \approx 28{,}000$ cubic feet. The assumption that the masonry and mortar mix used in the construction of the dome weighs 110 pounds per cubic foot provides the estimate of

$$27,581 \times 110 \approx 3,000,000 \text{ pounds}$$

for the weight of the shell of the dome above the row of windows. Assuming the even distribution of this weight over the 40 supporting ribs, we get a weight of about 75,000 pounds per rib. We saw in "The Hagia Sophia" section in Chapter 3 that this generates a horizontal thrust of about 27,000 pounds on each of the 40 buttresses near the base of the dome.

The fact that the shell of the dome of the Hagia Sophia is determined by two *concentric* spheres made it relatively simple to estimate its volume. The fact that its density is essentially constant made it easy to derive an estimate of the weight of the shell from the estimate of its volume. The dome of the Roman Pantheon is more complicated on both counts. The spheres that bound the inner and outer surfaces of its shell are not concentric and the density of the concrete of the shell varies.

Weighing the Dome of the Pantheon. We'll start with the cross section of the shell of the Pantheon provided by Figure 2.45. In Figure 7.13, the circles that form the outer and inner boundary of the cross section and the centers of these circles have been added. The lower boundary of what we will call the cap of the shell is indicated by the dashed horizontal line in the figure. The cap of the shell will be the focus of our discussion.

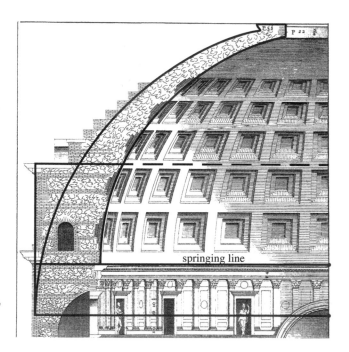

Figure 7.13. From Palladio's *I Quattro Libri*. Marquand Library of Art and Architecture, Princeton University

Figure 7.14 is an abstraction of Figure 7.13. It places half of the cross section of the dome into an *xy*-coordinate plane. The *x*-axis is determined by the horizontal diameter of the circle of the outer boundary of the cross section of the dome. The *y*-axis lies on the dome's vertical central axis. The constants R, r, D, E, a, b, and c have the following meaning:

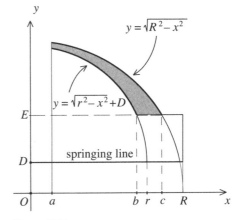

Figure 7.14

$R =$ the radius of the circle that determines the outer boundary of the cross section of the shell. The center of this circle is the origin O.

$r =$ the radius of the circle that determines the inner boundary of the cross section of the shell.

$D =$ the *y*-coordinate of the center of the inner circle (its *x*-coordinate is 0).

$E =$ the *y*-coordinate of the lower boundary (the dashed line) of the cap.

$a =$ the *x*-coordinate of the boundary of the oculus (the circular hole in the dome).

$b =$ the *x*-coordinate of the intersection of the lower boundary of the cap with the inner circle.

$c =$ the *x*-coordinate of the intersection of the lower boundary of the cap with the outer circle.

The equation of the outer circle is $x^2 + y^2 = R^2$. Solving for y gives us $y = \pm\sqrt{R^2 - x^2}$. Because only the upper half of the circle is being considered, the equation $y = \sqrt{R^2 - x^2}$ applies. The equation of the inner circle is $x^2 - (y - D)^2 = r^2$. So $y - D = \pm\sqrt{r^2 - x^2}$. We will consider only that part of the inner circle that lies above the line $y = E$. Therefore, $y \geq D$, hence $y - D = \sqrt{r^2 - x^2}$, so that the relevant equation is $y = \sqrt{r^2 - x^2} + D$. Recall from Chapter 2 that the radius of the inner surface of the shell of the Pantheon is 71 feet and that its oculus has a diameter of 24 feet. These data and Figure 2.45 can be used to obtain the estimates

$$r = 71, R = 92, D = 16, E = 48, a = 12, b = 64, \text{ and } c = 78,$$

all in feet. The computations of both the volume and the weight of the cap of the shell of the Pantheon that follow will ignore both the steprings and the coffering. We will assume that the two significant figures with which the data above are listed are reliable. The answers will be rounded off accordingly once the computations are complete.

We will now proceed as in the definition of the definite integral in the preceding section. Let n be a huge positive number and divide the interval $[a, b]$ (with a and b as listed above) into n equal pieces, each of length $dx = \frac{b-a}{n}$. Let x be a typical subdivision point and let dx be the distance to the next one. Now refer to Figure 7.15 and take a vertical segment that extends from the outer circle to the inner circle. Let its left boundary be at x and let the segment be dx thick. Revolve the segment around the *y*-axis as shown in the

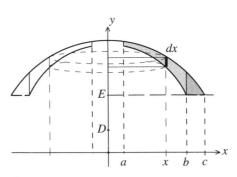

Figure 7.15

figure and notice that the segment traces out a circular ring-shaped band. The thickness of the band is dx. The radius of the circle is x, so the length of the circular band is the circumference $2\pi x$ of this circle. The height of the band is the difference $\sqrt{R^2 - x^2} - (\sqrt{r^2 - x^2} + D)$ between the y-coordinates of the two circles. Therefore the volume of this ring-shaped band is

$$2\pi x\left(\sqrt{R^2 - x^2} - \left(\sqrt{r^2 - x^2} + D\right)\right)dx.$$

(Because $2\pi x$ is the inner circumference of the band, this expression for the volume is only an approximation. But the approximation improves with smaller dx and tightens to an equality in the limit.) Take all these ring-shaped bands as x varies from a to $b - dx$ and notice that together they fill out the part V_1 of the volume of the cap of the shell (more precisely, that of the abstract form of the cap) that falls inside the vertical line $x = b$ (more precisely, inside the cylinder that the vertical line $x = b$ generates). Going to the limit, it follows that

$$V_1 = \int_a^b 2\pi x\left(\sqrt{R^2 - x^2} - \left(\sqrt{r^2 - x^2} + D\right)\right)dx.$$

To find the rest of the volume of the cap of the shell (again its abstract form) repeat for the interval $[b, c]$ what was just done for the interval $[a, b]$. Take any x with $b \le x \le c - dx$ and consider a vertical segment at x that extends from the outer circle to the lower boundary (the dashed line) of the cap. Again, let dx be the thickness of the segment. Rotate the segment as before and check that it traces out a circular band with volume

$$2\pi x\left(\sqrt{R^2 - x^2} - E\right)dx.$$

Summing up all the volumes of these circular bands and going to the limit gives us the remaining part V_2 of the volume of the cap of the shell as

$$V_2 = \int_b^c 2\pi x\left(\sqrt{R^2 - x^2} - E\right)dx.$$

The volume V of the abstract form of the cap of the shell is the sum $V = V_1 + V_2$ of the two definite integrals. To compute V_1, check that $-\frac{2}{3}\pi(R^2 - x^2)^{\frac{3}{2}} + \frac{2}{3}\pi(r^2 - x^2)^{\frac{3}{2}} - \pi Dx^2$ is an antiderivative of $2\pi x(\sqrt{R^2 - x^2} - (\sqrt{r^2 - x^2} + D))$, and then apply the Fundamental Theorem of Calculus to get

$$V_1 = \tfrac{2}{3}\pi\left((R^2 - a^2)^{\frac{3}{2}} - (R^2 - b^2)^{\frac{3}{2}}\right) - \tfrac{2}{3}\pi\left((r^2 - a^2)^{\frac{3}{2}} - (r^2 - b^2)^{\frac{3}{2}}\right) - \pi D(b^2 - a^2).$$

Plugging in $R = 92$, $a = 12$, $b = 64$, $r = 71$, $D = 16$, tells us that

$$V_1 \approx 984{,}818 - 656{,}875 - 198{,}649 = 129{,}294 \text{ cubic feet.}$$

By a similar calculation,

$$V_2 = \int_b^c 2\pi x\left(\sqrt{R^2-x^2} - E\right)dx = \frac{2}{3}\pi\left((R^2-b^2)^{\frac{3}{2}} - (R^2-c^2)^{\frac{3}{2}}\right) - \pi E(c^2-b^2).$$

Plugging in the previous values as well as $E = 48$ and $c = 78$, we get

$$V_2 \approx 361{,}442 - 299{,}783 = 61{,}659 \text{ cubic feet.}$$

Therefore an estimate for the volume of the cap of the shell of the Pantheon is

$$V \approx 129{,}294 + 61{,}659 = 190{,}953 \text{ cubic feet.}$$

What about the weight of the cap of the shell? Refer to the cross section of the dome in Figure 7.13. The density of the concrete of the shell is estimated to be 81 pounds per cubic foot from the top down to (and including) the third step ring, 94 pounds per cubic foot from there down to (and including) the fifth step ring, and 100 pounds per cubic foot from there down to the springing line of the inner dome. Taking the estimate of 191,000 cubic feet for the volume of the cap tells us that the weight W of the cap of the shell satisfies the estimate (in pounds)

$$15{,}000{,}000 \le 191{,}000 \times 81 \le W \le 190{,}000 \times 100 \le 19{,}100{,}000.$$

Recall that the volume and weight of the shell of the dome of the Hagia Sophia are about 27,500 cubic feet and 3,000,000 pounds, respectively. Is it reasonable that the volume and weight of the cap of the shell of the Pantheon should be so much greater? Yes, because a comparison of their shapes and sizes tells us that the shell of the dome of the Pantheon is *much* larger. (Problem 13 considers the matter.)

Having seen calculus in action in the study of domes, we will turn our attention to arches. You may recall (from Chapter 6, "Hanging Chains and Rising Domes") Robert Hooke's penetrating insight "as hangs the flexible line, so but inverted will stand the rigid arch." Our experience tells us that the shape of a hanging flexible string, cord, or chain is generally parabolic. But is it a parabola in precise mathematical terms or only in appearance? The discussion that follows will consider the shape of an arch obtained by inverting the hanging string and show that it is not a parabola, but the graph of a function defined by a combination of exponential terms.

The Shape of an Ideal Arch

This section studies a stable arch that satisfies the following conditions: (i) the only load on the arch is the weight of the arch, (ii) the only external support of the arch is at its base, and (iii) the gravitational forces on the arch are balanced perfectly by its reaction to the compressions that these forces generate. (The third condition parallels the reaction of the hanging flexible

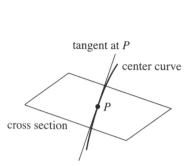

Figure 7.16

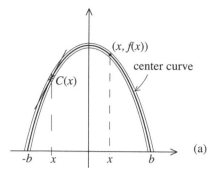

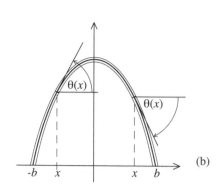

Figure 7.17

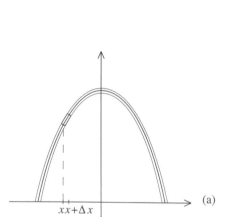

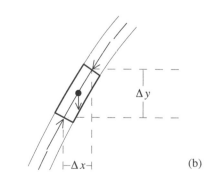

Figure 7.18

string by tension alone to the gravitational forces that act on it.) The central question will be: What is the exact shape of such an idealized arch?

The answer begins with the concept of the center curve of an arch. This curve is determined by the following property: a point P is on the center curve if P is the center of mass of the cross section of the arch with the plane perpendicular to the tangent of the curve at P. Figure 7.16 depicts what is involved. The center curve can be thought of as the line of thrust within the arch. We will assume that the arch is made of a homogeneous material and pursue the shape of the center curve of the arch.

The study that follows will use basic facts about vectors from "Dealing with Forces" in Chapter 2. Figure 7.17 shows the arch with its boundaries and center curve. We will let the center curve be the graph of a function $y = f(x)$ and determine the explicit form of this function. Let $(x, f(x))$, with $-b \leq x \leq b$, be any point on the center curve. Let $C(x)$ be the compression within the arch at that point (see Figure 7.17a) and let $\theta(x)$ be the angle that the tangent of the center curve makes with the horizontal at that point (as illustrated in Figure 7.17b). Finally, let w be the weight per unit length of the arch. We will assume that w is a constant.

Let Δx be a small positive quantity and consider the segment of the arch over $[x, x + \Delta x]$ shown in Figure 7.18a. The upper and lower boundaries of this segment are determined by the lines perpendicular to the tangents of the center curve corresponding to the coordinates x and $x + \Delta x$. The weight of the segment is approximately $w\Delta s$, where Δs is the length of the center curve falling within the segment. Focus on the enlarged version of the segment in Figure 7.18b. In line with our basic assumption, the gravitational force on the segment is counterbalanced by the difference between the compression $C(x)$ pushing up on it from below and the compression $C(x + \Delta x)$ pushing down on it from above. By balancing the vertical forces on the segment, we get $C(x) \sin \theta(x) \approx C(x + \Delta x) \sin \theta(x + \Delta x) + w\Delta s$. It follows that

$$C(x + \Delta x) \sin\theta(x + \Delta x) - C(x) \sin\theta(x) \approx -w\Delta s.$$

The smaller the Δx, the more closely the three forces act at the same point, the better the approximation. Now let $\Delta y = f(x + \Delta x) - f(x)$. By applying the Pythagorean Theorem to Figure 7.18b, we see that $(\Delta s)^2 \approx (\Delta x)^2 + (\Delta y)^2$. So

$$\Delta s \approx \sqrt{(\Delta x)^2 + (\Delta y)^2} = \sqrt{\left(1 + \frac{(\Delta y)^2}{(\Delta x)^2}\right)(\Delta x)^2} = \sqrt{1 + \left(\frac{\Delta y}{\Delta x}\right)^2}\,\Delta x.$$

Therefore,

$$\frac{C(x + \Delta x) \sin\theta(x + \Delta x) - C(x) \sin\theta(x)}{\Delta x} \approx -w\sqrt{1 + \left(\frac{\Delta y}{\Delta x}\right)^2}.$$

Now push Δx to zero and observe that three things happen simultaneously. The term on the left becomes the derivative of the function $C(x) \sin\theta(x)$, the square root term becomes $\sqrt{1 + f'(x)^2}$, and the approximation in the middle snaps to an equality. So

$$\frac{d}{dx} C(x) \sin\theta(x) = -w\sqrt{1 + \left(\frac{dy}{dx}\right)^2} = -w\sqrt{1 + f'(x)^2}.$$

By taking antiderivatives of both sides (refer back to the overview of Integral Calculus) we get

$$C(x) \sin\theta(x) = -w\int_{-b}^{x} \sqrt{1 + f'(t)^2}\,dt + \text{constant}.$$

Let's turn to the horizontal components of the forces involved. Let m be the mass per unit length of the arch. As weight is equal to mass times gravitational acceleration, $w = mg$. Because the weight of the segment is approximately $w\Delta s$, the mass of the segment is approximately $m\Delta s$. The difference between the two horizontal components of the compression is a force that will produce a horizontal acceleration a in the segment of the arch. Therefore by Newton's law of force,

$$C(x + \Delta x) \cos\theta(x + \Delta x) - C(x) \cos\theta(x) \approx (m\Delta s)a \approx am\sqrt{1 + \left(\frac{\Delta y}{\Delta x}\right)^2}\,\Delta x.$$

(This is only an approximation because the two horizontal forces do not act at precisely the same point and $m\Delta s$ is an approximation of the mass of the segment.) It follows that

$$\frac{C(x + \Delta x) \cos\theta(x + \Delta x) - C(x) \cos\theta(x)}{\Delta x} \approx am\sqrt{1 + \left(\frac{\Delta y}{\Delta x}\right)^2}.$$

Again push Δx to zero. Because the arch is stable, the segment does not move. So the acceleration is zero, and we get

$$\frac{d}{dx} C(x) \cos\theta(x) = 0.$$

Therefore $C(x)\cos\theta(x)$ is constant. Setting $x = 0$, we get that

$$C(x)\cos\theta(x) = C_0,$$

where $C_0 = C(0)\cos 0 = C(0)$ is the compression in the very top of the arch. By combining the two main conclusions that have now been reached,

$$\frac{C(x)\sin\theta(x)}{C(x)\cos\theta(x)} = -\frac{w}{C_0}\int_{-b}^{x} \sqrt{1 + f'(t)^2}\, dt + \text{constant}.$$

It follows that $\tan\theta(x) = -\frac{w}{C_0}\int_{-b}^{x} \sqrt{1 + f'(t)^2}\, dt + \text{constant}$. Because $\tan\theta(x)$ and $\frac{dy}{dx} = f'(x)$ are both equal to the slope of the tangent of the center curve at the point $(x, f(x))$,

$$\frac{dy}{dx} = -\frac{w}{C_0}\int_{-b}^{x} \sqrt{1 + f'(t)^2}\, dt + \text{constant}.$$

After taking derivatives of both sides,

$$\frac{d^2y}{dx^2} = -\frac{w}{C_0}\sqrt{1 + \left(\frac{dy}{dx}\right)^2}.$$

(It is common to denote the derivative of $f'(x)$ by $f''(x)$ or $\frac{d^2y}{dx^2}$.) It turns out that this equation—such an equation is called a differential equation—determines the function $y = f(x)$ explicitly.

Consider the exponential function $g(x) = e^x$ and define the hyperbolic sine and hyperbolic cosine functions respectively by

$$\sinh x = \frac{e^x - e^{-x}}{2} \quad \text{and} \quad \cosh x = \frac{e^x + e^{-x}}{2}.$$

(The label *hyperbolic* for these functions comes from the fact that they are related to the hyperbola in much the same way that the trigonometric functions are related to the circle.) Their graphs are depicted in Figure 7.19. For large positive values of x, the quantity $e^{-x} = \frac{1}{e^x}$ is small, so that both $\sinh x$ and $\cosh x$ are approximately equal to $\frac{1}{2}e^x$. For large negative values of x, the term $e^{-x} = \frac{1}{e^x}$ dominates. It is easy to check that $(\cosh x)^2 - (\sinh x)^2 = 1$, and because $g'(x) = e^x$, that $\frac{d}{dx}\sinh x = \cosh x$ and $\frac{d}{dx}\cosh x = \sinh x$.

Now consider the function

$$y = -\frac{C_0}{w}\cosh\left(\frac{w}{C_0}x\right) + D,$$

where w and C_0 are given above, and D is a constant (as yet unspecified). By the chain rule, $\frac{dy}{dx} = -\sinh(\frac{w}{C_0}x)$. So $\left(\frac{dy}{dx}\right)^2 = \left(\sinh(\frac{w}{C_0}x)\right)^2$, hence $1 + \left(\frac{dy}{dx}\right)^2 = 1 + (\sinh(\frac{w}{C_0}x))^2 = (\cosh(\frac{w}{C_0}x))^2$, and therefore $\sqrt{1 + \left(\frac{dy}{dx}\right)^2} = \cosh(\frac{w}{C_0}x)$. By applying the chain rule to $\frac{dy}{dx} = -\sinh(\frac{w}{C_0}x)$, we get $\frac{d^2y}{dx^2} = -\frac{w}{C_0}\cosh(\frac{w}{C_0}x)$. Combining this

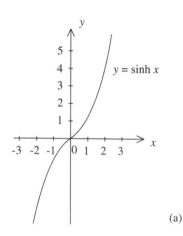

$y = \sinh x$

(a)

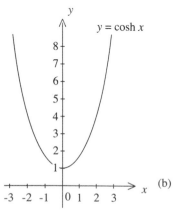

$y = \cosh x$

(b)

Figure 7.19

information tells us that the function $y = -\frac{C_0}{w} \cosh(\frac{w}{C_0}x) + D$ satisfies the same differential equation that the function $y = f(x)$ of the center curve satisfies. Because this differential equation has in essence only one solution (this fact belongs to the theory of differential equations), it follows that $y = f(x) = -\frac{C_0}{w} \cosh(\frac{w}{C_0}x) + D$ for a particular constant D. Let h be the maximal height that the center curve attains. Then $h = f(0) = -\frac{C_0}{w} \cosh(0) + D = -\frac{C_0}{w} + D$, so that $D = h + \frac{C_0}{w}$. Therefore, the center curve of the arch is the graph of the function

$$y = f(x) = -\frac{C_0}{w} \cosh\left(\frac{w}{C_0}x\right) + \left(h + \frac{C_0}{w}\right).$$

This graph provides the precise shape of an arch made of a homogeneous material in which the gravitational forces on the arch are balanced by the reactions to the compressions that these forces produce. This shape is an example of a catenary curve (in Latin, *catena* = chain).

The most impressive example of an arch with a geometry closely related to the catenary curve is the Gateway Arch in St. Louis. It was designed by Eero Saarinen (1910–1961) as the symbolic point of entry to the American West. The construction of the Gateway Arch was carried out from February 1963 to October 1965 by the firm Saarinen & Associates after the architect's death.

The Gateway Arch is 630 feet high and 630 feet wide at its base. The equation of the center curve of the arch was provided by Saarinen's structural engineers. It is expressed in blueprints as

Figure 7.20. The Gateway Arch in St. Louis. Photo by Bev Sykes

$$y = -A \cosh\left(\frac{B}{b}x\right) + (h + A),$$

where $h = 625.0925$ feet is the maximum height of the center curve, $b = 299.2239$ feet is one-half the distance between the two endpoints of the center curve at the base, $A = \frac{h}{Q_b/Q_t - 1} = 68.7672$, and $B = \cosh^{-1}\frac{Q_b}{Q_t} = 3.0022$ (meaning that $\cosh B = \frac{Q_b}{Q_t}$) where $Q_b = 1{,}262.6651$ and $Q_t = 125.1406$ are the cross-sectional areas (both in square feet) of the arch at the base and top, respectively.

The cross sections of the arch are best understood as follows. Consider a point P on the center curve and the plane through P perpendicular to the center curve. The cross section of the arch in this plane is an equilateral triangle with centroid at P *and* one vertex in the vertical plane of the center curve. The lengths of the sides of the triangles vary from 54 feet at the base of the arch to 17 feet at the top.

To see how the geometry of the Gateway Arch is related to the catenary, let's start with the curve given by the equation

$$y = -\frac{b}{B} \cosh\left(\frac{B}{b}x\right) + \left(y_0 + \frac{b}{B}\right),$$

Gateway Arch

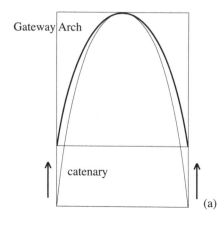

catenary

(a)

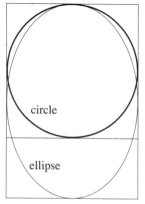

circle

ellipse

(b)

Figure 7.21

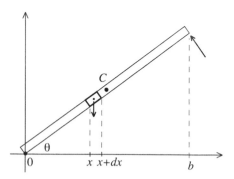

Figure 7.22

where b and B are as given above and $y_0 = y(0)$ is the height of the curve. To see that this curve is a catenary of the type discussed above, just arrange things so that $\frac{b}{B} = \frac{C_0}{w}$. Now multiply the y-coordinate of this catenary by the constant $\frac{AB}{b} = \frac{(68.7672)(3.0022)}{299.2239} \approx 0.69$. This multiplication compresses the catenary by this factor along its vertical dimension (as depicted in Figure 7.21a) and results in the function

$$f(x) = -\frac{AB}{b}\frac{b}{B}\cosh\left(\frac{B}{b}x\right) + \frac{AB}{b}\left(y_0 + \frac{b}{B}\right) = -A\cosh\left(\frac{B}{b}x\right) + (h + A).$$

So the center curve of the Gateway Arch is a compressed form of a catenary. The compression is exactly the same as the one that pushes an ellipse to a circle. Refer to Figure 7.21b. Why did Saarinen compress a catenary in this way in his design of the Gateway Arch? Perhaps because of the aesthetic consequence: the arch so compressed fits neatly into a 630 by 630 foot square.

The final section of this chapter applies the methods of calculus to study moments of force, centers of mass, and Coulomb's approach to the stability of an arch. These applications build on the account of "Analyzing Structures: Statics and Materials" in Chapter 6.

The Calculus of Moments and Centers of Mass

We begin with the important insight first discovered by Archimedes: When computing the moment of force that the weight of a component of a structure generates about some point of the structure, the entire weight of the component can be assumed to be concentrated at its center of mass. This principle is illustrated in a simple case and then applied to locate the center of mass of a segment of a semicircular arch.

Consider a thin, homogeneous beam. Figure 7.22 shows it in place in an xy-plane with one end fixed at the origin $x = 0$ and the other held in place at a point with x-coordinate b. The beam makes an angle θ with the horizontal. The beam has length L and weighs w units per one unit of length. So the total weight of the beam is wL. The point C indicates the position of the beam's center of mass. Because the beam is homogeneous, C is the geometric center of its rectangular cross section. Check, by using similar triangles, that the x-coordinate of C is $\frac{b}{2}$. Assume that the weight of the beam is concentrated at C. So the distance from the line of action of the gravitational force on the beam to the origin is $\frac{b}{2}$, and it follows from Archimedes's insight that the moment of force of the beam about the origin is $wL \cdot \frac{b}{2}$. Is this the same result that is obtained by regarding the beam to be divided into tiny pieces and adding up all the moments that they generate about the origin? Calculus will tell us that the answer is yes!

Proceeding as in the discussion Integral Calculus earlier in this chapter, let n be a huge positive number and divide the interval $[0, b]$ into n intervals each of length $dx = \frac{b}{n}$. For a typical subdivision point x, $x + dx$ is the very next one. These two points determine the small segment of the beam shown in Figure 7.22. Let l be the length of this segment. Observe that $\cos\theta = \frac{dx}{l}$. So $l = \frac{dx}{\cos\theta}$ and the weight of the small segment is $wl = w\frac{dx}{\cos\theta}$. The distance between the action of the weight of the segment and the origin $x = 0$ is very nearly equal to x, so that the moment of the segment about the origin 0 is very nearly $w\frac{dx}{\cos\theta} \cdot x = \frac{w}{\cos\theta} x\,dx$. By adding up all these "little" moments as x goes from 0 to $b - dx$ and going to the limit, we get

$$\int_0^b \frac{w}{\cos\theta} x\,dx.$$

By the Fundamental Theorem of Calculus this is equal to $[\frac{w}{\cos\theta}\frac{1}{2}x^2]\big|_0^b = \frac{w}{\cos\theta} \cdot \frac{b^2}{2} = w\frac{b}{\cos\theta} \cdot \frac{b}{2}$. Notice that $\cos\theta = \frac{b}{L}$. So $L = \frac{b}{\cos\theta}$, and it follows that the sum of all these little moments is equal to $wL \cdot \frac{b}{2}$, the result of the earlier computation. This confirms—at least in the special case of a thin, homogeneous beam—that a moment computation can be carried out by regarding the weight of an object to be concentrated at its center of mass. This turns out to be true in general.

We now return to Coulomb's analysis of the arch. Figure 7.23 shows the segment $ABba$ of the arch in Figure 6.30 in an xy-plane and collects relevant information. The assumption that the arch is circular was not needed in the "Analyzing Structures: Statics and Materials" section but will be in place now.

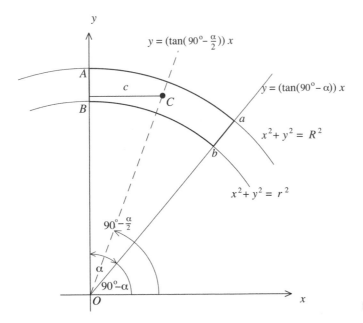

Figure 7.23

The inner and outer boundaries are given by circles with center the origin and radii R and r, respectively. We will assume that the depth of the arch (the dimension perpendicular to the xy-plane) is 1 unit and that it is made of a homogeneous material of density w. The point C is the center of mass of the arch segment and c is the distance from C to the y-axis. The figure shows the line through the boundary ab. It is determined by its angle $90° - \alpha$ with the x-axis. Because the material of the arch is homogeneous, C lies on the line through O that the angle $90° - \frac{\alpha}{2}$ determines. The fact that the slopes of the two lines are the respective tangents of the angles $90° - \alpha$ and $90° - \frac{\alpha}{2}$ determines their equations $y = (\tan(90° - \alpha))x$ and $y = (\tan(90° - \frac{\alpha}{2}))x$.

It is the goal of the study that follows next to determine the location of the center of mass C. Because the arch is homogeneous (with a depth of 1 unit) it suffices to locate the position of C in the xy-plane. Because C lies on the line $y = (\tan(90° - \frac{\alpha}{2}))x$, the x-coordinate c determines its location. We will compute the moment of force M_y of the arch segment around the y-axis in two ways. First, we consider the segment to be divided into small strips and compute M_y by adding all the "little" moments of the strips. Then M_y is computed again, this time under the assumption that the entire weight of the segment is concentrated at C. Setting the two results equal to each other determines the x-coordinate c and therefore the location of C. The understanding of the discussion that provides the details is facilitated by regarding the xy-plane of Figure 7.23 to be horizontal, the force of gravity as acting downward perpendicular to this plane, and the y-axis as the axis of rotation.

Figure 7.24 depicts a thin vertical strip through the arch segment. It is in typical position with its left edge at x, and it is dx thick. The weight of the strip is equal to its area times its density w. The moment of force of the strip around the y-axis is the product of its weight times its distance x to the y-axis. Because the thickness of the strip is dx and its length is determined by the upper and lower boundaries of the segment of the arch, it follows that the moment of the strip is $x \cdot w(\sqrt{R^2 - x^2} - \sqrt{r^2 - x^2})dx$ when $0 \leq x \leq r\sin\alpha$, and $x \cdot w(\sqrt{R^2 - x^2} - \tan(90° - \alpha)x)dx$ when $r\sin\alpha \leq x \leq R\sin\alpha$. Integral calculus informs us that the sum of the moments of all these thin strips is

$$M_y = \int_0^{r\sin\alpha} x \cdot w\left(\sqrt{R^2 - x^2} - \sqrt{r^2 - x^2}\right)dx$$
$$+ \int_{r\sin\alpha}^{R\sin\alpha} x \cdot w\left(\sqrt{R^2 - x^2} - \tan(90° - \alpha)x\right)dx.$$

Use the chain rule to show that $-\frac{1}{3}(R^2 - x^2)^{\frac{3}{2}}$ is an antiderivative of $x\sqrt{R^2 - x^2} = x(R^2 - x^2)^{\frac{1}{2}}$, and similarly that $-\frac{1}{3}(r^2 - x^2)^{\frac{3}{2}}$ is an antiderivative of $x\sqrt{r^2 - x^2} = x(r^2 - x^2)^{\frac{1}{2}}$. So by the Fundamental Theorem of Calculus,

$$M_y = \left[-\frac{1}{3}w(R^2 - x^2)^{\frac{3}{2}} + \frac{1}{3}w(r^2 - x^2)^{\frac{3}{2}}\right]\Bigg|_0^{r\sin\alpha}$$
$$+ \left[-\frac{1}{3}w(R^2 - x^2)^{\frac{3}{2}} - \frac{1}{3}w\tan(90° - \alpha)x^3\right]\Bigg|_{r\sin\alpha}^{R\sin\alpha}$$

$$= -\frac{1}{3}w(R^2 - r^2\sin^2\alpha)^{\frac{3}{2}} + \frac{1}{3}w(r^2 - r^2\sin^2\alpha)^{\frac{3}{2}} + \frac{1}{3}wR^3 - \frac{1}{3}wr^3$$

$$-\frac{1}{3}w(R^2 - R^2\sin^2\alpha)^{\frac{3}{2}} - \frac{1}{3}w\tan(90° - \alpha)\cdot R^3\sin^3\alpha$$

$$+\frac{1}{3}w(R^2 - r^2\sin^2\alpha)^{\frac{3}{2}} + \frac{1}{3}w\tan(90° - \alpha)\cdot r^3\sin^3\alpha$$

$$= \frac{1}{3}wr^3(1 - \sin^2\alpha)^{\frac{3}{2}} + \frac{1}{3}w(R^3 - r^3) - \frac{1}{3}wR^3(1 - \sin^2\alpha)^{\frac{3}{2}}$$

$$-\frac{1}{3}w(R^3 - r^3)\tan(90° - \alpha)\sin^3\alpha$$

$$= \frac{1}{3}wr^3\cos^3\alpha + \frac{1}{3}w(R^3 - r^3) - \frac{1}{3}wR^3\cos^3\alpha$$

$$-\frac{1}{3}w(R^3 - r^3)\cdot\frac{\sin(90° - \alpha)}{\cos(90° - \alpha)}\cdot\sin^3\alpha.$$

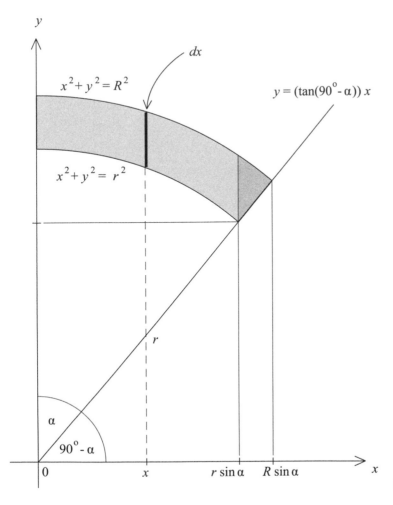

Figure 7.24

The next to last step made use of the identity $\sin^2\alpha + \cos^2\alpha = 1$ from Chapter 2, "Measuring Triangles." After inserting two more identities from that section, namely $\sin(90° - \alpha) = \cos\alpha$ and $\cos(90° - \alpha) = \sin\alpha$, we get

$$
\begin{aligned}
M_y &= \frac{1}{3}w(R^3 - r^3)\left(1 - \cos^3\alpha - \frac{\cos\alpha}{\sin\alpha} \cdot \sin^3\alpha\right) \\
&= \frac{1}{3}w(R^3 - r^3)\left(1 - \cos^3\alpha - \cos\alpha \cdot \sin^2\alpha\right) \\
&= \frac{1}{3}w(R^3 - r^3)(1 - \cos^3\alpha - \cos\alpha \cdot (1 - \cos^2\alpha)) = \frac{1}{3}w(R^3 - r^3)(1 - \cos\alpha).
\end{aligned}
$$

The result $M_y = \frac{1}{3}w(R^3 - r^3)(1 - \cos\alpha)$ for the moment of force of the arch segment obtained by summing up all the moments of the thin strips completes the first computation of M_y.

For the second computation of M_y, note first that the area between the two circles in Figure 7.23 is $\pi R^2 - \pi r^2 = \pi(R^2 - r^2)$. The area of the segment $ABba$ is the fraction $\frac{\alpha}{360°}$ of this. Because the segment is 1 unit in depth and has a density of w, it follows that the weight of the segment is $W_\alpha = w\pi(R^2 - r^2) \cdot \frac{\alpha}{360°}$. With the entire weight of the segment at C, we get

$$
M_y = c \cdot w\pi(R^2 - r^2) \cdot \frac{\alpha}{360°}.
$$

By equating the two results for M_y, we finally obtain that

$$
c = \frac{\frac{1}{3}(R^3 - r^3)(1 - \cos\alpha)}{\pi(R^2 - r^2)(\frac{\alpha}{360°})}.
$$

The calculus of trigonometric functions is greatly facilitated by using radian measures of angles instead of degrees. For example, with x in radians, the derivatives of the functions $\sin x$ and $\cos x$ are $\cos x$ and $-\sin x$, respectively. With degrees, this is not quite as simple. The conversion from degrees to radians is provided by the circle of radius 1. The angle of 360° at the center of this circle spans the entire circumference of 2π. So the angle 360° corresponds to the length 2π. Therefore, 1° corresponds to the length $\frac{2\pi}{360} = \frac{\pi}{180}$ along the circumference, and any angle α in degrees to the length $\alpha \cdot \frac{\pi}{180}$. This length is the radian measure of the angle α. The ratio $\frac{\alpha}{360°}$ of angles both in degrees is equal to the ratio $\frac{\alpha}{2\pi}$ of the angles both in radian measure. It follows that the substitution $\frac{\alpha}{2\pi}$ for $\frac{\alpha}{360°}$ converts the earlier expressions for W_α and c to

$$
W_\alpha = \frac{1}{2}w(R^2 - r^2)\alpha \quad \text{and} \quad c = \frac{\frac{2}{3}(R^3 - r^3)(1 - \cos\alpha)}{(R^2 - r^2)\alpha},
$$

where α is in radian measure.

Calculating Coulomb's Arch. This section explores Coulomb's analysis for the stability of an arch with the methods of calculus. Recall that Coulomb determines, in terms of the data of Figures 6.30 and 6.32, that an arch will not experience hinging failure if the horizontal compression H at the top of the arch satisfies the bounds $G_0 \leq H \leq G_1$, where G_0 is the *maximum value* achieved by $\frac{W_\alpha x_0}{y_0}$ and G_1 is the *minimum value* of $\frac{W_\alpha x_1}{y_1}$ with α ranging over the values $0° \leq \alpha \leq 90°$ in either case. Given that sliding failure of an arch is less likely than hinging failure, the inequality $G_0 \leq H \leq G_1$ is Coulomb's principal criterion for the stability of an arch.

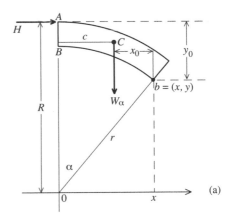

(a)

The study that follows will use the conclusions reached earlier in this section to express $\frac{W_\alpha x_0}{y_0}$ explicitly as a function of α and then consider the application of calculus to the problem of finding its maximum value G_0. (The analogous problem for the function $\frac{W_\alpha x_1}{y_1}$ and its minimum value G_1 is taken up in Discussion 7.3.) Figures 7.25a and 7.25b present the relevant information. The discussion of G_0 is based on Figure 7.25a and incorporates Coulomb's instruction to the readers of his *Essay on Problems of Statics* that H must be assumed to act at A, so as to make y_0 as large as possible and hence G_0 as small as possible. The analogous assumption for G_1 is included in Figure 7.25b. Letting H act at B makes y_1 as small as possible and hence G_1 as large as possible. (Discussion 7.3 considers the consequence of different locations for the action of H.)

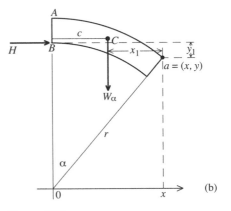

(b)

Figure 7.25

We begin by expressing $\frac{W_\alpha x_0}{y_0}$ as a function of α with α in radian measure. Figure 7.25a tells us that the coordinates of the point $b = (x, y)$ are

$$x = r \sin\alpha \quad \text{and} \quad y = r \cos\alpha.$$

From Figure 7.25a,

$$x_0 = x - c = r\sin\alpha - \frac{\frac{2}{3}(R^3 - r^3)(1 - \cos\alpha)}{(R^2 - r^2)\alpha} \quad \text{and} \quad y_0 = R - y = R - r\cos\alpha.$$

It follows that

$$
\begin{aligned}
W_\alpha x_0 &= \frac{1}{2}w(R^2 - r^2)\alpha\left(r\sin\alpha - \frac{\frac{2}{3}(R^3 - r^3)(1 - \cos\alpha)}{(R^2 - r^2)\alpha}\right) \\
&= \frac{1}{2}w(R^2 - r^2)r\alpha \cdot \sin\alpha - \frac{1}{3}w(R^3 - r^3)(1 - \cos\alpha)
\end{aligned}
$$

and therefore that

$$\frac{W_\alpha x_0}{y_0} = \frac{\frac{1}{2}wr(R^2 - r^2)\alpha\sin\alpha - \frac{1}{3}w(R^3 - r^3)(1 - \cos\alpha)}{R - r\cos\alpha}.$$

At this point stop to review (if necessary) the basic facts about the sine and cosine functions and their derivatives. The function $g_0(\alpha) = \frac{W_\alpha x_0}{y_0}$ is differentiable over the interval $0 \leq \alpha \leq \frac{\pi}{2}$ because the sine and cosine functions are differentiable and the denominator $R - r\cos\alpha > 0$ (because $R > r$). So the maximum G_0 of $g_0(\alpha)$ can be obtained by setting the derivative of $g_0(\alpha)$ equal to 0, solving for α, and sifting through the values obtained. Check by applying the quotient rule (and the product rule along the way) that the derivative of $g_0(\alpha)$ has numerator

$$\left[\frac{1}{2}wr(R^2 - r^2)(\sin\alpha + \alpha\cos\alpha) - \frac{1}{3}w(R^3 - r^3)\sin\alpha\right](R - r\cos\alpha)$$

$$-\left[\frac{1}{2}wr(R^2 - r^2)\alpha\sin\alpha - \frac{1}{3}w(R^3 - r^3)(1 - \cos\alpha)\right]r\sin\alpha$$

and denominator $[R - r\cos\alpha]^2$. Verify that after some algebra,

$$g_0'(\alpha) = \frac{\frac{1}{2}wr(R^2 - r^2)[R(\sin\alpha + \alpha\cos\alpha) - r(\sin\alpha\cos\alpha + \alpha)] - \frac{1}{3}w(R^3 - r^3)(R - r)\sin\alpha}{[R - r\cos\alpha]^2}.$$

To determine G_0, it remains (after factoring $R^2 - r^2$ as $(R + r)(R - r)$ and canceling the term $w(R - r)$) to set

$$\frac{1}{2}r(R + r)[R(\sin\alpha + \alpha\cos\alpha) - r(\sin\alpha\cos\alpha + \alpha)] - \frac{1}{3}(R^3 - r^3)\sin\alpha = 0$$

and to solve for α with $0 \leq \alpha \leq \frac{\pi}{2}$. There are ways of doing this—for specific values for r and R—both "hard" (with successive approximation techniques such as Newton's method) and "soft" (with computer applications such as Maple, Mathematica, and MATLAB). The fact that all approaches to the solutions are beyond the scope of this text confirms the validity of Coulomb's comment that it is always easier to find G_0 by trial and error than by the "exact methods" of calculus. Discussion 7.3 takes up Coulomb's suggestion (see one of the paragraphs that concludes "Analyzing Structure" in Chapter 6) that G_0 "will easily be found" by starting with $\alpha = \frac{\pi}{4}$ (or $\alpha = 45°$), calculating $\frac{W_\alpha x_0}{y_0}$, and by repeating this calculation with successively smaller α (moving in the direction of the keystone in the process).

We will close with a final reflection. Even though Coulomb's investigation of the stability of an arch (as presented in Chapter 6) brings in parameters (such as the coefficients τ, σ, and μ) that reflect the strength of the materials, the assumptions that he subsequently makes reduce this investigation to strictly geometric considerations. In particular, the maximum value G_0 of $\frac{W_\alpha x_0}{y_0}$ depends only on R and r and has nothing to do (except for w) with the materials the arch is made of. At the end, Coulomb's analysis is consistent with the conclusion of Heyman's Safe Theorem (of "Hanging Chains and Rising Domes" in Chapter 6) that the stability of a masonry arch is primarily determined by its geometry (and the assumption that the building material is able to resist compression).

Problems and Discussions

All the problems that follow deal with mathematical issues that have been the focus of this chapter.

Problem 1. Let $f(x) = x^2$. Use the limit definition

$$\lim_{\Delta x \to 0} \frac{f(x + \Delta x) - f(x)}{\Delta x}$$

to show that $f'(x) = 2x$. Then use it to find the derivative of $g(x) = \frac{1}{x} = x^{-1}$. [Hint: In each case, rewrite the algebraic expression in such a way that Δx can be canceled before the limit is taken.]

Problem 2. Use the chain rule to find the derivative of the function $f(x) = (4 - x^2)^{\frac{3}{2}}$.

Problem 3. What is the domain of the function $f(x) = (x^2 - 1)^3$? This function has three critical numbers that divide the number line into four intervals. Find the critical numbers and determine whether the function is increasing or decreasing over each of these intervals. Locate the x-coordinates of the high points of the graph (the maximum values of the function) and the low points of the graph (the minimum values of the function). Do the same thing for the functions $g(x) = (x^2 - 4)^{\frac{1}{3}}$ and $h(x) = (x^2 - 9)^{\frac{2}{3}}$.

Problem 4. Let $f(x)$ be a function and let $F(x)$ be an antiderivative of $f(x)$. Recall that for any small number dx the product $f(x)\,dx$ is approximated by $F(x + dx) - F(x)$. Let $f(x) = 3x^2$ and notice that $F(x) = x^3$ is an antiderivative. Take $x = 3$. Let $dx = 0.01$ and check how good the approximations is. Then check again for $dx = 0.0001$. Is the approximation tighter?

Problem 5. Consider the function $f(x) = 5 - x^2$ with $0 \le x \le 2$.

i. Take $n = 4$ and compute the sum that arises in the definition of the integral $\int_0^2 (5 - x^2)\,dx$. Do so with two decimal place accuracy.
ii. Take $n = 6$ and repeat the computation of the sum arising in $\int_0^2 (5 - x^2)\,dx$ again with two decimal place accuracy.
iii. Use the Fundamental Theorem of Calculus to find the precise value of this integral. Why are the results of both (i) and (ii) only very rough approximations of this value?

Problem 6. Consider the parabola with equation $y = x^2 + 1$. Sketch its graph, highlight the area that the integral $\int_{-3}^3 (x^2 + 1)\,dx$ represents, and find the area by evaluating the integral. Now cut the parabola with the horizontal line $y = 10$. Use Archimedes's area formula from the section "Remarkable Curves and Remarkable Maps" of Chapter 4 to compute the area of the parabolic section that the cut determines. Check the value of the integral by subtracting the area of the parabolic section from a rectangle.

Problem 7. Interpret the definite integral $\int_{-4}^4 \sqrt{16 - x^2}\,dx$ as the area under a curve and use this interpretation to evaluate the integral.

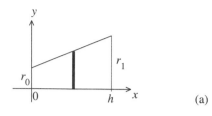

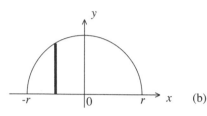

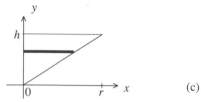

Figure 7.26

(a)

(b)

(c)

Problem 8. Show that $\int_2^5 \left(4 + \sqrt{9 - (x - 5)^2}\right) dx = 12 + \frac{9}{4}\pi$ by interpreting the integral as the area under a circular arc.

Problem 9. Study the discussion "Volumes of Rotation and Lengths of Curves" earlier in this chapter. Then use Figure 7.26 to express the volumes of the solids specified below as definite integrals. Evaluate each of them by using the Fundamental Theorem of Calculus.

   i. A cut-off cone of height h and circular boundaries of radius r_0 and r_1, respectively.
   ii. A sphere of radius r.
   iii. A cone with height h and circular base of radius r. In this case set up the integral in the variable y.

Problem 10. Review the basic properties of the exponential function e^x and study the definitions of the hyperbolic functions $y = \sinh x$ and $y = \cosh x$ that arise in the section "The Shape of and Ideal Arch."

   i. Show that $(\cosh x)^2 - (\sinh x)^2 = 1$. Consider a uv-coordinate plane. Note that the points $(\cos x, \sin x)$ lie on the circle $u^2 + v^2 = 1$ and that the points $(\cosh x, \sinh x)$ lie on the hyperbola $u^2 - v^2 = 1$.
   ii. Use the fact that $\frac{d}{dx} e^x = e^x$ to show that $\frac{d}{dx} \sinh x = \cosh x$ and $\frac{d}{dx} \cosh x = \sinh x$.

Problem 11. Compute the lengths of the two graphs in Figure 7.27 between the given points by using the length of a curve formula. In the case of Figure 7.27a check your answer with the distance formula. In the case of Figure 7.27b refer to Problem 10.

Problem 12. Turn to the section "Problems and Discussions" of Chapter 3. Refer to the conclusions of Problems 4 and 5 about the original dome of the Hagia Sophia and the speculative narrative that they rely on. Use this information and the discussion "Weighing the Dome of the Hagia Sophia" in this chapter to derive the estimate of 23,300 pounds for the weight of this original dome.

Problem 13. Use Figure 7.14 and the data $r = 71$, $R = 92$, $a = 12$, and $b = 64$ to estimate the minimal and maximal thickness of the shell of the cap of the dome of the Pantheon. Your conclusions should confirm that its shell is much larger than the spherical shell of the Hagia Sophia with its 50-foot inner radius and 2.5-foot thickness.

Problem 14. Study the computation of the volume of the cap of the shell of the Pantheon and in particular Figures 7.13, 7.14, and 7.15. By adding two more definite integrals to $V_1 = \int_a^r 2\pi x(\sqrt{R^2 - x^2} - (\sqrt{r^2 - x^2} + D))dx$, express

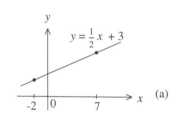

(a)

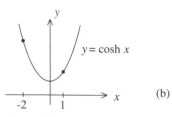

(b)

Figure 7.27

the volume of the shell of the dome *above the springing line* as a sum of three definite integrals. Evaluate the three integrals to obtain an estimate of this volume. Then estimate the weight of that part of the shell.

Problem 15. Consider the abstract cross section of the cap of the shell of the Pantheon depicted Figure 7.14. Flip and rotate the cross section so that the center of the outer circle of the cross section remains at O, the center of the inner circle is the point (D, a), the springing line is on the vertical line $x = D$, and the vertical line $x = E$ is a boundary of the cap. Draw a careful diagram of the cross section of the cap of the shell in this new position in the *xy*-plane. With the cross section in the new position, the volume of the cap of the shell is a volume of rotation about the *x*-axis. Use this observation to express the volume of the cap of the shell as a sum of two definite integrals. Pay attention to the limits of integration (but there is no need to evaluate the integrals).

Problem 16. Let the *z*-axis of an *xyz*-coordinate system represent the central vertical axis of a dome. Suppose that the top of the drum from which the dome rises lies in the *xy*-plane and that the exterior height of the dome above the drum is h. Let $A(z)$ be the cross-sectional area of the shell at a height z with $0 \le z \le h$. Explain by appealing to a carefully drawn diagram why the volume of the shell of the dome is given by the definite integral $\int_0^h A(z)\,dz$.

Problem 17. Refer to Figures 4.29 and 4.32 and discuss how the conclusion of Problem 16 might be applied to provide an estimate of the volume of the double shell of the cathedral Santa Maria del Fiore of Florence. Refer to Figure 5.37 and Problems 8 and 9 in Chapter 5 and describe how Problem 16 might be used to estimate the volume of the double shell of the basilica of St. Peter's.

The next several problems arise from the discussion in "The Shape of an Ideal Arch." In this regard, the angle θ of Figure 7.17b is negative when measured in the clockwise direction and positive when measured counterclockwise. In particular, the angle $\theta(-b)$ is positive and $\theta(b) = -\theta(-b)$ is negative.

Problem 18. Use the fact that $\sin(-\theta) = -\sin\theta$ to show that $C(b)\sin\theta(b) = -C(-b)\sin\theta(-b)$. Consider Figure 7.17a and let $L(x) = \int_{-b}^x \sqrt{1 + f'(t)^2}\, dt$ be the length of the center curve of the arch from $(-b, 0)$ to $(x, f(x))$. Then $L = L(b)$ is the full length of the center curve. Recall the formula

$$C(x)\sin\theta(x) = -w\int_{-b}^x \sqrt{1 + f'(t)^2}\, dt + \text{constant}.$$

Use it with $x = b$ and $x = -b$ to show that the constant is $\frac{wL}{2}$. Conclude that $C(x)\sin\theta(x) = w(\frac{L}{2} - L(x))$. Show that the compression in the arch satisfies $C(x) = \sqrt{w^2(\frac{L}{2} - L(x))^2 + C_0^2}$.

Problem 19. Check that the coordinates of the points $(0, h)$ and $(b, 0)$ satisfy the equation $y = -A\cosh(\frac{B}{b}x) + (h + A)$ of the center curve of the Gateway Arch.

Problem 20. Recall that the graph of the equation $y = -A\cosh(\frac{B}{b}x) + (h + A)$ of the center curve of the Gateway Arch is gotten by compressing a catenary, but that it is not a catenary. This means that the analysis of the ideal arch does not apply to it. Review the basic assumptions of this analysis. Which of them are satisfied by the Gateway Arch and which are not?

Problem 21. Our analysis has shown that the center line of an ideal arch is the graph of a function of the form $y = -H\cosh(\frac{x}{H}) + \text{constant}$, with H a constant, and that this function satisfies the differential equation $\frac{d^2y}{dx^2} = -\frac{1}{H}\sqrt{1 + (\frac{dy}{dx})^2}$. The center line of the Gateway Arch was seen to be the graph of a function of the form $y = -KH\cosh(\frac{x}{H}) + \text{constant}$, where H and K are constants. Show that this function satisfies the differential equation $\frac{d^2y}{dx^2} = -\frac{1}{H}\sqrt{K^2 + (\frac{dy}{dx})^2}$.

Problem 22. Because a hanging chain (that is uniform, perfectly flexible, and does not lengthen when stretched) supports its own weight exclusively by tensile forces, the situation of the hanging chain is analogous to that of the ideal arch. Derive the mathematical shape of the hanging chain by adapting the study of the ideal arch. Your conclusion (if correct) will provide a verification of Robert Hooke's observation "as hangs the flexible line, so but inverted will stand the rigid arch."

The next two problems turn to section "The Calculus of Moments and Centers of Mass," in particular to Figure 7.23 and the value $c = \frac{\frac{2}{3}(R^3 - r^3)(1 - \cos\alpha)}{(R^2 - r^2)\alpha}$ for the x-coordinate of the center of mass C of the segment of the arch. Since the arch is assumed to be homogeneous, C is the centroid of the segment. Because it is on the line $y = \tan(\frac{\pi}{2} - \frac{\alpha}{2})x$, C is the point $(c, \tan(\frac{\pi}{2} - \frac{\alpha}{2})c)$.

Problem 23. Suppose that $\alpha = \frac{\pi}{2}$ and that $r = \frac{1}{2}R$. Show that C is the point $(\frac{14}{9\pi}R, \frac{14}{9\pi}R)$. Because $\frac{14}{9\pi} \approx 0.495$, this is very close to the point $(\frac{1}{2}R, \frac{1}{2}R)$. Sketch the segment of the arch and assess this answer.

Problem 24. Suppose that $\alpha = \frac{\pi}{2}$ and that $r = \frac{5}{6}R$. Show that $C = (\frac{182}{99\pi}R, \frac{182}{99\pi}R)$ and that the distance $CO < 0.828R$. Notice that $0.833R < r$. This implies that C lies outside the segment of the arch. Is this conclusion problematic?

Discussion 7.1. A Theorem of Pappus and Guldin. Paul Guldin (1577–1643), mathematician and astronomer, was born into a Jewish family in St. Gallen, Switzerland. He later converted to Catholicism and became a Jesuit priest. He taught mathematics at a Jesuit college in Rome and at the

University of Vienna. The second volume of Guldin's work *De centro gravitatis* contains the statement:

> If any plane figure revolves about an external axis in its plane, the volume of the solid so generated is equal to the product of the area of the figure and the distance traveled by the centroid of the figure.

Recall that for a homogeneous region, the centroid is the center of mass. Guldin seemed not to know that his insight had already been formulated in the *Mathematical Collection* of Pappus of Alexandria, one of the last great Greek mathematicians (who lived around A.D. 300). Whether Guldin was guilty of plagiarism is a matter of debate by historians.

Let's check Guldin's statement, now called the Theorem of Pappus and Guldin, in the case of Figure 7.23, where the plane figure is the segment *ABba* of the arch and the external axis is the *y*-axis. We know that the centroid *C* of this segment of the arch is a distance $c = \frac{\frac{2}{3}(R^3 - r^3)(1 - \cos\alpha)}{(R^2 - r^2)\alpha}$ (with α in radian measure) from the *y*-axis. So it needs to be verified that $V = A \times 2\pi c$, where *A* is the area of the segment and *V* is the volume obtained by rotating the segment one complete revolution around the *y*-axis.

Problem 25. Show that the area of the arch segment is $A = \frac{1}{2}(R^2 - r^2)\alpha$. [Hint: Find the difference between the areas of the full circles, and take an appropriate portion of it.]

Problem 26. Verify that the volume obtained by rotating the sector of Figure 7.28a one complete revolution about the *y*-axis is $\frac{2}{3}\pi R^3(1 - \cos\alpha)$. [Hint: Refer to the computation (in "Volumes of Spherical Domes") of the volume of the cap of the shell of the Pantheon for the approach to the solution, and to the computation (in "The Calculus of Moments and Centers of Mass") of the moment of the segment of the arch for some of the particulars.]

Problem 27. By considering the conclusion of Problem 26 and referring to Figure 7.28b, explain why the volume *V* obtained by rotating the segment of the arch one complete revolution around the *y*-axis is equal to $V = \frac{2}{3}\pi(R^3 - r^3)(1 - \cos\alpha)$.

Problem 28. Use information already developed to check that the Theorem of Pappus and Guldin holds for the rotation of the arch segment *ABba*.

Discussion 7.2. A Thin-Shelled Vault. Excursions often end (and begin) at airports, so it is appropriate that our *Mathematical Excursions to the World's Great Buildings* near their end with a discussion of an airport terminal. Along with Ove Arup (and other prominent architects and structural engineers), Eero Saarinen contributed to the design of thin-shell concrete vaults in the 1950s. Not surprisingly, therefore, Saarinen was called to serve on the jury

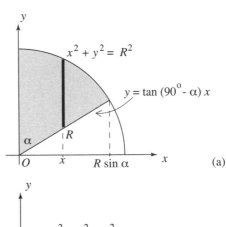

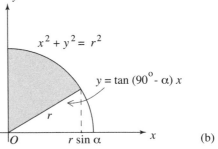

Figure 7.28

that ultimately awarded the Sydney Opera commission to Jørn Utzon. His positive assessment of Utzon's design was influential and the famous Australian landmark owes its existence in good measure to him.

Perhaps Saarinen's most acclaimed structure (other than the Gateway Arch) was the TWA Flight Center at the John F. Kennedy Airport in New York City. Begun in 1956, it was finished in 1962 by Saarinen & Associates after the architect died in 1961. The building quickly became a New York City landmark and icon of modern architecture. Perhaps in part because he had studied sculpture and wanted to react to the boxy shapes of the International Style, Saarinen created a dynamic, soaring building that reflects the excitement of travel. The structure consists of four interacting vaulted domes supported by four Y-shaped columns. Together, the vaults form a large, umbrellalike shell that curves 50 feet high and 315 feet long over the passenger areas. The vaults are made of sculptured concrete braced within by invisible webs of reinforcing steel. Seen from a distance, the structure has the shape of a bird extending its wings. The upward-moving curves and the bands of skylights at the interfaces between the vaults give the interior a sense of lightness and airiness. Saarinen referred to the terminal as "a building in which the architecture itself would express the drama and specialness and excitement of travel . . . , a place of movement and transition. . . . The shapes were deliberately chosen in order to emphasize an upward-soaring quality of line. We wanted an uplift." The building was completely refurbished before it became part of Jet Blue's Terminal 5 in 2008.

Problem 29. The years from 1956 to 1962 were also the years during which Utzon and Arup wrestled with the vault design for the Sydney Opera. (This was discussed in Chapter 6.) Compare Saarinen's thin-shell concrete vaults of Figure 7.29 with those of Utzon in Plate 24. What are the similarities? Why do you think that the challenges posed by Utzon's vaults were greater?

Figure 7.29. Saarinen's terminal at JFK showing three of the four vaults and two of the Y-shaped columns. Photo by Pheezy

Discussion 7.3. Studying Coulomb's Criterion. This discussion explores aspects of Coulomb's stability criterion $G_0 \leq H \leq G_1$ developed in the section "Analyzing Structures: Statics and Materials" of Chapter 6 and examined in "The Calculus of Moments and Centers of Mass" of this chapter.

Problem 30. Assume that $R = 5$ and $r = 4$, both in feet, and $w = 150$ pounds per cubic foot. Use the "trial and error" method that Coulomb proposes to find the maximum value of the function $g_0(\alpha) = \frac{W_\alpha x_0}{y_0}$ over $0 \leq \alpha \leq \frac{\pi}{2}$. [Hint: Start with $\alpha = \frac{\pi}{4}$ or $\alpha = 45°$. First change α by increments of $\frac{\pi}{20}$ or $9°$ (both positive and negative) to gain a general sense of the value(s) of α that provide the maximum. Then refine your search by using increments of $\frac{\pi}{40}$ or $4.5°$ and later $\frac{\pi}{180}$ or $1°$.]

Problem 31. Use Figure 7.25b to express $\frac{W_\alpha x_1}{y_1}$ as a function $g_1(\alpha)$ with α in radian measure ranging over the interval $0 \leq \alpha \leq \frac{\pi}{2}$. Then show that

$$g_1'(\alpha) = \frac{\frac{1}{2} wR(R^2 - r^2)[r(\sin\alpha + \alpha\cos\alpha) - R(\sin\alpha\cos\alpha + \alpha)] + \frac{1}{3} w(R^3 - r^3)(R - r)\sin\alpha}{[r - R\cos\alpha]^2}$$

What difficulties does the use of $g_1'(\alpha)$ in the determination of the minimum value G_1 of the function $g_1(\alpha)$ need to deal with?

The final problems deal with the question of the location of the action of the horizontal force H in Figures 7.25a and 7.25b. We will see that difficulties arise if H is considered to be acting at B in Figure 7.25a and at A in Figure 7.25b.

Problem 32. Suppose that the force H acts at point B in Figure 7.25a (in the context of G_0) and at point A in Figure 7.25b (in the context of G_1). Show that

$$\frac{W_\alpha x_0}{y_0} = \frac{1}{2} w(R^2 - r^2)\left(\frac{\alpha\sin\alpha}{1 - \cos\alpha}\right) - \frac{1}{3}\frac{w}{r}(R^3 - r^3) \quad \text{and}$$

$$\frac{W_\alpha x_1}{y_1} = \frac{1}{2} w(R^2 - r^2)\left(\frac{\alpha\sin\alpha}{1 - \cos\alpha}\right) - \frac{1}{3}\frac{w}{R}(R^3 - r^3).$$

Problem 33. Consider the function $f(\alpha) = \frac{\alpha\sin\alpha}{1 - \cos\alpha}$. Use the quotient and product rules to show that $f'(\alpha) = \frac{\sin\alpha - \alpha}{1 - \cos\alpha}$. Notice that neither $f(\alpha)$ nor $f'(\alpha)$ is defined at $\alpha = 0$. Use basic facts about the sine and cosine to show that the function f is decreasing over the interval $0 < \alpha \leq \frac{\pi}{2}$.

Problem 34. Continue to consider $f(\alpha) = \frac{\alpha\sin\alpha}{1 - \cos\alpha}$. Review L'Hospital's rule and apply it twice to show that $\lim_{\alpha \to 0} f(\alpha) = \lim_{\alpha \to 0} \frac{\alpha\sin\alpha}{1 - \cos\alpha} = 2$. Since $f(\frac{\pi}{2}) = \frac{\pi}{2}$, it follows from a conclusion of Problem 33 that $\frac{\pi}{2} \leq f(\alpha) < 2$ over $0 < \alpha \leq \frac{\pi}{2}$.

Problem 35. Combine information provided by Problems 32 and 34 to conclude that $\frac{W_\alpha x_0}{y_0}$ has no maximum value over $0 < \alpha \leq \frac{\pi}{2}$, but that it makes sense to take $G_0 = w[(R^2 - r^2) - \frac{1}{3r}(R^3 - r^3)]$. Conclude also that the minimum value of $\frac{W_\alpha x_1}{y_1}$ is $G_1 = w[\frac{\pi}{4}(R^2 - r^2) - \frac{1}{3R}(R^3 - r^3)]$ over $0 < \alpha \leq \frac{\pi}{2}$.

Problem 36. Consider the case $R = 5$, $r = 4$, both in feet, and $w = 150$ pounds per cubic foot. Check that substitution into the equations of Problem 35 provides the values $G_0 = 150(9 - \frac{61}{12}) \approx 588$ pounds and $G_1 = 150(\frac{9\pi}{4} - \frac{61}{15}) \approx 450$ pounds. Given this, how can the horizontal force H satisfy $G_0 \leq H \leq G_1$?

Glossary of Architectural Terms

This glossary contains the relevant architectural terms that arise in the text. The mathematical terms are not listed. They are defined in the text when they are first introduced. Terms in *italics* also appear in the glossary.

acropolis The citadel, or high fortified area of an ancient Greek city. The Acropolis is often understood to be the acropolis of the city of Athens.

aggregate Hard materials such as sand, gravel, pebbles, stone or masonry fragments added to *cement* to make *concrete*.

aisle A division in a church parallel to the *nave*. It is often separated from the nave by an *arcade*.

amphitheater An arena or stadium with a central oval stage for contests or performances surrounded by rows of seats arranged in parallel oval tiers. A classical Greek or Roman theater is semicircular in *plan*. Seating is arranged in tiers that rise in expanding semicircles from a semicircular stage. An amphitheater (the prefix amphi is Greek for "on both sides") is in effect a combination of two such semicircular structures.

apse A vaulted recess in a church at the end of the *nave*, or, more accurately, at the end of the *chancel*. It is usually semicircular or polygonal in *plan*.

arcade A sequence of *arches* supported by *columns* or *piers*, or a covered walk enclosed by a line of such arches on one or both sides.

arch A vertical structural element capable of spanning a horizontal gap. It is usually curved and consists of wedged segments called *voussoirs*. It can either be load-bearing or decorative. A Roman arch is semicircular. A Gothic arch is pointed.

Islamic architecture makes use of the pointed arch, the double arch, the horseshoe arch, and the ogee arch. The ogee arch is pointed and distinguished by the way it changes curvature from concave to convex.

ashlar A block of masonry cut so that its faces and sides are rectangular. Also, masonry in which all stones are squared, giving a uniform pattern of vertical and horizontal joints.

attic A horizontal structure below the roof of a building or below the *springing* of a dome. It follows the contours of the wall structure below it and may have windows or be decorated with *relief* sculpture.

balustrade A railing made with contoured or tapered posts.

Baroque architecture An architecture that makes use of classical and *Renaissance* forms. It is distinguished by its rich ornamentation, an integration of painting and sculpture, an interplay of different surfaces, and the dramatic use of light. Sometimes, in contrast with the clear and rational forms of the Renaissance style, it is very elaborate, ornate, and cluttered. If taken to the extreme, it is referred to as Rococo architecture.

barrel vault A cylindrical, continuous ceiling or roof structure usually made of masonry and having the cross section of an arch that is often semicircular or pointed.

basilica A church designed with a long and high central space called *nave*.

The nave is flanked by one or two *aisles* that are lower and separated from the nave by *arcades*. Another common feature is a section perpendicular to the nave, called *transept*, that divides the nave in two, giving the *plan* of the church the shape of a cross. The shorter segment is the *chancel*. It includes the altar, an area for clergy and musicians called a *choir*, and a recess called an *apse*. The basilica design is based on a standard building form used by the Romans for administrative purposes. In its religious meaning, the term basilica is a distinction given to more important churches that confers on them a number of ceremonial privileges.

basilica plan A *plan* for a church that reflects the features of a *basilica*.

bay A compartment or division of the interior or exterior of a building, normally marked by its vertical supports.

beam A *structural element* often in horizontal or slanting position with uniform square or rectangular cross section. Often supported by vertical columns or piers.

buttress A *structural element* designed to resist an outward *thrust* and to provide lateral stability. Often massive and inclined to the structure.

Byzantine architecture An architecture based on Roman forms. *Arcades, colonnades,* and *domes* are often organized with dominating vertical or horizontal axes. It is an architecture of both aisled *basilicas* and domed churches with *central plan*. Vaults are ever present. The murals, mosaics, and marbles that cover their curved surfaces contribute an illusion of weightlessness and an intense, jewel-like effect.

capital The part of a column just above the shaft. It is often ornately decorated.

capomaestro In the Middle Ages and the *Renaissance*, a master mason or master builder, usually the lead architect of a project.

catenary The shape of a hanging cable, chain, or rope that is uniformly loaded and perfectly flexible. The center curve of an arch that has its only external support at its base and responds to the forces of gravity only by its reaction to compression has the shape of a catenary.

cathedral A major church that is the seat of a bishop. The word comes from the Greek "cathedra," meaning seat or bench.

cement A powder made by heating a mixture of clay and limestone. When mixed with an *aggregate* and water it hardens to form *concrete*. The Romans discovered a volcanic powder that has the properties of cement.

centering A temporary bracing, usually of timber, to support arches, vaults, or domes during construction. It can either extend upward from the ground or be placed higher up, for example, near the *springing*.

central plan A design for a church that features a central *dome* and a structure that radiates symmetrically from its center. Often in the shape of a *Greek cross*.

chancel In a church in *basilica plan*, the shorter part of the subdivision created by the *transept*. It includes the altar, the *choir*, and the *apse*.

choir The part of the *chancel* used during religious service by the clergy and accompanying singers and musicians.

clerestory The highest section of the wall of the *nave* of a church in *basilica* design, usually pierced by windows, called clerestory windows.

coffer A recessed or sunken panel in a ceiling or *vault* in the shape of a square, hexagon, or other polygon. Coffering is a regular array of *coffers*.

colonnade A series of *columns* in line.

column A vertical, usually load-bearing structural element usually with a circular horizontal cross section that is relatively small compared to the height. Classically, it consists of a *capital* at the top, a shaft as its main part, and a base at the bottom. It is usually made of stone and often tapers as it rises. In the *Doric order*, the capital is plain. In the *Ionic order*, the capital is decorated with scrolls. In the *Corinthian order*, the capital is decorated with leaf-like formations.

column and beam The basic structure consisting of two *columns* supporting a horizontal *beam*.

compression A force pushing against a *structural element* putting the element under pressure.

concrete Artificial stone made with *cement*, water, and *aggregates* such as crushed stone, *masonry*, or sand. Reinforced concrete is concrete that is strengthened by embedded iron mesh or rods. Prestressed concrete refers to a concrete slab that is compressed and strengthened by cables that run under great tension inside the slab along its length. Roman concrete was neither reinforced nor prestressed.

Corinthian order Pertaining to a classical Greek configuration of *columns* and what they support. The columns are relatively delicate, in a design said to have been based on the proportions of a slender female figure. It features columns with capitals that are ornately decorated with leaflike formations.

course A line of stone blocks or bricks.

crossing The space at the intersection of the *nave* and *transept* of a church. It often has a *dome* or tower over it.

crossing piers Large piers at the *crossing* of a church, usually four in number, that serve as the primary support of a *dome* or tower over the *crossing*.

cupola Usually a *dome* of smaller size.

deformation A change of shape of a *structural element* due to the action of a *load*.

dome A curved roof or ceiling structure that spans and encloses the space underneath it. A dome is usually symmetric about a central vertical axis and has a circular or polygonal base. Domes are often spherical in shape.

Doric order Pertaining to a classical Greek configuration of *columns* and what they support. The columns are relatively sturdy, in a design said to have been based on the proportions of a male figure. It often features columns with a simple capital and an array of vertical grooves or *flutes* along the shaft.

drum A vertical cylindrical or polygonal wall from which a *dome* or *lantern* rises. It is often supported by *pendentives* or *squinches*. Alternatively, a drum is one of the stacked cylindrical segments that columns are often made with.

duomo The Italian word for a cathedral church, one that is (or was) the seat of a bishop. The German and French equivalents are "Dom" and "dôme," respectively. The words derive from the Latin "domus," meaning house, as in "domus Dei," house of God. The term *dome* has the same origin.

elevation The vertical face of a building, external or internal, or a scale drawing of it.

equilibrium A state of balance. A situation in which forces act on a structure, but the resultant of the acting forces is zero, so that no motion results. A condition of equilibrium is stable if the structure returns to its state of balance on its own if it is temporarily disturbed, and it is unstable if it does not.

flutes Decorative vertical grooves on a column.

flying buttress An arched *buttress*, usually made of masonry, that extends outside the structure from the *springing* of a *vault* to a freestanding outer *pier*.

footing A layer of load-bearing material at the bottom of a wall or *column* so as to distribute its weight more widely and securely over the *foundation*.

force A push or pull on or by a *structural element*. An external force is one that is applied on the element from outside it. The most prevalent external forces are gravitational. An internal force or reaction is generated by a structural element in response to an external force.

foundation That part of a structure that meets the ground and through which *loads* are transferred to the ground.

free body diagram or **force diagram** A diagram in which the *forces* acting on a structural element (often a single point of the element) are represented as *vectors*. Each vector points in the direction of the force it represents and is of a length proportional to the magnitude of the force.

fresco A painting on a wall with water-based colors executed while the plaster is still wet.

frieze The part of a structure above the horizontal slabs that the *columns* support. Often decorated with sculpture in low *relief*.

fulcrum A support about which free rotation is possible.

funicular polygon A force diagram derived from a *funicular structure*.

funicular structure A structure composed of ropes or cables carrying suspended weights, or more generally a structure in which all operative forces are tensions or compressions.

Gothic arch An arch in a shape obtained by joining two opposed circular arcs so that they meet at a point at the top.

Gothic architecture Referring to a soaring style of architecture that features *Gothic arches*, high *ribbed vaults*, richly colored stained glass windows large in proportion to the wall spaces they penetrate, curving exterior *flying buttresses*, and tall spires and towers.

granite The most durable stone used in construction. Its hardness and the difficulty (and cost) of cutting and shaping it restrict its use to structures exposed to severe conditions and stresses. It varies in weight from 160 to 190 pounds per cubic foot. Its crushing resistance varies from 600 to 1200 *tons* per square foot.

Greek cross A cross with four arms of equal length, often used in the design of early Christian churches and later Byzantine churches.

groin vault A continuous ceiling or roof structure, usually made of masonry, formed by the perpendicular intersection of a *barrel vault* with one or more smaller *barrel vaults*.

hemicycle A semicircular structure or part of such a structure. Often the rounded termination of a church.

hoop A horizontal closed curve, most often a circle, obtained by intersecting a domed or cylindrical structure with a horizontal plane.

hoop stress *Tensions* generated along *hoops* in a domed structure by the weight of the *dome*. These *tensile forces* will tend to stretch the *shell* of the dome along the hoops and lead to cracks along the *meridians*.

hypostyle Having a flat roof that rests on a rectangular arrangement of rows of *columns*.

Ionic order Pertaining to a classical Greek configuration of *columns* and what they support. The columns are relatively delicate, in a design said to have been based on the proportions of a female figure. The columns have *capitals* that are decorated with scroll-like formations. Their shafts often feature an array of vertical grooves or *flutes*.

Ionic volute The scrolls that decorate the *capitals* of columns in the *Ionic order*. The scrolls are shaped by spiral curves.

Islamic architecture An architecture featuring *arches*, *squinches*, *vaults*, *domes*, and towers. These *structural elements* have classical origins, but the need for large spaces for communal prayer in mosques led to the development of new forms and ingenious techniques of construction. There are regional variations, but the extensive use of decoration, such as carved, inlaid, painted and gilded stone, wood, plaster, glazed brick, *terra cotta*, or tile is a common theme. Geometric, floral, and calligraphic motifs predominate.

keystone The wedge-shaped block, or *voussoir*, at the top of a masonry arch.

lantern A small windowed *cupola* or tower on top of a *dome* admitting light into the interior space below.

lever A rigid bar supported by and free to rotate about a fixed point (called a *fulcrum*) or axis. A *force* applied on one side of the point or axis will effect a rotation, or bending *moment* on the other side.

limestone Any stone having carbonate of lime as principal constituent. There are many different types. They make good building materials, but fine even-grained stone containing no cracks and vents needs to be selected. It is easily worked and forms a good, even-colored surface. Its weight varies from 120 to 170 pounds per cubic foot. It has a crushing resistance from 90 to 500 *tons* per square foot. It is less durable than *granite* or *sandstone* and absorbs relatively much water.

lintel A common name for a single block of stone in horizontal position that is supported at its ends by posts. *Post* and *lintel* are other terms for a *column* and *beam*, respectively.

load An *external force* on a *structural element* including the weight of the element.

loggia A part of a building where one or more sides are open to the air, the opening being *colonnaded* or *arcaded*. Often a porch or large open arcaded or colonnaded recess at ground floor level.

marble A form of recrystallized *limestone*.

masonry Building material such as stone, brick, or sun-dried mud. Individual units are usually joined together with *mortar*.

mass The quantity of matter in a body or object. The mass of an object is the ratio W/g, where W is the weight of the object and g is the gravitational constant ($g \approx 32$ feet/second^2 in the units feet and seconds).

meridian A curve on the surface of a *dome* that is determined by a plane through the vertical central axis.

mihrab A prayer niche in the wall of a mosque oriented toward Mecca.

minaret A tall, slender tower attached to or near a mosque used for the call to prayer.

moment of force or **bending moment** refers to the capacity of a *force* to produce a rotation or bending in a *structural element*. Quantitatively, it is the product of the magnitude of the force times the distance to the axis of rotation.

mortar A mixture made with lime and or cement, sand, and water, used as a bonding agent between bricks or stones.

nave The long, high central section of a church generally reserved for the laity. It is often flanked by lower aisles on each side, and separated from these by arcades.

obelisk Obelisks are monumental pillars cut from a single stone, usually granite, that often stood in pairs to mark and enhance the entrances of Egyptian temples. Obelisks are square in cross section and taper to a pyramid at the top. Their height is about nine or ten times the diameter of the base. Historical records in Egyptian hieroglyphic script are often cut into its four sides.

order One of several carefully proportioned classical arrangements of *columns* and the elements that they support. The *Corinthian, Doric,* and *Ionic* orders are preeminent examples.

pediment A classical architectural element consisting of the triangular section that often crowns a *portico* or facade. It usually rests on a horizontal structure carried by columns. Pediment also refers to a triangular or rounded decorative element over a door, window, or niche.

pendentive A *structural element* often serving as transition between vertical walls or *arches* meeting at an angle and the part of the base of a *drum* or *dome* that rises above them. Pendentives often merge into vertical *piers* that support them from below. The inner surface of a pendentive has the shape of a curving triangle.

peristyle A circular arrangement of *columns* often in support of a structure above it.

perspective The way in which objects appear to the eye. Also a set of techniques for depicting, on a flat surface, the objects, shapes, and spatial relationships that are seen.

pier A solid support or pillar, usually square or rectangular in cross section, often the bottom section of a *column*. A sturdy upright structure of masonry acting mainly to support a vertical load. A *crossing pier* is one of usually four main piers at the *crossing* of a church that serve as the primary support of a *dome* or tower.

pilaster A flat rectangular *column* or *pier* projecting from a wall. Often nonstructural, but including a *capital*, shaft, and base. Often with the features prescribed by a classical *order*.

pile A large beam driven into the soil in arrays so as to provide firm support for the foundations of a building or other structure.

pinnacle A relatively small upright structure capping a *column, buttress,* or *pier*. The weight with which a pinnacle pushes down on the structural element below it adds *stiffness* to the element.

plan A drawing made to scale to represent the basic horizontal section of a structure.

portico A porch structure, often at the entrance of a building, and often with a roof supported by column elements arranged in one of the classical *orders*. Also a walkway with a roofed structure supported by an *arcade* or *colonnade*.

post and lintel Another name for a basic *column* and *beam* configuration.

reaction An *internal force* generated by a component of a structure in response to an *external force*.

relief A form of sculpture in which the represented objects are not freestanding but project from a flat surface. In high relief the representations rise substantially from the flat background. In low relief, best known as bas-relief, they may rise only a little above the flat surface.

Renaissance architecture A style of architecture that uses classical Greek and Roman elements such as *arches, columns, colonnades,* and *porticos* in clear, rational, and well-proportioned compositions.

resultant The combined effect of any number of *forces* acting at the same point.

rib An embedded and projecting band on and in a *vault* or *dome*, usually of masonry, often forming a part of a primary structural frame.

ribbed vault A continuous ceiling or roof structure, usually made of masonry, having the cross section of an arch and having a network of structurally critical arching ribs embedded in it and protruding from it. The webbing of a ribbed vault refers to the masonry or stone that fills in the regions between the ribs.

Romanesque architecture An architectural style, often of churches in *basilica plan*, featuring heavy masonry construction with thick walls, small windows, the use of semicircular Roman *arches* and *arcades*, as well as *barrel* and *groin vaults*.

rotunda A building having a circular *plan* and a *dome*. Also, a large circular, or cylindrical room.

rubble Rough fragments of broken stone or bricks used in masonry construction. *Piers* are sometimes made with a rubble core surrounded by *ashlar* facing.

safety margin or **safety factor** A measure of the structural capacity of a system, such as a bridge or building, beyond the loads that it actually carries. It is the ratio of the estimated strength of the system, divided by the strength it needs to have to carry the loads that it is subjected to.

sandstone A stone formed by grains of sand (usually the minerals quartz or feldspar) bound under pressure by binding materials such as clays or silica. The quality of the stone depends largely on the nature of these cementing materials because sand is very durable. Sandstone is used for the best *ashlar* work. Sandstone weighs in the range of 120 to 145 pounds per cubic foot. It has a crushing resistance from 200 to 900 *tons* per square foot.

section In architecture, a diagrammatic drawing to scale in which the elements of a building are represented as if cut by a plane, often a vertical plane. In mathematics, a figure obtained from another by a cut. The concepts conic section and section of a circle or sphere are examples.

shaft The main vertical element of a *column*. More precisely, the part of a column between the base and the *capital*.

shear A *force* acting transverse to the axis of a *structural element* such as a beam. If too severe, shear can cause sliding failure of the element.

shell The curving masonry or concrete part of a *dome* or *vault*. In other words, the solid, structural part of a *dome* or *vault*.

slab A flat stone or concrete element often spanning a horizontal gap. A slab is usually much wider than a *beam*, in other words, it often extends significantly in a second horizontal direction.

span The horizontal distance between the vertical supports of an *arch*, *beam*, *vault*, or *dome*.

springing The position from where an *arch*, *vault*, or *dome* begins to curve inward from its vertical supports.

squinch An *arch* or configuration of arches, usually of masonry, anchored in a corner of a vertical wall and slanting upward to support a superstructure such as a *dome*.

stability The capacity of a structure to remain in stable *equilibrium* under the action of applied forces.

statics The study of architectural and other structures that are in *equilibrium*, especially the loads on such structures, the thrusts generated by them, how they respond, and why they fail.

stiffness A measure of a structure's resistance to *deformation*.

stoa An ancient Greek *portico*, usually walled at the back with a colonnade at the front, designed to provide a sheltered promenade.

stress A measure of the local intensity of a *force* on the material of a structure. This intensity is commonly expressed as the magnitude of the force divided by the square area of the surface on which it acts.

structural component A fully formed unit of construction that may be as small as a brick, or as large as a prefabricated concrete section of a wall. It is identified in terms of the fabrication or construction process rather than the structural role.

structural element A basic unit of an architectural structure expected to be capable of carrying its own weight and the other loads that it is intended to support. An *arch*, *dome*, *shell*, and *vault* are examples, as are the constituent *beams*, *columns*, *piers*, and *slabs*.

structural form The external geometric configuration of a *structural element* or a larger structure.

structure A system of *structural elements*. In a building, often the complete system that plays the primary load-bearing role.

tension or **tensile force** A *force* along a longitudinal structural member tending to stretch it.

thrust A lateral or slanting *force*. Commonly, the force exerted by an *arch* or *dome* on its supports, especially the outward horizontal component of such a force.

tie-beam A main horizontal member of a structure that acts in tension to prevent the structure from spreading. In particular, a beam in a timber roof that serves this purpose.

ton A unit of weight or force equal to 2000 pounds. The metric version, the **tonne**, is a unit of mass equal to 1000 kilograms, or equivalently to approximately 2204 pounds.

tracery A dividing segment of stone within an ornamental window that provides strength and separates glass panels.

transept The structure of a church in *basilica plan* perpendicular to the nave that supplements the nave to give the church a cross-shaped configuration.

truss An assembly of interconnected, often triangular *structural elements* often made of wood or metal, designed to broadly act as a *beam*. It is generally load-bearing. A **simple truss** consists of a single, rigid structural triangle.

vault A continuous ceiling or roof structure having the cross section of an arch. It is usually made of masonry or concrete. The *barrel vault, groin vault*, and *ribbed vault* are three major types.

vector The representation of a *force* as an arrow in which the direction of the arrow represents the direction of the force and the length of the arrow is proportional to the magnitude of the force.

voussoir A stone worked into the shape of a wedge with curving edges and used as a building block of a masonry arch or vault. The voussoir at the top of an arch is the *keystone*.

These references are the important sources of information for the text. They will also be useful to a reader who wishes to explore its topics further. The entries in the first section of historical references are relevant to two or more chapters. They are followed by references for the specific topics discussed in the chapters.

History of Civilization, Architecture, and Mathematics

C. Boyer and U. Merzbach, *A History of Mathematics*, John Wiley & Sons, New York, 1991.

David Burton, *The History of Mathematics: An Introduction*, 7th ed., McGraw-Hill, New York, 2010.

Kenneth Clark, *Civilization: A Personal View*, Harper & Row, New York, 1969.

Will and Ariel Durant, *The Story of Civilization*, 11 vols. Simon and Schuster, New York, 1935–1975 (especially vol. 2, *The Life of Greece*; vol. 3, *Caesar and Christ*; vol. 4, *The Age of Faith*; and vol. 5, *The Renaissance*).

P. Goldberger, *Why Architecture Matters*, Yale University Press, New Haven, 2009.

T. Heath, *A History of Greek Mathematics*. Vol. 1, *From Thales to Euclid*; vol. 2, *From Aristarchus to Diophantus*, Dover Publications, New York, 1981.

David Jacobs, *Architecture*, Newsweek Books, New York, 1974.

Victor Katz, *A History of Mathematics*, 3rd ed., Addison Wesley, Boston, 2008.

Morris Kline, *Mathematical Thought from Ancient to Modern Times*, Oxford University Press, New York, 1972.

David Macauley, *Building Big*, Houghton Mifflin, Boston, 2000. Companion to the PBS Series.

Rowland Mainstone, *Developments in Structural Form*, 2nd ed., Architectural Press, Oxford, 2001.

Robert Mark, *Light, Wind, and Structure: The Mystery of the Master Builders*, MIT Press, Cambridge, 1990.

Robert Mark and David P. Billington, Structural Imperative and the Origin of New Form, *Technology and Culture*, vol. 30, no. 2 (1989), 300–329. Special Issue: Essays in Honor of Carl W. Condit.

Marian Moffett, Michael Fazio, and Lawrence Wodehouse, *Buildings across Time: An Introduction to World Architecture*, McGraw-Hill, Boston, 2004.

James Neal, *Architecture: A Visual History*, PRC Publishing, London, 1999.

H. Schlimme, editor, *Practice and Science in Early Modern Italian Building: Towards an Epistemic History of Architecture*, Electa, Milan, 2006.

Dirk Struik, *A Concise History of Mathematics*, Dover Publications, New York, 1948.

Time-Life Books, *The Great Ages of Man: A History of the World's Cultures*, 20 vols. Time, Inc., New York, 1966 (especially *Ancient Egypt*, *Classical Greece*, *Imperial Rome*, *Byzantium*, *Age of Faith*, *Renaissance*, *Age of Enlightenment*, and *Twentieth Century*).

B. L. Van der Waerden, *Science Awakening*, Oxford University Press, New York, 1961.

Chapter 1. Humanity Awakening: Sensing Form and Creating Structures

Norman Crowe, *Nature and the Idea of a Man-Made World*, MIT Press, Cambridge, 1997.

Downs Dunham, Building an Egyptian Pyramid, *Archaeology*, vol. 9, no. 3 (1956), 159–165.

Chapter 2. Greek Geometry and Roman Engineering

Marcus Frings, The Golden Section in Architectural Theory, *Nexus Network Journal, Architecture and Mathematics*, vol. 4, no. 1 (2002), 1–18.

Mark Wilson Jones, *Principles of Roman Architecture*, Yale University Press, New Haven, 2000.

Rowland Mainstone and Robert Mack, On the Structure of the Roman Pantheon, *Art Bulletin*, vol. 68, no. 4 (1986), 673–674.

Robert Mark and Paul Hutchinson, On the Structure of the Roman Pantheon, *Art Bulletin*, vol. 68, no. 1 (1986), 24–34.

Michael J. Ostwald, Under Siege: The Golden Mean in Architecture, *Nexus Network Journal, Architecture and Mathematics*, vol. 2, nos. 1–2 (2000), 75–82.

W. K. West, Problems in the Cultural History of the Ellipse, *Technology and Culture*, vol. 19, no. 4 (1978), 709–712.

Chapter 3. Architecture Inspired by Faith

James S. Ackerman, "Ars Sine Scientia Nihil Est" Gothic Theory of Architecture at the Cathedral of Milan, *Art Bulletin*, vol 31. no 2 (1949), 84–111.

G. Binding, *Medieval Building Techniques*, Tempus Publishing, Stroud, Gloucestershire, Great Britain, 2004.

J. M. Bloom, On the Transmission of Designs in Early Islamic Architecture, *Muqarnas*, vol. 10 (1993), 21–28. Essays in Honor of Oleg Grabar.

Martin S. Briggs, Gothic Architecture and Persian Origins, *Burlington Magazine for the Connoisseurs*, vol. 62, no. 361 (1933), 183–189.

Ahmet S. Cakmak, Rabun M. Taylor, and Eser Durukal, The Structural Configuration of the First Dome of Justinian's Hagia Sophia (A.D. 537–558): An Investigation Based on Structural and Literary Analysis, *Soil Dynamics and Earthquake Engineering*, vol. 29, no. 4 (2009), 693–698.

K.A.C. Creswell, The Origin of the Persian Double Dome, *Burlington Magazine for Connoisseurs*, vol. 24, no. 128 (1913), 94–99, and vol. 24, no. 129 (1913), 152–156.

William Emerson and Robert L. van Nice, Hagia Sophia, Istanbul: Preliminary Report of a Recent Examination of the Structure, *American Journal of Archeology*, vol. 47, no. 4 (1943), 403–436.

Helen C. Evans, Byzantium Revisited: The Mosaics of Hagia Sophia in the Twentieth Century, *The 4th Annual Pallas Lecture*, Modern Greek Program, University of Michigan, February 9, 2006.

Deborah Howard, *The Architectural History of Venice*, Yale University Press, New Haven, 2002.

Deborah Howard, Venice and Islam in the Middle Ages: Some Observations of Architectural Influence, *Architectural History*, vol. 34 (1991), 59–74.

William MacDonald, Design and Technology in Hagia Sophia, *Perspecta*, vol. 4 (1957), 20–27.

H. McCague, A Mathematical Look at a Medieval Cathedral, *Math Horizons*, American Mathematical Society, Providence, RI, 2003.

Arthur U. Pope, Gothic Architecture and Persian Origins, *Burlington Magazine for the Connoisseurs*, vol. 62, no. 363 (1933), 293–294.

Christine Smith, East or West in 11th-Century Pisan Culture: The Dome of the Cathedral and Its Western Counterparts, *Journal of the Society of Architectural Historians*, vol. 43, no. 3 (1984), 195–208. In Memoriam: Kenneth J. Conant.

Emerson H. Swift, The Latins at Hagia Sophia, *American Journal of Archeology*, vol. 39, no. 4 (1935), 458–474.

Rabun Taylor, A Literary and Structural Analysis of the First Dome on Justinian's Hagia Sophia, Constantinople, *Journal of the Society of Architectural Historians*, vol. 55, no. 1 (1996), 66–78.

Hermann Weyl, *Symmetry*, Princeton University Press, Princeton, 1952.

Chapter 4. Transmission of Mathematics and Transition in Architecture

Martin S. Briggs, Architectural Models-I, *Burlington Magazine for Connoisseurs*, vol. 54, no. 313 (1929), 174–175, 178–181, 183.

Giovanni Fanelli and Michele Fanelli, *Brunelleschi's Cupola: Past and Present of an Architectural Masterpiece*, Mandragora, Florence, 2004.

F. Prager and G. Scaglia, *Brunelleschi: Studies of His Technology and Inventions*, MIT Press, Cambridge, 1970.

Howard Saalman, *Filippo Brunelleschi: The Cupola of Santa Maria del Fiore*, A. Zwemmer, London, 1980.

Howard Saalman, Giovanni di Gherardo da Prato's Designs Concerning the Cupola of Santa Maria del Fiore in Florence, *Journal of the Society of Architectural Historians*, vol. 18, no. 1 (1959), 11–20.

Hans Samelson, Letter in the *Journal of the Society of Architectural Historians*, vol. 57, no. 3 (1998), 360–361.

Franklin K. B. Toker, Florence Cathedral: The Design Stage, *Art Bulletin*, vol. 60, no. 2 (1978), 214–231.

Chapter 5. The Renaissance: Architecture and the Human Spirit

James Ackerman, *The Architecture of Michelangelo*, University of Chicago Press, Chicago; Penguin Books, New York, 1986.

James Ackerman, *Palladio*, Penguin Books, New York, 1966.

Richard J. Betts, Structural Innovation and Structural Design in Renaissance Architecture, *Journal of the Society of Architectural Historians*, vol. 52, no. 1 (1993), 5–25.

Suzanne Boorsch, The Building of the Vatican: The Papacy and Architecture, *Metropolitan Museum of Art Bulletin*, new series, vol. 40, no. 3 (1982–83), 1–64.

Fritjof Capra, *The Science of Leonardo*, Doubleday, New York, 2007.

Deborah Howard, Venice between East and West: Marc'Antonio Barbaro and Palladio's Church of the Redentore, *Journal of the Society of Architectural Historians*, vol. 62, no. 3 (2003), 306–325.

James Lees-Milne, *Saint Peter's: The Story of Saint Peter's Basilica in Rome*, Little, Brown and Company, Boston, 1967.

R. Mainstone, The Dome of St. Peter's: Structural Aspects of Its Design and Construction, and Inquiries into Its Stability, *Architectural Association Files*, vol. 39 (1999), 21–39.

Charles B. McClendon, The History of the Site of St. Peter's Basilica, Rome, *Perspecta*, vol. 25 (1989), 32–65.

H. A. Millon and V. M. Lampugnani, editors, *The Renaissance from Brunelleschi to Michelangelo: The Representation of Architecture*, Rizzoli, New York, 1994.

Henry A. Millon and Craig Hugh Smyth, *Michelangelo Architect: The Facade of San Lorenzo and the Drum and Dome of St. Peter's*, Companion to an exhibition at the National Gallery of Art, 9 October–11 December 1988, organized by the National Gallery of Art, Washington, in association with the Casa Buonarroti, Florence, and Olivetti. Olivetti, Milan, 1988.

Branco Mitrovic and Ivana Djordjevic, Palladio's Theory of Proportions and the Second Book of the "Quattro Libri dell'Architettura," *Journal of the Society of Architectural Historians*, vol. 49, no 3 (1990), 279–292.

Earl Rosenthal, The Antecedents of Bramante's Tempietto, *Journal of the Society of Architectural Historians*, vol. 23, no. 2 (1964), 55–74.

Howard Saalman, Michelangelo: S. Maria del Fiore and St. Peter's, *Art Bulletin*, vol. 57, no. 3 (1975), 374–409.

Giorgio Vasari, *The Lives of the Most Excellent Painters, Sculptors and Architects*, Florence, 1550.

Stephen R. Wassel, Andrea Palladio (1508–1580), *Nexus Network Journal, Architecture and Mathematics*, vol. 10, no. 2 (2008), 213–225.

Rudolf Wittkower, *Architectural Principles in the Age of Humanism*, W. W. Norton & Company, New York, 1971.

Chapter 6. New Architecture: Materials, Structural Analysis, Computers, and Design

O. Arup and R. S. Jenkins, The Evolution and Design of the Concourse at the Sydney Opera, *Proceedings: Institution of Civil Engineers*, vol. 39 (1968), 541–565.

Turpin C. Bannister, The Genealogy of the Dome of the United States Capitol, *Journal of the Society of Architectural Historians*, vol. 7, no. 1/2 (1948), pp. 1–19, 31, with two appendixes by August Schoenborn, pp. 20–30.

Antonio Becchi, Massimo Corradi, Federico Foce, and Orietta Pedemonte, editors, *Essays on the History of Mechanics: In Memory of Clifford Ambrose Truesdell and Edoardo Benvenuto*, Birkhäuser Verlag, Basel, 2003.

J. A. Bennett, *The Mathematical Science of Christopher Wren*, Cambridge University Press, Cambridge, 1982.

C. Blasi, E. Cosson, and I. Iori, The Fractures of the French Panthéon: Survey and Structural Analysis, *Engineering Fracture Mechanics*, vol. 75, iss. 3–4 (2008), 379–388.

P. Block, M. DeJong, J. Ochsendorf, As Hangs the Flexible Line: Equilibrium of Masonry Arches, *Nexus Network Journal, Architecture and Mathematics*, vol. 8, no. 2 (2006), 9–19.

Glenn Brown, The Work of Thomas U. Walter, Architect, pp. 317–389, in William B. Bushong, editor, *Brown's History of the United States Capitol*, House Document No. 108-240 (1901–1903), United States Capitol Preservation Commission, 1998.

James W. P. Campbell and Robert Bowles, The Construction of the New Cathedral, pp. 207–219 in Derek Keene, Arthur Burns, and Andrew Saint, editors, *St. Paul's The Cathedral Church in London*, Yale University Press, New Haven, 2004.

J.-A. de Coulomb, Essai sur une application des règles de maximis & minimis à quelques problème de statique, relatifs à l'architecture, *Mémoires de Mathématiques & Physique*, vol. 7, 1773, pp. 343–382, Paris 1776.

Kerry Downes, Wren and the New Cathedral, pp. 191–206 in Derek Keene, Arthur Burns, and Andrew Saint, editors, *St. Paul's The Cathedral Church in London*, Yale University Press, New Haven, 2004.

S. B. Hamilton, The Place of Sir Christopher Wren in the History of Structural Engineering, *Transactions—Newcomen Society for the Study of the History of Structural Engineering*, vol. 14 (1933–34), 27–42.

J. Heyman, *Coulomb's Memoir on Statics*, Cambridge University Press, Cambridge, 1972.

J. Heyman, *The Masonry Arch*, Ellis Horwood, Chichester, England; Halsted Press, John Wiley & Sons, New York, 1982.

J. Heyman, *The Stone Skeleton: Structural Engineering of Masonry Architecture*, Cambridge University Press, Cambridge, 1995.

Derek Keene, Arthur Burns, and Andrew Saint, editors, *St. Paul's The Cathedral Church in London*, Yale University Press, New Haven, 2004.

Nicoletta Marconi, Technicians and Master Builders for the Dome of St. Peter's in Vatican in the Eighteenth Century: The Contribution of Nicola Zabaglia (1664–1750), *Proceedings of the Third International Congress on Construction History*, Cottbus, Germany, May 2009, 991–999.

J. L. Meriam and L. G. Kraige, *Engineering Mechanics: Statics*, 5th ed., John Wiley & Sons, New York, 2002.

Y. Mikami, *Utzon's Sphere: Sydney's Opera House—How It Was Designed and Built*, Shokokusha Publishing Co., Tokyo, 2001.

William J. Mitchell, Roll Over Euclid: How Frank Gehry Designs and Builds, pp. 352–353 in J. Fiona Ragheb, editor, *Frank Gehry, Architect*, Guggenheim Museum Publications, New York, 2001.

William J. Mitchell, A Tale of Two Cities: Architecture and the Digital Revolution, *Science*, vol. 285 (1999), 839–841.

Ricard de Montferrand, *Eglise cathedrale de Saint Isaac*, Paris 1845.

John Nutt, *Constructing a Legacy: Technological Innovation and Achievements*, pp. 103–121 in Anne Watson, editor, *Building a Masterpiece: The Sydney Opera House*, Powerhouse Publishing, Sydney; Lund Humphreys Publishing, London, 2006.

H. W. Robinson, Robert Hook as Surveyor and Architect, *Notes and Records of the Royal Society of London*, vol. 6, no. 1 (1948), 48–55.

J. Rondelet, Memoire historique sur le dôme du Panthéon Francaise, Paris, *Traite theorique et practique de l'art et batir*, vol. 4, part 2, Paris, 1797.

J. Summerson, J. H. Mansart, Sir Christopher Wren and the Dome of St. Paul's Cathedral, *Burlington Magazine*, vol. 132, no. 1042 (1990), 32–36.

Sydney Opera House, *Nomination by the Government of Australia for Inscription on the World Heritage List 2006*, Australian Government Department of the Environment and Heritage, 2006.

David Taff, *Computers and the Opera House: Pioneering a New Technology*, pp. 84–101 in Anne Watson, editor, *Building a Masterpiece: The Sydney Opera House*, Powerhouse Publishing, Sydney; Lund Humphreys Publishing, London, 2006.

S. Timoshenko, *History of Strength of Materials*, Dover Publications, New York, 1983.

Jørn Utzon, *The Sydney Opera House: Utzon's Design Principles*, Sydney Opera House Trust, Sydney, 2002.

Pierre Varignon, *Nouvelle Méchanique*, C. Jombert, Paris, 1725.

Anne Watson, editor, *Building a Masterpiece: The Sydney Opera House*, Powerhouse Publishing, Sydney; Lund Humphreys Publishing, London, 2006.

S. Wren, *Parentalia: or Memoirs of the family of the Wrens*, London 1750, rpt. Gregg Press, 1965.

W. Zalewski and E. Allen, *Shaping Structures: Statics*, John Wiley & Sons, New York, 1998.

Chapter 7. Basic Calculus and Its Application to the Analysis of Structures

Mario Como and Antonio Grimaldi, Large Structures Behaviour: The Past and the Future, pp. 47–79 in Michel Frédmont and Franco Maceri, editors, *Novel Approaches in Civil Engineering*, Springer-Verlag, Berlin, 2004.

Alexander J. Hahn, *Basic Calculus: From Archimedes to Newton to Its Role in Science*, Springer-Verlag, New York, 1998.

Robert Osserman, How the Gateway Arch Got Its Shape, *Nexus Network Journal, Architecture and Mathematics*, 12 (2010).

Robert Osserman, Mathematics of the Gateway Arch, *Notices of the AMS*, Providence, RI, 2010, 220–229.

James Stewart, *Single Variable Calculus: Concepts and Contexts*, Brooks/Cole, Pacific Grove, CA, 2005.

Note: Page numbers in *italics* refer to images or diagrams. References to images in the color plate section are also *italicized*.

The sources of visual material other than those indicated in the captions are as follows (numerals refer to image numbers).

answers.com (2.2)

backtoclassics.com (Plate 21)

bible-history.com (2.35)

Reprinted with the permission of Cambridge University Press. Copyright © 1995 Cambridge University Press (6.2)

commons.wikimedia.org (1.1, 1.9, 1.13, 3.2, 3.19, 3.29, 3.45, 5.1, 5.4, 5.40, 5.51, 5.71, 6.1c, 6.6, 7.20, 7,29, Plate 2, Plate 3, Plate 4, Plate 12, Plate 16, Plate 17a–d, Plate 20)

commons.wikimedia.org, © Creative Commons Attribution (1.3, 1.5, 1.6, 3.11, 3.12, 3.30, 3.31, 3.36, 5.16, 6.1d, Plate 8)

commons.wikimedia.org, © GNU Free Documentation License and Creative Commons Attribution (2.1, 2.3, 2.33, 2.36, 2.37, 2.61, 2.62, 2.64, 3.23, 3.37, 3.44, 4.34, 5.2, 5.3, 5.5, 5.14, 5.15, 5.18, 5.20, 5.23, 5.32, 6.1a, 6.1b, 6.49, Plate 10, Plate 13, Plate 24)

counterlightsrantsandblather1.blogspot .com/2009/11/renaissance-saint-peters -michelangelo.html (5.33)

c2r4a1.blogspot.com/2007/03granada -spain-21-22-feb-2006-alhambra.html (Plate 17e)

en.wikipedia.org (3.22, 5.31, 5.44, 5.72, Plate 18)

flickr.com/photos (2.65, 3.9, 3.21)

Fotolia (5.30)

fotothing.com, © Creative Commons Attribution (6.54)

gaudiclub.com/ingles/i_vida/colonia .asp (6.53)

intranet.arc.miami.edu/rjohn/images/ Brunelleschi/S.%20Maria%20del%20 Fiore%20Floor%20Plan.jpg (4.27)

jstor.org/stable (3.6, 5.34, 5.36, 5.39, 6.8)

mcah.columbia.edu/dbcourses/public portfolio.cgi?view=1526# (5.35)

Used with permission of the Princeton University Library (2.38, 2.45, 3.27, 3.28, 5.17, 5.19, 5.21, 5.69, 5.70, 6.35, 6.39, 6.40, 6.43, 6.44, 6.45, 6.47, 7.13, Plate 15)

Used with permission of the Princeton University Library, © IGDA: Pubbli-aer-foto (info@pubbliaerfoto.it) (2.38)

Used with permission of the Princeton University Library, image courtesy of John Blazejewski/Princeton University Library (3.34, 3.46, 5.64, 5.73, 6.4, 6.5, 6.12, 6.13, 6.14, 6.48, Plate 7)

Réunion des Musées Nationaux/Art Resource, NY, © RMN (Institut de France)/Art Resource (5.26)

shutterstock.com (2.25, 3.7, 3.10, 4.33, Plate 6)

www.thecultureconcept.com (2.12)

t771unit3.pbworks.com/w/page/ 6766569/Chapter%209-%20The%20 Early%20Middle%20Ages (Plate 9)

© The Warburg Institute, University of London (2.42)

wikimediafoundation.org, © GNU Free Documentation License and Creative Commons Attribution (3.24, Plate 14)